BASIC CIVIL ENGINEERING

Dr. Ing. Alamirew M. Tola

Toronto Academic Press

BASIC CIVIL ENGINEERING

Dr. Ing. Alamirew M. Tola

Toronto Academic Press
4164 Lakeshore Road
Burlington ON L7L 1A4
Canada
www.tap-books.com
Email: orders@arclereducation.com

ISBN: 978-1-77956-306-4

ABOUT THE AUTHOR

Dr.-Ing. Alamirew M. Tola is a distinguished civil engineer with a Ph.D. in Highway and Transportation Engineering from the prestigious University of Rostock, Germany. He is an assistant professor at the Kombolcha Institute of Technology, Wollo University, Ethiopia, where he brings a wealth of knowledge and experience to his students and colleagues. Dr.-Ing. Tola's expertise spans a wide array of areas within civil engineering, including Building Information Modeling (BIM), traffic safety, advanced highway design, smart transportation, traffic management and optimization, urban transportation, and machine learning in civil engineering. His dedication to the field is evident through his innovative teaching methods and his commitment to groundbreaking research. Passionate about pushing the boundaries of his field, Dr. Tola continuously seeks to integrate new technologies and methodologies into his work, aiming to improve the sustainability and functionality of civil engineering projects.

Table of Contents

5 GEOTECHNICAL ENGINEERING 127

6 CONSTRUCTION TECHNOLOGY 165

7 TRANSPORTATION ENGINEERING 205

List of Figures

List of Tables

PREFACE

Civil engineers design, construct, supervise, operate, and maintain large construction projects and systems, including roads, buildings, airports, tunnels, dams, bridges, and systems for water supply and sewage treatment. Many civil engineers work in design, construction, research, and education. Civil engineers are rightfully proud of their legacy. During the past century, clean water supplies have extended general life expectancies. Transportation systems serve as an economic and social engine. New bridges, blend strength and beauty, speed transport, and bring communities closer together. Public and private construction, for which engineers provide the essential underpinnings of design and project oversight, produces hundreds of thousands of jobs and drives community development.

Basic Civil Engineering provides the reader with the necessary fundamental concepts of structural behavior, dealing with the design, construction, and maintenance of the physical and naturally built environment. The global civil engineering profession has increasingly recognized the reality of shrinking resources, the desire for sustainable practices and design, and the need for social equity in the consumption of resources. Civil engineering is changing and needs to change more. This book attempts to put the evolution of technology and the development of engineering into its historical context and with it the emergence of problems whose solutions extend beyond technology alone if they are to achieve systems-level solutions. The balancing of economic, social, and environmental objectives, both now and into the future, has become known as sustainable development and this is increasingly a key driver at a range of organizational and spatial scales. This book is designed to enrich the preliminary conceptual knowledge about civil engineering to the students of civil branches or non-civil branches of engineering.

Contents/Structure of the Book

The textbook is structured into ten comprehensive chapters that cover essential aspects of civil engineering, providing a foundational understanding and practical insights into various disciplines within the field.

Chapter 1 serves as a gateway to the world of civil engineering, offering an overview of its history, scope, and fundamental principles. It introduces students and professionals to the diverse roles civil engineers play in shaping infrastructure and society. Chapter 2 is focused on the properties, characteristics, and applications of materials commonly used in civil engineering projects. It covers essential topics such as aggregates, cement, concrete, and other construction materials vital for structural integrity and durability.

Chapter 3 focuses on the principles of structural analysis and mechanics, essential for designing and assessing the stability and strength of various civil engineering structures. It includes topics such as stress, strain, structural components, and load analysis.

Surveying and leveling techniques are crucial for accurately mapping and measuring land, and essential for planning, designing, and constructing infrastructure projects. Chapter 4 covers surveying instruments, methods, and applications in civil engineering.

Chapter 5 explores the properties and behavior of soils and rocks, critical for designing foundations, slopes, and earthworks. Topics include soil mechanics, site investigations, and geotechnical testing methods. Chapter 6 discusses construction techniques, processes, and technologies used in civil engineering projects. It covers construction methods, equipment, safety considerations, and project management practices essential for successful project execution.

Chapter 7 focuses on planning, designing, and maintaining transportation systems. It covers topics such as highway design, traffic engineering, transportation planning, and sustainable transportation solutions. Chapter 8 addresses the impact of civil engineering projects on the environment and explores sustainable practices. Topics include water and wastewater treatment, air pollution control, and environmental regulations.

Chapter 9 deals with the flow and management of water. It covers topics such as open channel flow, hydrology, flood control, irrigation systems, and hydraulic structures like dams and levees. Chapter 10 focuses on project management principles specific to civil engineering projects. It covers project planning, scheduling, cost estimation, risk management, and quality control, crucial for delivering projects on time and within budget.

Each chapter in this book is designed to provide a thorough understanding of its respective topic, making it an essential resource for students, educators, and professionals in the field of civil engineering. Whether studying foundational principles or advancing knowledge in specialized areas, the textbook equips readers with the knowledge and skills necessary for successful careers in civil engineering and infrastructure development.

Audience

This book targets a diverse audience ranging from students and professionals in civil engineering to enthusiasts seeking foundational knowledge in the field. It serves as an essential resource for undergraduate students pursuing degrees in civil engineering, offering comprehensive coverage of core principles and practices essential for their academic journey and future careers.

For aspiring engineers, the book provides a structured introduction to key concepts such as structural analysis, construction materials, surveying, and project management. It aims to build a solid understanding of foundational theories and practical applications, preparing students to tackle real-world challenges in infrastructure development and construction projects.

Professionals in related fields, such as architecture or urban planning, will also find value in the book's detailed explanations and case studies, enhancing their interdisciplinary understanding of civil engineering principles and practices. It serves as a refresher for practitioners looking to reinforce their knowledge base or explore new areas within civil engineering.

Moreover, individuals with a general interest in engineering and construction will benefit from the book's accessible language and illustrative examples, making complex engineering concepts understandable and engaging. Whether preparing for exams, exploring career opportunities, or simply expanding one's knowledge, this book aims to empower its readers with a comprehensive and practical understanding of the fundamental principles that underpin civil engineering projects and infrastructure development globally.

Learning Features

- Start-of-chapter learning objectives
- Key terms in each chapter
- Summary Points at the end of each chapter
- Review Questions at the end of each chapter
- Case Studies and Activities in each chapter for solving real problems
- References at the end of each chapter

CHAPTER 1

Introduction to Civil Engineering

LEARNING OBJECTIVES

After studying this chapter, you will be able to:

- Understand the concept of civil engineering;
- Identify branches of civil engineering;
- Know the scope of different fields of civil engineering; and
- Focus on career opportunities in civil engineering.

KEY TERMS FROM THIS CHAPTER

Airports
Buoyancy
Contractor
Flood control systems
Natural disasters
Sewerage systems
Transportation engineers
Bridges
Canals
Dams
Military engineering
Pipelines
Sustainability
Turnkey

INTRODUCTION

Civil engineering is a field of professional engineering that focuses on designing, building, and managing the infrastructure of the physical and naturally occurring environment. This includes public structures like roads, bridges, canals, dams, airports, sewer systems, pipelines, building components, and railways.

Civil engineering is typically divided into various sub-disciplines and is known as the second-oldest engineering field after military engineering. The purpose of civil engineering is to differentiate between non-military engineering and military engineering. This field can be found in the public sector, ranging from municipal public works departments to federal government agencies, as well as in the private sector, from small local firms to large global Fortune 500 companies.

1.1. CONCEPT OF CIVIL ENGINEERING

Civil engineering, possibly the most ancient engineering field, focuses on the built environment. Its origins can be traced back to the moment when individuals began sheltering themselves with roofs or laying down tree trunks to ford rivers. The built environment is key to defining modern civilization, with buildings and bridges being some of the most prominent structures that we associate with civil engineering, particularly within the sub-discipline of structural engineering.

Transportation engineers design roads, railroads, subway systems, and airports, while civil engineers are responsible for less visible creations as well. For instance, every time you turn on a water faucet, you can expect water to come out thanks to the work of civil engineers. New York City, for example, has an impressive water supply system that receives billions of gallons of high-quality water from the Catskills, located over one hundred miles away.

In the same way, there is a lack of concern among people about the fate of water once it has been used. The traditional field of sanitary engineering has transformed into modern environmental engineering, becoming so important that many academic departments have adjusted their names to reflect this change, now offering civil and environmental engineering programs. Civil engineering 2D drafting involves creating detailed drawings for structures made of steel, wood, R.C.C., and more, including plans, sections, elevations, and other technical drawings for buildings.

1.1.1. Definition and Scope of Civil Engineering

Civil engineering involves harnessing natural power sources to benefit man through planning, designing, and constructing structures for environmental control, development of resources, buildings, transportation, and other necessary facilities for the well-being and enjoyment of humanity.

Civil engineering focuses on the planning, design, estimation, supervision, management, and maintenance of various structures such as buildings, roads, bridges, dams, and more.

1.1.2. History

1.1.2.1. Civil Engineering as a Discipline

Civil engineering is the utilization of physical and scientific principles to address society's challenges. Its roots can be traced back to advancements in physics and mathematics. As a multifaceted profession, civil engineering encompasses various specialized sub-disciplines, such as structures, materials science, geography, geology, soils, hydrology, environmental science, mechanics, project management, and more. During ancient and medieval times, artisans like stonemasons and carpenters were responsible for most architectural design and construction, eventually becoming master builders.

Knowledge was passed down through guilds and seldom replaced by new developments. The buildings, roads, and infrastructure created during this period were often similar, with any advancements being made on a small scale.

In the 3rd century BC, Archimedes laid the foundation for a scientific approach to physical and mathematical problems in civil engineering with his work, including Archimedes' principle and the practical application of Archimedes' screw. In the 7th century AD, Indian mathematician Brahmagupta utilized arithmetic with Hindu-Arabic numerals for volume computations in excavation work.

1.1.2.2. Civil Engineering Profession

Engineering has been a crucial part of human history dating back to the earliest civilizations such as ancient Egypt, the Indus Valley civilization, and Mesopotamia. It is believed that civil engineering practices first emerged between 4000 and 2000 BC as people transitioned from nomadic lifestyles to settled communities, creating a demand for shelter. This era also saw advancements in transportation, with the invention of the wheel and sailing becoming vital technologies.

Historically, there was no clear separation between civil engineering and architecture, with the terms engineer and architect often being used interchangeably. The construction of the pyramids in Egypt around 2700–2500 BC and the Qanat water management system in modern-day Iran, which is over 3,000 years old and over 71 kilometers long, are examples of early large structure constructions in history (Mays, 2010), Iktinos built the Parthenon in Ancient Greece between 447–438 BC, Roman engineers constructed the Appian Way around 312 BC, General Meng T'ien oversaw the building of the Great Wall of China on orders from Ch'in Emperor Shih Huang Ti around 220 BC (Oakes William, Leone Les & Gunn Craig 2001), and ancient Sri Lanka saw the construction of stupas like the Jetavanaramaya and intricate irrigation systems in Anuradhapura. Romans also created various civil structures across their empire, including aqueducts, insulae, harbors, bridges, dams, and roads (Figure 1.1).

In the 18th century, the term civil engineering was created to encompass non-military engineering practices. The first school for civil engineering, École Nationale des Ponts et Chaussées, was established in France in 1747, with similar institutions appearing in other European countries like Spain. John Smeaton is considered the first civil engineer for designing the Eddystone Lighthouse. In 1771, Smeaton and his colleagues founded the Smeatonian Society of Civil Engineers, a group that gathered for social dinners rather than technical discussions.

The Institution of Civil Engineers was established in London in 1818, with renowned engineer Thomas Telford serving as its inaugural president in 1820. The institution was granted a Royal charter in 1828, officially acknowledging civil engineering as a distinct profession. The charter outlined civil engineering as:

The skill of harnessing the immense power of nature for the benefit of mankind, to facilitate production and trade within states, including external and internal commerce. This is achieved through the construction of various infrastructures such as roads, bridges, aqueducts, canals, and river navigation systems for efficient transportation and exchange.

Figure 1.1. *Chichen Itza was a large pre-Columbian city in Mexico built by the Maya people of the Post Classic. The northeast column temple also covers a channel that funnels all the rainwater from the complex some 40 meters (130 ft) away to a rejollada, a former cenote.*

Source: By Daniel Schwen – Own work, CC BY-SA 4.0, https://commons.wikimedia.org/w/index.php?curid=7647000.

Additionally, this art is utilized in creating ports, harbors, moles, breakwaters, and lighthouses to enhance maritime activities, as well as in developing navigation techniques using artificial power for commercial purposes. Furthermore, it involves the use of machinery for various applications and the implementation of drainage systems in urban areas.

1.1.2.3. Civil Engineering Education

Norwich University, established in 1819 by Captain Alden Partridge, was the inaugural private college in the United States to offer civil engineering instruction. Rensselaer Polytechnic Institute granted the first degree in civil engineering in the U.S. in 1835. Cornell University awarded the first civil engineering degree to a woman, Nora Stanton Blatch, in 1905. In the early 19th century in the UK, the distinction between civil engineering and military engineering, overseen by the Royal Military Academy in Woolwich, led to the development of new engineering education programs. This included the founding of the Class of Civil Engineering and Mining at King's College London in 1838 to address the expanding railway system's need for skilled engineers. Additionally, the College for Civil Engineers in Putney was established in 1839, and the University of Glasgow created the UK's first Chair of Engineering in 1840.

1.1.3. Function of Civil Engineering

The responsibilities of civil engineers can be categorized into three groups: pre-construction tasks such as feasibility studies, site investigations, and design; construction-related tasks like communication with clients, consulting engineers, and contractors; and post-construction tasks such as maintenance and research.

1.1.3.1. Feasibility Studies

Before beginning any significant project, a thorough assessment of the goal and preliminary exploration of potential plans is essential. This includes considering different possible schemes, along with potential alternatives. Feasibility studies may examine various methods, such as choosing between a bridge or tunnel for a water crossing, or determining the best route once a method is chosen. It is crucial to address both economic and engineering challenges throughout the planning process.

1.1.3.2. Site Investigations

An initial site inspection is included in the feasibility study; however, a more thorough investigation is typically necessary once a plan has been approved. Investing in a comprehensive analysis of the ground and substructure can lead to significant cost savings in the future by avoiding expensive remedial work or construction changes.

Soil mechanics, a crucial aspect of large-scale construction projects, did not see serious study until the mid-1930s. Karl von Terzaghi, the pioneer of science, marks its official beginning in 1936 with the First International Conference on Soil Mechanics and Foundation Engineering at Harvard University. Since then, specialist societies and journals have emerged worldwide, and most civil engineering faculties in universities offer courses in soil mechanics.

Construction is a broad term that refers to the art and science of creating objects, systems, or organizations. The word originates from Latin "construction" and Old French "construction."

1.1.3.3. Design

The design of engineering projects often involves the integration of design principles from various fields such as hydraulics, thermodynamics, and nuclear physics. Research in structural analysis and material technology has led to more efficient designs, innovative concepts, and optimized material usage. The evolution of structural theory and material science has enabled more precise stress analysis and systematic testing. With advanced theories, abundant design data, and the ability to conduct rigorous computer analysis, modern designers can create more sophisticated and optimized structural designs.

1.1.3.4. Construction

Private clients may start the promotion of civil engineering works, but the majority of projects are typically carried out for big corporations, government bodies, and public boards. While some of these organizations have in-house engineering teams, consulting engineers are often brought in for larger, specialized projects.

The consulting engineer may need to conduct feasibility studies, propose a project plan, and provide an estimated cost. This engineer is responsible for designing the project, creating specifications and drawings, and preparing legal documents for competitive bidding. After evaluating bids, the engineer selects the most suitable one. While not a party to the contract, the engineer's responsibilities are outlined within it, including overseeing construction and certifying project completion. All actions must align with client interests, and professional organizations enforce ethical standards. The resident engineer serves as the consulting engineer's senior representative on the construction site.

In recent years, there has been a rise in the popularity of turnkey or package contracts. Under these agreements, the contractor is responsible for financing, designing, specifying, constructing, and commissioning the entire project. The consulting engineer is hired by the contractor rather than the client in these cases. Typically, the contractor is an incorporated company that wins the contract based on the consulting engineer's specifications and general drawings. The consulting engineer must approve any changes made and review the detailed drawings.

1.1.3.5. Maintenance

The contractor is responsible for maintaining the works to meet the consulting engineer's standards. This includes ancillary and temporary works that are part of the construction project. The contractor is required to carry out a period of maintenance after construction is completed, and the final payment of the contract price is withheld until approved by the consulting engineer. Government engineering and public works departments focus mainly on maintenance and employ direct labor for this purpose.

1.1.3.6. Research

Research in the field of civil engineering is conducted by various entities, including government agencies, industrial foundations, universities, and other institutions. In most countries, government-controlled agencies like the United States Bureau of Standards and the National Physical Laboratory of Great Britain are involved in a wide range of research activities related to building research, roads, highways, hydraulic research, water pollution, and more. While many of these agencies receive government funding, they also rely on income from research projects initiated by the industry.

1.1.4. Importance of Civil Engineering in Society

The importance of civil engineering in today's world cannot be overstated. From skyscrapers to bridges, it plays a critical role in shaping our modern infrastructure. But its impact goes far beyond just physical structures; civil engineering serves as the cornerstone for societal progress. So, what makes civil engineering so vital to our society?

Here are the main reasons:

- **Infrastructure Development:** Civil engineering is crucial for constructing and maintaining the essential infrastructure that sustains contemporary society. Various civil engineering structures such as ports, roads, bridges, parks, airports, and public transportation systems are crucial for facilitating the movement of people and goods, supplying clean water, managing waste, driving economic progress, and promoting connectivity.
- **Safe and Sustainable Living:** Civil engineers are responsible for designing structures and systems that protect communities and promote their well-being. They develop flood control systems and implement sustainable urban planning strategies to minimize the effects of natural disasters and support the long-term sustainability of communities.
- **Clean Water and Sanitation:** Having access to clean water and proper sanitary facilities is a basic human right. Civil engineers play a crucial role in designing water treatment plants, managing sewage

systems, and maintaining pipelines to ensure that communities have safe and reliable access to these essential resources. Their efforts not only improve public health but also enhance the overall quality of life.

- **Environmental Conservation:** They lead the way in environmental conservation, incorporating eco-conscious designs in their projects that promote renewable energy, sustainable building methods, and responsible waste disposal to minimize their environmental footprint.
- **Economic Growth:** Civil engineering projects stimulate economic growth by creating job opportunities, attracting investments, and fostering development. Major construction projects play a key role in boosting local economies and providing employment options. Furthermore, civil engineering serves as a catalyst for prosperity.
- **Innovation and Technological Advancements:** Civil engineering is constantly evolving with innovative technologies like smart infrastructure, 3D printing in construction, and advanced materials pushing boundaries in the field. These advancements not only improve efficiency but also help to reduce environmental impact.

Civil engineers are the masterminds responsible for designing and maintaining the intricate infrastructure that keeps urban areas running smoothly. Without their expertise, cities would struggle to function with their essential networks of roads, airports, and public transportation systems facilitating the seamless flow of people and goods.

The planners of transportation systems meticulously layout and create mobility solutions. They aim for maximum efficiency, safety, and inclusivity. Their efforts extend beyond mere building projects, aiming to unite communities, alleviate traffic, and stimulate economic development.

Additionally, civil engineering has acknowledged its responsibility in environmental preservation during a time characterized by heightened environmental awareness. Sustainable infrastructure projects are leading the way in this endeavor.

Furthermore, civil engineers are at the forefront of creating environmentally sustainable infrastructure that aims to minimize energy usage and implement green energy solutions that utilize renewable sources. These endeavors play a vital role in moving towards a more eco-friendly future and demonstrate a dedication to ethical construction techniques. By prioritizing advancement, civil engineering serves as a protector of both human health and the preservation of our environment.

1.2. BRANCHES OF CIVIL ENGINEERING

Civil Engineering focuses on enhancing production and transportation methods within states, for both domestic and international trade. It involves the planning and maintenance of infrastructure such as roads, bridges, railroads, aqueducts, canals, river navigation, docks, and warehouses, to facilitate internal communication and trade. Additionally, it is involved in the construction of ports, harbors, breakwaters, lighthouses, and the implementation of artificial power for commercial navigation.

Civil engineering is used to safeguard property from natural forces, such as building embankments to protect land from sea erosion or river flooding. It also involves utilizing streams and rivers for various purposes like powering machines, providing water to cities, towns, and irrigation, and removing harmful waste by draining towns and areas to maintain public health.

In today's world, civil engineering encompasses various aspects such as energy production and distribution, aircraft and airport development, chemical process plant and nuclear power station construction, as well as water desalination. These areas of focus can be categorized into construction, transportation, maritime and hydraulic engineering, power systems, and public health.

1.2.1. Construction

Most civil engineering contracts involve construction work, and the use of steel and concrete as building materials has shifted the design responsibility from architects to civil engineers. Engineers analyze building problems based on function and economics to determine the structural design of a building.

1.2.2. Transportation

The construction of Roman roads and bridges was carried out by military engineers, while the pavements developed by McAdam and bridges by Perronet were the result of civil engineering. The civil engineer also played a major role in the building of canals in the 18th century and railways in the 19th, which greatly contributed to the Industrial Revolution by offering efficient and fast transportation of goods. Today, civil engineers are involved in a wide range of transportation projects, including traffic studies, designing road, rail, and air systems, as well as constructing pavements, embankments, bridges, and tunnels.

1.2.3. Maritime and Hydraulic Engineering

The art of harbor construction and shipbuilding dates back to ancient times. Many developing nations now see the creation of a large, efficient harbor as a priority in order to facilitate the import of industrial machinery and raw materials, as well as the export of finished products. In developed countries, the growth of global trade, the use of larger ships, and the increase in overall tonnage demand faster and more effective handling. The civil engineer is tasked with creating deeper berths, installing alongside-handling equipment (such as for ore), and making navigation improvements to meet these demands.

Water supply has been a crucial aspect of early civilizations, and the need for water has only increased over time. In developed nations, the demand for water is primarily for industrial and household use. However, in many regions such as the Indus basin, large-scale projects are being implemented to improve irrigation systems in order to meet the growing food demand. These projects often include the construction of hydroelectric power plants to support industrial growth.

Large dams are now considered one of the biggest construction projects worldwide, with organizations such as the International Commission on Large Dams playing a key role in promoting design advancements. Building large impounding dams near populated areas necessitates meticulous attention to safety engineering, focusing on soil mechanics and stress analysis. Regulatory bodies typically oversee the qualifications and licensing of engineers responsible for designing and inspecting dams.

1.2.4. Power

Throughout history, civil engineers have been essential in the mining industry for extracting coal and metals, with tunnel construction being a recurring duty across various civil engineering sectors. In the 20th Century, the development and building of power stations evolved alongside the increasing need for electricity, leading to the introduction of nuclear power stations which introduced a new realm of design and construction, including the use of pre-stressed concrete pressure vessels for the reactor. The discovery of natural gas in significant quantities, along with the exploitation of oil fields, has sparked a revolution in gas production. Innovative developments such as shipping in liquid form from the Sahara and piping from the North Sea bed have emerged as key advancements in the industry.

1.2.5. Public Health

The management of drainage and liquid-waste disposal is interconnected with measures to reduce pollution and promote water reuse. Changes in urban development within water catchment areas can impact runoff patterns, while river training and regulation can lead to floods, necessitating flood prevention and control measures.

The production of durable goods in large quantities, such as automobiles and refrigerators, has led to challenges in disposing of solid waste in modern society. Additionally, the increasing use of non-degradable packaging materials has compounded this issue. Civil engineers are crucial in protecting the environment by designing infrastructure that minimizes harm and pollution, focusing on enhancing the environment rather than causing damage.

1.3. SCOPE OF DIFFERENT FIELDS OF CIVIL ENGINEERING

Civil engineering, the first branch of engineering, has been evolving since ancient civilizations. The American Society of Civil Engineering describes it as a profession that utilizes knowledge of mathematics and physical sciences to develop economical ways of utilizing natural materials and forces for the betterment of humanity through study, experience, and practice.

1.3.1. Surveying

Surveying is the art of creating maps. Before beginning any development project in a particular area, it is essential to determine the precise positions of various objects in both the horizontal and vertical directions. This is achieved through the process of surveying. In the past, surveyors used traditional instruments such as chains, tapes, and leveling tools. However, in today's electronic age, more advanced equipment like distance meters and total stations are employed to obtain highly accurate results with ease. Furthermore, modern technologies like photogrammetry and remote sensing have revolutionized the field of surveying, making it more efficient and effective.

Surveying is a method used to obtain precise measurements of the Earth's surfaces on a larger scale. This process includes collecting measurement data, analyzing and interpreting the data to make it usable, as well as determining the relative position and size based on specific measurement requirements. Surveying serves two contrasting functions – determining current horizontal and vertical positions for mapping purposes and setting markers for construction control or land boundary indication. Surveying, a crucial element in the development of human environments for centuries, is often overlooked in terms of importance. It is a necessary step in the planning and carrying out of all types of construction projects. Surveying has been important since the beginning of history, and without it, many significant scientific discoveries would not have been possible. In modern times, surveying is used primarily in transportation, construction, land division, and communication projects.

Surveying techniques are generally consistent worldwide, with minor variations in methodology and the use of a few handheld tools. Instruments used in surveying are primarily produced in Switzerland, Austria, Great Britain, the United States, Japan, and Germany. Japanese-made instruments closely resemble those from Western countries.

1.3.2. Building Materials

Shelter is an essential requirement for civilization. Throughout history, efforts have been made to provide good shelter using a variety of building materials such as stones, bricks, timber, lime, cement, sand, jellies, and tiles. Advancements in technology have led to the use of steel, aluminum, glass, glazed tiles, plaster of Paris, paints, and varnishes, which have enhanced the quality of buildings. Mortar, a mixture of binding materials like lime and cement with sand, and concrete, a mixture of cement, sand, jelly, and water, are commonly used in construction. The combination of concrete with steel bars, known as reinforced cement concrete or RCC, has enabled the construction of strong and durable tall structures. Civil engineers must have knowledge of the properties of building materials to ensure their proper use. As new and improved building materials become available, a skilled civil engineer will be quick to incorporate them into their projects.

There are a variety of options for selecting used building materials, including reclaimed, second-hand, recycled, vintage, or reused items. Before deciding on which type of used building materials to purchase, it is important to consider factors such as the type of project, quality, style, and price of each item.

The initial consideration in the evaluation process is the nature of the building project in question. For instance, if the goal is to construct or replace a roof for a house, it is not advisable to use reclaimed materials from a previous structure. Instead, opting for recycled roofing materials like rubber or tin would be a more suitable choice.

There are many options for adding crown molding to a home. Homeowners can consider visiting companies that specialize in selling reclaimed or vintage building materials. Sometimes, individuals renovating or demolishing older properties may offer others the opportunity to browse and choose from crown molding and other used building materials. For example, one might find an original hardwood floor in an old home that the owner is willing to sell. It is recommended to carefully inspect the floor and conduct a polish test on a small area to ensure it can be refinished and restored.

When selecting used building materials, it's important to take style into consideration. Vintage, traditional, and older styles can be found in abundance at places like older properties or companies that salvage and resell items. However, if you're looking for modern materials, you may have to seek out recycled building materials made from used items.

When considering purchasing items, price is an important factor to take into account. Recycled and reclaimed items generally have a higher price compared to brand new manufactured items. This is mainly because the process of reclaiming or recycling old items into new ones can be expensive. Additionally, vintage or antique used building materials are often unique and one-of-a-kind, making them more valuable due to the original materials and detailed workmanship that may not be found in a standard building materials store.

1.3.3. Construction Technology

Civil engineering focuses primarily on construction, an activity that is constantly evolving and improving. With land costs on the rise, there is a growing need for tall structures in urban areas, while in rural areas the focus is on building low-cost constructions. It is essential to harness technology that utilizes locally available materials in order to meet these demands.

Construction Technology is a practical construction program that offers specialized training in both residential and commercial construction. Students will gain skills in project planning, organization, problem-solving, and communication to effectively manage building projects.

Construction is the process of erecting or assembling large structures. Although it is often used interchangeably with building, it typically refers to major projects like buildings, ships, and aircraft, as well as public works including roads, dams, and bridges.

1.3.4. Structural Engineering

The load on a structure is transferred to the ground, causing internal stresses in various components. For example, in a building, a slab transfers its load to the ground through beams, columns, and footings. Analyzing internal stresses in a structure is called Structural Analysis while determining the appropriate size of structural components is known as structure design. Structures can be made of masonry, RCC, or steel. Classical analysis improved until the mid-1960s, after which numerical methods and computer-aided design became common. Matrix Method and Finite Elements Analysis are used to analyze complex structures. Civil engineers must provide both safe and cost-effective structures, using mathematical optimization techniques to find economical solutions. The importance of earthquake-resistant structures has grown due to frequent earthquakes, leading to significant research in this field. All of these aspects are part of the field of structural engineering.

1.3.5. Geotechnical Engineering

Soil characteristics differ from location to location and can vary from one depth to another even in the same location. The properties of soil may change with the seasons due to fluctuations in moisture levels. It is important to ensure that the load from a structure is securely transferred to the soil by accurately assessing its safe bearing capacity. The field of Civil Engineering that focuses on this aspect is known as Geotechnical Engineering.

In addition to determining the safe bearing capacity for building foundations, geotechnical engineering encompasses a range of studies necessary for designing pavements, tunnels, earthen dams, canals, and earth retaining structures. This field also involves researching ground improvement techniques.

Geotechnical engineering, a division of civil engineering, focuses on the characteristics and properties of various earth materials. Experts in this field analyze the composition of soil and rock layers to understand their physical and chemical attributes. With this knowledge, they specialize in designing foundations and structures for buildings, roads, and various other projects.

Geotechnical engineering aims to create soil stabilization systems that safeguard individuals. This involves guaranteeing the structural integrity of buildings on shaky ground, as well as protecting vital infrastructure from the effects of earthquakes and landslides. Geotechnical engineers are also tasked with addressing soil-related issues in underwater environments, such as those surrounding marinas and offshore platforms.

The process of geotechnical design commences with a subsurface exploration, during which soil samples are extracted through test pits or boreholes. Geotechnical engineers subsequently analyze the soil's characteristics, such as its stability, the existence of air or rock voids, and the chemical composition of the ground. This examination enables them to assess the potential environmental repercussions of soil disturbance and devise appropriate measures to ready the area for construction.

1.3.6. Hydraulics

Water is an essential requirement for all organisms. The study of the mechanics of water and its flow properties is a crucial area in Civil Engineering, known as hydraulics.

Hydraulics is a mechanical process that utilizes the power of liquid pressure. In systems that use hydraulics, controlled movement is achieved by pressurized liquid being pumped through cylinders to move pistons. Hydraulics is an integral part of mechatronics, which involves the integration of mechanical, electronic, and software engineering in the development and production of products and processes.

Basic hydraulic systems such as aqueducts and irrigation systems rely on gravity to generate water pressure for delivery. These systems harness water's innate properties to facilitate self-delivery. In comparison, more intricate hydraulic systems utilize pumps to intensify the pressure of liquids, often oils, to propel a piston within a cylinder. Additionally, valves are utilized to regulate the flow of oil within the system.

Hydraulic systems function similarly to pneumatic systems by using fluids, but they use liquids instead of gases. Unlike pneumatics, hydraulics can handle higher pressures, up to 10,000 psi compared to about 100 psi in pneumatic systems. This is because liquids are incompressible, allowing for more efficient power transfer without energy loss from compression, except when air enters the hydraulic lines. Hydraulic fluids have the ability to lubricate, cool, and transmit power in hydraulic systems. In contrast, pneumatic systems require separate oil lubrication, which can be messy when combined with air pressure. Pneumatics, while less complex, are easier to control and maintain, as well as safer due to their lower risk of fire. Additionally, the compressibility of the gas in pneumatic systems can absorb shock and protect the mechanism, making them more reliable in certain situations.

1.3.7. Water Resources and Irrigation Engineering

Water needs to be provided for agricultural fields and drinking purposes. Therefore, it is important to locate appropriate water sources and store the water. The process of identifying, planning, and constructing water retaining structures such as tanks and dams, and transporting stored water to fields is referred to as water resources and irrigation engineering.

Water resources are essential sources of water that are beneficial to humans. They are necessary for the existence of life as we know it. Water is used for a variety of purposes such as agriculture, industry, household activities, recreation, and environmental conservation. Fresh water is required for almost all of these human activities.

Irrigation involves providing water to dry land using artificial methods like pipes, hoses, or ditches. This land typically has crops, grass, or vegetation that don't receive sufficient water from regular rainfall or natural sources. The need for irrigation may arise during dry seasons with low rainfall, or due to the land's inherent inability to retain enough water for fertility. Water for irrigation can be sourced from nearby lakes, reservoirs, rivers, or wells.

1.3.8. Transportation Engineering

Transportation infrastructure is a crucial requirement, with civil engineers responsible for designing and creating high-quality and cost-effective roads. This involves planning base courses, selecting suitable surface materials, constructing cross-drainage structures, road intersections, culverts, bridges, tunnels, and more. Railways are also an important mode of long-distance transportation, with transportation engineering encompassing the design, construction, and maintenance of railway lines and signal systems.

Additionally, the development of airports and harbors is essential for efficient transportation. To properly plan these facilities, traffic surveys must be conducted. This process includes carrying out traffic surveys, as well as designing, constructing, and maintaining roads, bridges, railways, harbors, and airports, all falling under the umbrella of transportation engineering.

Transportation engineering is a specialized field within civil engineering that specifically deals with the transportation of people and goods. Professionals in this field work towards making transportation as efficient, convenient, and environmentally friendly as possible. Given the significance of transportation in modern society, transportation engineering is a crucial aspect of civil engineering. It includes various subsets such as aerospace, waterway, air transportation, urban transportation, and highway engineering.

Transportation engineering plays a significant role in the pre-planning stage of a project before it even begins. During this phase, the structural engineer conducts a thorough analysis of the current transportation system and assesses the potential impact of the planned project on the surrounding area. The engineer must compile an inventory or database of vital information, including population size, land usage, existing travel patterns, community expectations, budget constraints, and relevant laws and regulations.

Highway engineering focuses on the design, building, and maintenance of highways and local roads. Highway engineers may be tasked with improving a busy and hazardous intersection by implementing new signage, construction, traffic flow, or lighting. They must consider various factors such as driving habits, types of vehicles using the road, and the current road conditions to determine the best course of action for enhancing the intersection.

1.4. CAREER OPPORTUNITIES IN CIVIL ENGINEERING

Civil engineers play a vital role in shaping communities through the design and construction of various structures like highways, airports, and dams. To equip individuals for careers in this field, civil engineering programs provide training in construction techniques, advanced math concepts, and research. Civil engineers are responsible for planning and developing structures to meet the public's needs in sectors like transportation, coastal construction, and environmental engineering. They may work on projects that involve sustainable buildings, safe roads and bridges, and aesthetically pleasing yet resilient structures.

Choosing a career in civil engineering enables individuals to make a positive difference in their communities, all while pursuing their specific interests. To excel in this field, civil engineers must possess keen attention to detail, advanced mathematical abilities, strong organizational skills, and a knack for problem-solving. Effective team leadership and communication skills are also crucial, as civil engineers are required to collaborate with clients and supervise construction teams.

1.4.1. Civil Engineering Career Outlook

The salaries for professionals with a civil engineering degree can differ based on their location and the industry they work in. The Bureau of Labor Statistics (BLS) reports that the median income for civil engineers is $87,060 nationwide. However, those working in the oil and gas extraction industry have an average salary of $129,870. In California, civil engineers earn an annual average of $110,480.

These factors can also influence the availability of career opportunities. In particular, the fields of architecture, engineering, and related services provide the most jobs for professionals in these industries. California, Texas, and New York have the highest number of civil engineering jobs. Additionally, experience plays a significant role in income, as shown in the table below. Salaries for civil, structural, and environmental engineers increase with years of experience. Professionals with at least 10 years of experience often earn $30,000 more annually than entry-level workers (Table 1.1).

Table 1.1. Median Annual Salary for Civil Engineering Careers

Job Title	Entry-Level (0–12 months)	Early Career (1–4 Years)	Midcareer (5–9 Years)	Experienced (10–19 Years)
Civil Engineer	$63,000	$69,000	$83,000	$98,000
Structural Engineer	$67,000	$72,000	$85,000	$99,000
Construction Manager	$62,000	$71,000	$83,000	$92,000

Source: PayScale, https://www.bestcolleges.com/careers/science-and-engineering/civil-engineering/.

1.4.2. Skills Gained With a Civil Engineering Degree

Civil engineering programs offer students a comprehensive understanding of both mathematics and physics and ways to implement them into construction projects. Additionally, students will acquire the ability to design plans, renderings, models, and blueprints to visually represent their ideas. Furthermore, they will cultivate a range of professional and leadership skills to enable them to make informed decisions, foster collaboration, and oversee projects. These skills may encompass the following:

- **Problem-Solving:** Solving problems is the foundation of engineering, especially for civil engineers who must tackle tasks like optimizing hydroelectric power extraction from rivers or constructing earthquake-resistant buildings. Engineers must possess the skills to analyze problems, create solutions, and weigh the pros and cons of different methods in order to effectively address these challenges.
- **Project Management:** Civil engineers are responsible for supervising the activities of surveyors, planners, drafters, construction managers, civil engineering technicians, and laborers. Skillful coordination of these teams necessitates proficiency in goal-setting, work plan creation, and accountability. In addition to technical skills, civil engineers must possess a range of interpersonal skills to inspire their workers and address conflicts effectively.
- **Mathematics:** Many civil engineering programs include in-depth classes in geometry, trigonometry, calculus, and applied statistics. Students may also be required to take specific courses tailored to their chosen career path within civil engineering. For instance, a student interested in becoming a geotechnical engineer may need to study structural and soil mechanics to successfully design foundations.
- **Organization:** Civil engineers who possess strong organizational skills are able to effectively manage their tasks and deadlines. These skills enable professionals to allocate resources efficiently and ensure compliance with local, state, and federal regulations. To further develop these skills, many engineering students enroll in business administration courses and gain practical experience through internships or supervised fieldwork.

- **Communication:** Civil engineers need to be skilled communicators in order to work on projects that involve public interest. They must effectively communicate with various audiences, such as elected officials and citizen advocacy groups, and also collaborate with expert colleagues like architects and urban planners. Engineering students improve their communication abilities through classroom discussions and group projects.

1.4.3. Civil Engineering Career Paths

Below are some of the best career opportunities available for individuals with a degree in civil engineering.

- **Surveyor:** Surveyors determine property boundaries and locations for building sites by measuring and recording them. They research land records and titles to ensure data accuracy, presenting their findings to authorities. Surveyors create maps to establish water and land boundaries based on legal documents such as deeds and leases.
- **CAD Technician:** These experts utilize computer-aided design software to effectively plan projects. They may review drawings and designs, conduct research to comply with building code standards, and work with both 2-D and 3-D modeling technologies. CAD technicians often have a background in drafting and architecture.
- **Building Engineer:** These experts offer maintenance and repair services for apartment buildings, offices, and medical facilities. They may also supervise maintenance for privately owned hotels or factories. Building engineers possess in-depth knowledge of heating and cooling systems, enabling them to conduct thorough inspections and repairs. They are responsible for ensuring that structures meet local fire and building codes as well as federal health and disability regulations.
- **Water Hygiene Engineer:** Water hygiene engineers are responsible for overseeing operations and treatment protocols at water or wastewater treatment plants. They conduct equipment inspections, monitor meters and gauges, and evaluate data to determine the necessary actions. They also prepare reports of their findings and offer recommendations for improvements.
- **Urban Planner:** Urban planners collaborate with public officials, developers, and local residents to strategize the organization and utilization of land. They collect and analyze data, market research, and economic trends to inform decisions on potential site developments. Urban planners stay informed on zoning regulations and building codes to ensure projects meet legal requirements. They also communicate and present proposals to communities, planning authorities, and committees.
- **Fire Engineer:** The fire engineer ensures the implementation of fire safety and protection measures. They collaborate with authorities and clients to adhere to fire safety regulations and codes, conducting research to determine

the appropriate placement and installation of fire suppression and safety systems. Additionally, engineers conduct equipment inspections to identify any faults or damages.

- **Transport Planner:** Transport planners, whether in the public or private sector, are responsible for creating transportation strategies that cater to the needs of all types of transport users, such as cyclists and pedestrians. In their development strategies, planners must take into account environmental impact, efficiency, and safety concerns. They also analyze data on traffic flows and congestion to provide insights and prepare reports on transportation issues.
- **Field Service Engineer:** Field service engineers, also known as engineering technicians, are mainly tasked with maintaining equipment, scheduling repairs, and installing electrical and engineering products. They work closely with clients who may rent, lease, or own such equipment, relying on logic and expertise to troubleshoot and resolve issues.
- **Construction Manager:** Construction managers are responsible for preparing cost estimates, creating budgets, and developing work timetables. They work closely with subcontractors, architects, and engineers to ensure successful collaboration on projects. Additionally, construction managers ensure compliance with state and federal laws, as well as building and safety codes. They oversee various types of projects, including road and bridge construction, as well as public, residential, commercial, and industrial structures.
- **Environmental Engineer:** Environmental engineers utilize engineering, soil science, and biology to create solutions for environmental challenges. They assess investigative reports and design projects that address issues such as recycling, waste management, and pollution control. Engineers also interpret scientific data to make recommendations and provide support for legal actions.
- **Construction Estimator:** Construction estimators are responsible for determining the various factors that impact the cost of a project, including production time, materials, and labor. They analyze blueprints and work closely with architects and clients in order to provide precise cost estimates. Throughout the project, construction estimators continuously monitor and adjust estimates and totals, proposing strategies to minimize expenses. They are meticulous in documenting estimated and actual project costs to ensure thorough record-keeping.
- **Design Engineer:** Collaborating with research and development teams, design engineers establish design parameters and fulfill project requirements. They commonly utilize CAD or CAE software to generate models and blueprints, modifying existing plans or creating prototypes as needed to ensure projects adhere to specified guidelines.
- **Geotechnical Engineer:** Geotechnical engineers evaluate the

soil and rock conditions of a given location to assess the structural reliability and long-term viability of a planned construction or mining project. They analyze the potential hazards such as erosion, landslides, earthquakes, and other geological events, providing recommendations to ensure the safety and durability of the site.

- **Water Resources Engineer:** Water resources engineers are responsible for designing innovative systems and equipment to effectively manage human water resources. They may develop water treatment facilities, underground wells, and utilize natural springs to ensure citizens have access to clean, uncontaminated water for various purposes. When creating or updating water systems, these engineers carefully consider factors such as costs, government regulations, and other relevant aspects to provide sustainable solutions.
- **Structural Engineer:** Structural engineers work closely with architects and construction managers as consultants to develop layouts for buildings, bridges, tunnels, and other infrastructure projects. They analyze environmental forces, such as snow and wind, to determine load requirements, and possess expertise in various construction materials like steel, concrete, masonry, and wood. Based on their calculations, structural engineers advise on the best material, structure design, and support systems for optimal project outcomes.
- **Consulting Engineer:** When buildings or structures need to improve performance, consulting engineers are brought in to help. They assist with all aspects of the design process, including analysis, construction details, and future structural support. Consulting engineers also offer guidance on potential design flaws, maintenance costs, and operational considerations. By working with different construction teams, these professionals help develop creative and efficient design strategies.
- **Nuclear Engineer:** Nuclear engineers are responsible for designing and creating nuclear equipment, including reactor cores and radiation shielding. They oversee the operation and upkeep of nuclear power plants to ensure they meet safety regulations. Additionally, nuclear engineers may create procedures for plant operations and the safe disposal of nuclear waste. In the event of nuclear accidents, engineers analyze the situation to suggest improvements for future safety measures and prevention strategies.
- **Planning Engineer:** Planning engineers are responsible for developing project plans and selecting appropriate engineering methods to meet project requirements. They establish deadlines, track progress through analysis and reporting, and collaborate with team members to allocate tasks and meet goals. By using historical and current data, planners formulate strategies for upcoming projects.
- **Engineering Project Manager:** Engineering project managers are responsible for supervising

all individuals involved in an engineering project, such as design and planning engineers, construction contractors, subcontractors, architects, and bureaucrats. They assign tasks, track deadlines and budget limits, and generate reports and documentation. Their goal is to create project strategies that comply with regulations and fulfill commitments.

- **City Engineer:** City engineers work for city governments to construct new structures, as well as repair or update existing infrastructure. They are responsible for designing roads, bridges, and other city infrastructure within set schedules and budget constraints. City engineers take into account environmental impact, health and safety concerns, and federal guidelines in their planning. They may also collaborate with city officials and the community to ensure that projects meet the needs of the population while upholding high-quality standards.

CASE STUDY

O2I PROVIDED CONSTRUCTION DRAWINGS FOR A TOP MANUFACTURING FIRM

The Client

Based out of Cleveland, Ohio, USA, the client has over 63 manufacturing locations and has operations & joint ventures in 23 countries; and reaches out to 160 countries with a network of distributors as well as sales offices. With a history of 123 years of rich history behind it, the client is a global manufacturer of the highest quality welding, joining, and cutting equipment.

As a market leader, the client caters to heavy fabrication, Pipeline, automotive/ transportation, maintenance & repair, etc. Known as a high-quality product manufacturer, the client works with the latest technologies and offers the best to its growing list of customers.

Requirement & Challenges

The client required 3D modeling and 2D drafting for a few of their new components and was looking for an experienced engineering service provider with expertise in PTC Creo 4.0. Since the components were new, those needed to be created with the minutest details in mind for the easy manufacturing process and to omit cost manufacturing error.

The final designs needed to incorporate the initial suggestions made by the manufacturing team. The project also required a team with an understanding of the standards and guidelines to design the component that can qualify in every safety and quality check.

Solution

Outsource2india has over a decade of experience in providing engineering support services like 3D part modeling, 2D drafting, 3D rendering, and CNC services, among a hoard of other services. For this project, the team consisted of four mechanical engineers and a quality analyst to ensure the models met all the specifications noted by the client. The team underwent training for three days and understood the nitty-gritty of the component to accommodate the changes as marked in the pdf files received from the client. The team also had daily review meetings to keep the client aware of all the changes that were made to the drawings. The team provided the 3D modeling for the two components in 30 days. The client was deeply pleased with the effort and workmanship the team offered. The output file was provided in PTC Creo and PDFs files.

The Outsource2india Advantage

Outsource2india is one of the top companies that offer mechanical engineering solutions to help analyzing, designing, manufacturing as well as maintenance of mechanical systems. O2I has helped hundreds of clients in the past 21 years.

CLASS ACTIVITY

THE STRUCTURE OF COMPOSITE MATERIALS

Materials

The macro/micro assessment of the structure of composite engineering materials.

- **Equipment**: Optical microscope, or hand lens with about 10 X power
- **Specimens**: Small examples of a number of different composite materials from the following categories;
    - Timber – hardwood, softwood, plywood, compressed fiberboard.
    - Cement, concrete
    - Asphalt
    - Laminates – glass
- **Process**: To examine the surface of each specimen with the naked eye, and appropriate magnifying equipment, to determine the physical structure of each sample. Surfaces showing differing features in color or texture should be photographed or sketched. Components of the specimen should be identified and structure identified, where it can be determined and their physical hardness should be quantified when compared with other features of the material.

Some components of materials can be elements, compounds, or amorphous in structure. Crystals in a structure should be identified and noted.

Data recorded from this process should be examined and research taken on to identify the particular composites and their use.

SUMMARY

- Civil engineering deals with the built environment and can be dated to the first time someone placed a roof over his or her head or laid a tree trunk across a river to make it easier to get across.
- The main scope of civil engineering is planning, designing, estimating, supervising, managing, and maintenance of structures like buildings, roads, bridges, dams, etc.
- Throughout ancient and medieval history most architectural design and construction was carried out by artisans, such as stonemasons and carpenters, rising to the role of master builder. Knowledge was retained in guilds and seldom supplanted by advances. Structures, roads, and infrastructure that existed were repetitive, and increases in scale were incremental.
- The functions of the civil engineer can be divided into three categories: those performed before construction (feasibility studies, site investigations, and design), those performed during construction (dealing with clients, consulting engineers, and contractors), and those performed after construction (maintenance and research).
- The most important object of Civil Engineering is to improve the means of production and traffic in states, both for external and internal trade. It is applied in the construction and management of roads, bridges, railroads, aqueducts, canals, river navigation, docks, and storehouses, for the convenience of internal intercourse and exchange; and in the construction of ports, harbors, moles, breakwaters, and lighthouses; and the navigation by artificial power for the purposes of commerce.
- Civil engineering careers, then, allow people to positively impact their communities while focusing on areas that interest them. To succeed, civil engineers should be detail-oriented individuals with strong mathematical, organizational, and problem-solving skills. They should also excel at team leadership and communication since they must discuss plans with clients and guide workers on building processes.
- Civil engineering programs provide students with a deep understanding of math and physics and how to apply them to building projects. Students will also learn to create plans, renderings, models, and blueprints to help visualize their designs.

REVIEW QUESTIONS

1. What is civil engineering and its scope?
2. What is the function of civil engineering?
3. How does civil engineering impact society?
4. What is the opportunity of civil engineering?
5. Which field of civil engineering is most in demand?

REFERENCES

1. Barbosa, H. J. C., & Bernardino, H. S. (2010). Genetic programming in civil, structural and environmental engineering. *Computational Technology Reviews, 4*, 115–145. http://dx.doi.org/10.4203/ctr.4.5.
2. Cardoso Teixeira, J. M. (2005). Construction site environmental impact in civil engineering education. *European Journal of Engineering Education, 30*(1), 51–58. http://dx.doi.org/10.1080/03043790512331313831.
3. Chong, K. P. (2005). Nanotechnology in civil engineering. *Advances in Structural Engineering, 8*(4), 325–331. http://dx.doi.org/10.1260/136943305774353151.
4. Coleman, S. (2000). Hydraulics in civil and environmental engineering. *Journal of Hydraulic Engineering, 126*(9), 724–725. http://dx.doi.org/10.1061/(ASCE)0733-9429(2000)126:9(724).
5. Diab, Y. G. (2002). The environmental quality of civil engineering infrastructures. *IABSE Symposium Report, 86*(17), 1–7. http://dx.doi.org/10.2749/222137802796335631.
6. Easa, S. M., Li, S., & Shi, Y. (1998). GIS technology for civil engineering education. *Journal of Professional Issues in Engineering Education and Practice, 124*(2), 40–47. http://dx.doi.org/10.1061/(ASCE)1052-3928(1998)124:2(40).
7. Findik, F., & Findik, F. (2021). Civil engineering materials. *Heritage and Sustainable Development, 3*(2), 154–172. http://dx.doi.org/10.37868/hsd.v3i2.74.
8. Grigg, N. S., Criswell, M. E., Fontane, D. G., & Siller, T. J. (2005). Information technology in civil engineering curriculum. *Journal of Professional Issues in Engineering Education and Practice, 131*(1), 26–31. http://dx.doi.org/10.1061/(ASCE)1052-3928(2005)131:1(26).
9. Gulliver, J. S. (2020). Climate change adaptation: Civil and environmental engineering challenge. *Journal of Hydraulic Engineering, 146*(12), 01820001. http://dx.doi.org/10.1061/(ASCE)HY.1943-7900.0001850.
10. Highter, W. H., Johnson, J. H., Phang, M. K., & Watwood, V. B. (1993). Bachelor of civil technology and civil engineering programs. *Journal of Professional Issues in Engineering Education and Practice, 119*(2), 118–125. http://dx.doi.org/10.1061/(ASCE)1052-3928(1993)119:2(118).
11. Janowski, A., Szulwic, J., & Tysi⌧c, P. (2021). Spatial modelling in environmental analysis and civil engineering. *Applied Sciences, 11*(9), 3945. http://dx.doi.org/10.3390/app11093945.
12. Mays, L. (2010). *Ancient Water Technologies*. Springer. ISBN 978-90-481-8631-0
13. Oakes, W. C., Leone, L. L., & Gunn, C. J. (2001). *Engineering Your Future*. Great Lakes Press. ISBN 978-1-881018-57-5
14. Pahl, P. J. (2002). Information technology for civil engineering. *IABSE Symposium Report, 86*(20), 7–14. http://dx.doi.org/10.2749/222137802796335154.
15. Petrella, A., & Notarnicola, M. (2022). Recycled materials in civil and environmental engineering. *Materials, 15*(11), 3955. http://dx.doi.org/10.3390/ma15113955.

16. Shamim, A., Ahmad, S., Khitab, A., Anwar, W., Khushnood, R. A., & Usman, M. (2018). Applications of nano technology in civil engineering. *International Journal of Strategic Engineering, 1*(1), 48–64. http://dx.doi.org/10.4018/IJOSE.2018010104.
17. Shankar, B. V. (2017). Nano technology in civil engineering. *International Journal for Research in Applied Science and Engineering Technology, V*(VIII), 2192–2194. http://dx.doi.org/10.22214/IJRASET.2017.8313.
18. She, Z. (2013). Information technology in civil engineering management. *International Journal on Advances in Information Sciences and Service Sciences,* 5(2), 632–641. http://dx.doi.org/10.4156/AISS.vol5.issue2.80.
19. Topping, B. H. V., & Tsompanakis, Y. (2011). *Civil and Structural Engineering Ccomputational Technology*. Stirlingshire, UK: Saxe-Coburg Publications. http://dx.doi.org/10.4203/csets.28.
20. Xu, H. (2014). Greening of civil engineering materials. *Applied Mechanics and Materials, 584–586*, 1768–1770. http://dx.doi.org/10.4028/www.scientific.net/AMM.584-586.1768.

CHAPTER 2

Civil Engineering Materials

LEARNING OBJECTIVES

After studying this chapter, you will be able to:

- Learn about the cementitious materials;
- Discuss on aggregates;
- Evaluate the admixtures for concrete;
- Know about mortars and concretes; and
- Give an overview of bricks.

KEY TERMS FROM THIS CHAPTER

Carpentry	Coal powder
Durability	Gypsum cements
Hydroxides	Insulation
Natural materials	Plumbing
Polymers	Roofing

INTRODUCTION

Engineering structures are built using a variety of materials, collectively known as engineering or building materials. Civil engineers must be familiar with the properties of these materials in order to ensure the structural integrity of their projects. Factors such as water resistance, strength, durability, temperature resistance, appearance, and permeability must be carefully considered when selecting the most suitable material for a specific construction purpose.

Construction materials are any substances utilized for building purposes. Various natural materials like clay, rocks, sand, wood, twigs, and leaves have historically been used in construction. In addition to natural materials, there are also man-made products, some more synthetic than others. The production of these materials is a well-established industry in many countries and their use is typically divided into specialized trades such as carpentry, insulation, plumbing, and roofing.

2.1. CEMENTITIOUS MATERIALS

Cementitious materials are various substances that are combined with water or another liquid to create a cement paste that can be shaped while still pliable but will harden into a solid form. Mixing sand with the paste produces mortar while adding a mixture of coarse and fine aggregate creates concrete.

2.1.1. Types of Cementitious Materials

There are a plethora of cement types and classification schemes. The chemical component that causes the cement to set or harden is one of the most basic ways to classify materials. Based on this, the most significant category of contemporary cement is made up of silicate and aluminate cement, in which calcium silicates and aluminates serve as the setting agents. This category includes natural, aluminous, and Portland cement.

Limes were once often utilized as the only cementitious material; however, their sluggish setting and hardening are incompatible with modern needs. Limes' hardening is attributed to the conversion of hydroxides to carbonates. As a result, their main purpose now is to give mortars and stucco more durability and to soften otherwise harsh cement. The use of limes is beneficial in that their slow setting promotes healing and the recementing of hairline cracks.

A different type of cement is made from calcined gypsum and its derivatives. Gypsum cements are commonly used for indoor plasterwork, as well as for making boards and blocks. However, the solubility of gypsum limits its use in construction projects in areas with high humidity levels. Oxychloride cements, on the other hand, have unique properties that make them a specialty choice. While they are more expensive than traditional cements, they excel in certain applications such as creating spark proof floors.

Masonry cement, mortar cement, and other cementitious materials like polymers, fly ash, and silica fume are commonly used in construction due to their convenience and cost-effectiveness. Polymers, which are long-chain plastics, can enhance the qualities of concrete compared to traditional materials. Silica fume, a byproduct of electric-arc furnaces, reacts with limes in concrete to create a cementitious material with significantly smaller particles than traditional cement.

2.1.2. Portland Cements

The most widely used modern cement is Portland cement, which is created by carefully combining particular raw components to create a final product that satisfies ASTM C150 specifications for one of eight different cement kinds. The raw ingredients are composed of two minor compounds, gypsum ($CaSO_4 \cdot 2H_2O$) and magnesia (MgO), and four important compounds, namely lime (CaO), iron (Fe_2O_3), silica (SiO_2), and alumina (Al_2O_3). Typically, shale, marl, calcite, or limestone are the sources of calcareous (CaO) minerals. The sources of argillaceous (SiO_2 and Al_2O_3) minerals are sand, clay, and shale. The location of the manufacturing facility and the availability of raw resources determine the materials utilized to make any particular cement. Numerous industrial by-products can be used to make Portland cement.

Cement is made by first mining the basic ingredients, grinding them into a powder, and then combining them in specific ratios. Burning coal powder, gas, or oil heats the upper end of a rotating kiln to 2,600–30,000F. This is where the blend is fed. Rewarmers and the use of alternative fuel sources, like discarded tires, are utilized to lower the cost of fuel because the manufacture of cement is an energy-intensive operation. Steel belts supply the iron component, and burning tires give the heat needed to create the clinker. The basic components are chemically fused together into hard nodules known as cement clinker by exposure to elevated temperature. After cooling, the clinker is passed through a ball mill and ground to a fineness where essentially all of it will pass a No. 200 sieve (75 m). Gypsum is added in small quantities during the grinding process to control temperature and regulate the setting time of the cement. The final product, known as Portland cement, is typically sold in 94 lb bags.

The most popular application for Portland cement is in concrete, a complicated mixture of cement, aggregates, water, and sometimes chemical and mineral admixtures. Portland cement is rarely used on its own; examples of applications are fine grout and a cement slurry for well-hole filling. Examining the relationship between the different Portland cement characteristics and their possible impact on the final concrete is crucial. Portland cement concrete's strength and durability make it a popular choice for structural applications. Strength is a generally directly proportional indicator of overall durability that is simple to assess. Although it is difficult to quantify, specific durability can be defined by adjusting the aggregate's characteristics and cement's chemistry.

2.1.3. Aluminous Cements

The process involves blending bauxite and limestone to create a fine powder, resulting in cement with rapid-hardening properties and high early-age strength. Strength tests on 4-inch cubes of 1:2:4 concrete show the varying strengths between normal Portland, high-early-strength Portland, and aluminous cement.

Due to the significant heat released quickly by aluminous cement during hydration, caution should be taken to avoid using the cement in areas where this heat cannot be properly dispersed. It is generally recommended to avoid constructing aluminous-cement concretes in layers exceeding 12 inches, as the resulting temperature increase may severely compromise the concrete's strength.

Compared to Portland cement, aluminous cement is far more resistant to the effects of sulfate waters. Additionally, compared to silicate cement, they seem to be far more resistant to assault from water containing aggressive carbon dioxide or weak mineral acids. Their main application is in concretes where their extremely high early strength or sulfate resistance can be utilized, and where the additional cement cost is not a significant consideration. Aluminous cements are also used in conjunction with firebrick to produce refractory concrete. When exposed to high temperatures, the hydration products dehydrate, leading to the formation of a ceramic bond with the aggregates.

2.1.4. Natural Cement

A naturally occurring mixture of calcareous and argillaceous materials is calcined at a temperature lower than that of sintering to create natural cement. According to ASTM C10, "Specification for Natural Cement," the temperature must not rise above what is required to extinguish carbonic acid gas. Natural cements differ greatly in terms of composition and qualities since they are manufactured from naturally occurring components without any special attempt to modify the composition. Certain natural cements have qualities that are nearly identical to those of Portland cement, while others are significantly weaker. The main applications for natural cement are as an additive in Portland cement concretes and as masonry mortars.

2.1.5. Limes

They are primarily composed of calcium oxide (CaO) and are found naturally in limestone, marble, chalk, coral, and shells. They are mainly used in mortars for construction purposes.

2.1.5.1. Hydraulic Limes

In order to provide enough free lime to allow hydration and enough unhydrated calcium silicates to give the dry powder its hydraulic properties, they are created by calcining limestone that contains silica and alumina to a temperature just below the point of incipient fusion. Hydraulic limes are relatively weak since they contain a high amount of lime and little silicate. Masonry mortars are where they are primarily used. A hydraulic lime that contains more than 10% silica will sink to the bottom.

2.1.5.1.1. Quicklimes

When limestone is heated to a temperature above 1700F, it loses its carbon dioxide content and becomes quicklime. Quicklime is primarily composed of calcium and magnesium oxides with impurities like silica, iron, and aluminum oxides, which are typically kept below 5%. If impurities exceed 10%, the product may be classified as hydraulic lime.

There are two recognized classes of quicklime: dolomitic and high-calcium. Magnesium oxide is often found in less than 5% of high-calcium quicklime. In a typical dolomitic quicklime, magnesium oxide content ranges from 35 to 40%. Magnesian limes are a small kind of quicklimes that have a magnesium oxide content of 5 to 35%. The capacity of quicklime to slake with water is one of its best qualities. A chemical reaction occurs when two to three times the weight of water is added to quicklime. When calcium oxide and water mix, calcium hydroxide is created, and enough heat is generated to bring the mixture to a boil.

The end result is a finely split suspension of calcium hydroxide in water, together with magnesium hydroxide or oxide in the case of dolomitic lime. When the semifluid mass cools, it solidifies into a putty that is manageable to handle with a shovel or a hod. Construction uses this slaked quicklime putty after it has cooled and, ideally, been screened. It is usually necessary to completely slake quicklime. The putty yield is typically between 70 and 100 feet per ton of quicklime, though it can vary based on the kind of quicklime, how much of it burns, and the slaking circumstances.

The main application of putty is in masonry mortars, where its high plasticity greatly improves the workability of the mortar. It is sometimes added to concrete as an admixture to enhance workability. In certain areas, putty is also used as a finishing coat for plaster, allowing for maximum utilization of its plasticity.

2.1.5.1.2. Mason's Hydrated Lime

Hydrated limes are created by adding a small amount of water to quicklime, resulting in a fine powder once the hydration process is complete. The product is then sorted to remove any large particles and packaged in 50-lb sacks. This type of hydrated lime is always produced in factories, while quicklime putty is typically made on-site. The hydrated limes used by masons are suitable for various construction applications, including mortars, base-coat plasters, and concrete. They maintain the composition of quicklime, with high-calcium hydrates containing mainly calcium hydroxide, and normal dolomitic hydrates containing calcium hydroxide and magnesium oxide.

The plasticity of mortars made from standard mason's hydrated lime (Type N) is satisfactory. It is superior to most cements but not as high as mortars made with an equal amount of slaked putty.

When dolomitic quicklime is hydrated normally at atmospheric pressure, only the calcium fraction is hydrated, leaving the magnesium oxide portion mostly unaltered chemically. Magnesium oxide is changed into magnesium hydroxide when pressure is applied to dolomitic quicklime. This leads to the so-called special hydrates (Type S), which exhibit a high degree of flexibility as soon as they are wetted with water, in addition to having their magnesia contents nearly entirely hydrated. Compared to mortars created from Type N hydrates, those made from Type S hydrates are more workable. Actually, Type S hydrates are almost as practical as quicklime putty that has been slaked. The user of this type of hydrate may therefore have the convenience of a bagged product and a high degree of workability without having the trouble and possible hazard of slaking quicklime.

2.1.5.1.3. Finishing Hydrated Limes

Finishing hydrated limes are ideal for use as a plaster finishing layer. They are distinguished by their high level of brightness and flexibility. Almost all finished hydrated limes are made in the Toledo region of Ohio from dolomitic limestone. The typical hydrate is made up of calcium hydroxide and magnesium oxide. When initially wet, it is no more plastic than Type N Mason's hydrates. It differs from the latter in that, after soaking overnight, the finishing hydrated lime acquires a very high degree of flexibility, whilst the mason's hydrate exhibits relatively little development in plasticity with soaking.

2.1.6. Low-Temperature Gypsum Derivatives

When gypsum rock ($CaSO_4$ $2H_2O$) is heated to a relatively low temperature, about 130°C, three-fourths of the water of crystallization is driven off. The finished product is called by a variety of names, including hemihydrate, calcined gypsum, and first settle stucco. Plaster of Paris is its more well-known moniker. It is often a fine, white powder. While it will set in water, it will lose strength and eventually dissolve if exposed to water for an extended period of time.

Plaster of Paris, with or without retarded set, is used as a molding or gauging plaster. Moulding plaster is used to create beautiful plaster items. The gauging plaster is used to finish hydrated lime and provide a smooth white-coat finish on plaster walls. Manufacturers utilize unretarded plaster of Paris to create gypsum blocks, tiles, and gypsum board (wallboard, lath, backerboard, coreboard, and so on).

When the plaster is combined with sisal fiber and a retardant, it is sold under the name of hardwall plaster or cement plaster, though the latter name is deceptive as it does not include Portland cement. Hardwall plaster is commonly used as a base-coat plaster when mixed with water and two to three parts of sand by weight. In some instances, wood fiber is used instead of sand to create a "wood-fibered" plaster.

Combining hardwall plaster with the appropriate aggregate produces unique effects. A lightweight plaster is created by mixing perlite or vermiculite material.

Gypsum plasters, in general, have a firm set, reach full strength when dry, exhibit no aberrant volume variations, and have good fire-resistant properties. They are not, however, well suited for usage in continuous damp or intermittent wet situations.

2.1.7. Oxychloride Cement

Magnesium oxychloride cement, also known as Sorel cement, is formed by mixing lightly calcined magnesium oxide with a solution of magnesium chloride. It is very effective in flooring compositions where it is combined with colored pebbles. Oxychloride cement flooring is spark-proof and more robust than concrete floors.

Oxychloride cement has exceptionally strong bonding power and, as a result, it may be utilized with larger amounts of aggregate than Portland cement. Oxychloride cement also binds effectively with wood and is used to create partition blocks or tiles with wood shavings or sawdust as the aggregate. It is generally water resistant, but should not be worn in continuously moist environments.

2.1.8. Masonry Cement

Many commercial masonry cement are made from a combination of Portland cement and pulverized limestone, with limestone content ranging from 50 to 60%. They are marketed in bags weighing 70 to 80 pounds, with each bag supposedly holding one cubic foot. The price per bag is often lower than that of Portland cement, but the cost per ton is greater due to the use of a lighter bag.

Because there are no restrictions on chemical composition or physical requirements, masonry cement standards are fairly flexible. Some manufacturers make significant changes to the composition based on competition, weather conditions, or material availability. The qualities of the mortars produced might vary greatly.

2.1.9. Fly Ashes

Fly ash that meets the standards of ASTM C618, "Specification for Fly Ash and Raw or Calcined Natural Pozzolan for Use as a Mineral additive in Portland Cement Concrete," is commonly used as a cementitious material and additive.

Natural pozzolans are formed from diatomaceous earths, opaline cherts, and shales, among other things. While they have the same ASTM classification as fly ash, they are not as widely available and hence do not elicit the same amount of interest or investigation.

Fly ashes are formed when coal is burned, often in an electricity-generating unit. The ash that would otherwise be expelled via the chimney is caught using a variety of methods, including electrostatic precipitators.

The fly ash can be sized before it is shipped to concrete suppliers.

All fly ashes have pozzolanic characteristics, which means they may react with calcium hydroxide at normal temperatures to generate cementitious compounds. When cement is combined with water, a chemical process called hydration occurs. This reaction produces calcium silicate hydrate (CSH) and calcium hydroxide $[Ca(OH)_2]$. Fly ashes have a high amount of silicon dioxide (SiO_2). In the presence of moisture, $Ca(OH)_2$ reacts with SiO_2 to generate another CSH.

Type F ashes are pozzolanic in nature and are produced by the combustion of anthracite or bituminous coal. They have been demonstrated in studies and practice to give enhanced sulfate resistance and lower alkali-aggregate growth. The combustion of lignite or subbituminous coals produces type C fly ashes. Because of the chemical features of coal, Type C fly ashes have cementitious qualities in addition to their pozzolanic properties. Type C fly ashes may affect the durability of concretes that include them.

2.1.10. Silica Fume (Microsilica)

Silica fume, also known as microsilica, is a condensed gas formed as a byproduct of metallic silicon or ferrosilicon alloys in electric arc furnaces. (While both words are accurate, microsilica (MS) is a less ambiguous moniker. The Canadian standard CAN/CSA-A23.5-M86, "Supplementary Cementing Materials," restricts amorphous SiO_2 to a maximum of 85% and large to 10%. Many MS have more than 90% SiO_2.

MS has an average diameter of 0.1 to 0.2 m and a particle size that is approximately 1% that of Portland cement. Due to its tiny size, MS cannot be used in its raw form. Manufacturers provide it in three forms: densified, slurry (with or without water-reducing admixtures), and pelletized. Densified or slurried MS can be used in concrete.

The pelletized materials are compressed to the point where they do not break down when mixed. Due to its minuscule size, MS enhances several beneficial characteristics in concrete. It significantly enhances the long-term strength and reacts efficiently with $Ca(OH)_2$ to produce a useful material instead of waste. MS is typically incorporated in concrete projects with a design strength exceeding 12,000 psi. It offers improved sulfate resistance and reduced permeability in concrete. Furthermore, its tiny size enables MS to fill microcracks and small openings.

2.2. AGGREGATES

The term "aggregate" refers to a wide range of materials, including air-cooled blast furnace slag, gravel, boulders, cobbles, manufactured and natural lightweight aggregates, and native and manufactured sands. The sizes of the aggregates can be used to further characterize them.

2.2.1. Normal-Weight Aggregates

These typically have specific gravities between 2.0 and 3.0. They are usually distinguished by size as follows:

Boulders	Larger than 6 in
Cobbles	6 to 3 in
Coarse aggregate	3 in to No. 4 sieve
Fine aggregate	No. 4 sieve to No. 200 sieve
Mineral filler	Material passing No. 200 sieve

Normal-weight aggregates, which are used in the majority of concrete construction projects, are extracted by mining and crushing formational material or by emptying riverbeds. Ordinarily, 144 lb/ft^3 of fine and coarse aggregate are used to make concrete. Typically, crushed boulders and cobbles are used to create different sizes of coarse aggregate, manufactured sand, and mineral filler instead of being used in their raw form as mined. Water and weathering act on glacial and river deposits to produce gravel and naturally occurring sand. These materials require little processing because of their smooth, rounded surfaces and uniform particle sizes. These materials are available in two aggregate sizes: coarse and fine. All of the material in fine aggregates passes through a 3/8-inch sieve. The majority of the material is retained on the No.4 sieve in coarse aggregates.

In asphaltic concrete, mortar, or Portland-cement concrete, aggregates make up the majority of the volume percentage. The coarse and fine aggregates make up between 60 and 75% of the total volume of a Portland cement concrete mix. Aggregates make up between 75 and 85% of the mix volume in asphaltic concrete. The aggregates are therefore not inert filler materials.

The service life and durability of the material system in which the aggregate is used, such as Portland-cement concrete, asphaltic concrete, mortar, or aggregate base, are clearly impacted by each aggregate property. A coarse or fine aggregate's suitability for use in mortar or concrete is determined by a number of factors, such as gradation and the quantity of fine material that passes the No. 200 sieve, organic impurities, potential alkali reactivity, hardness, soundness, particle shape, volume stability, and resistance to freezing and thawing. Property limits are specified in ASTM C33, "Specification for Concrete Aggregates," C637, "Specification for Aggregates for Radiation-Shielding Concrete," and C330, "Specification for Lightweight Aggregates for Structural Concrete," for aggregates used in general building construction. Various trade or governmental organizations have written standards for other types of construction, like airports and highways.

2.2.1.1. Gradation of Aggregates

The distribution of aggregate sizes in a concrete mix is crucial because it affects several factors, including workability (i.e., how much work is needed to place the mix in forms), durability in place, and overall economy. Granular and fine-aggregate grading limit ranges are provided by ASTM C33. The latter are arranged in size order from Size 1 (31\2 to 11½2 in) to Size 8 (3½8 to No. 8). A different gradation for manufactured sand is specified by the National Stone Association in C33 than for fine aggregate, primarily for the No. 200 and 100 mesh sieves. With higher percentages passing through each sieve, the NSA gradation is notably finer.

The fine materials that contribute to the workability of concrete are made up of angular particles called rock fines, as opposed to silts and clays found in natural sand. For aggregate production and quality-control testing, the different gradations offer standard sizes.

They support the creation of concrete with suitable characteristics. However, using standard individual grading limits should be done with caution. With appropriately graded aggregates, acceptable or cost-effective concrete might not be possible if the number of aggregate sizes is restricted or if there is insufficient overlap between aggregate sizes. This is because the combined gradation is gap-graded. To maximize the void content of the combined aggregates, a dense or well-graded size distribution is the ideal condition. Though individual aggregates may not meet the standard limits, it is still possible to produce acceptable concrete by combining them to create a dense gradation.

2.2.1.2. Amount of Fine Material Passing the No. 200 Sieve

The content that passes the No. Clay, silt, or a combination of the two make up a 200 sieve. It raises the aggregate's water requirement. Huge quantities of substances smaller than No. 200 could also be a sign that there are clay coatings on the coarse aggregate, which would weaken the aggregate's bond with the cement matrix. ASTM C117, "Materials Finer than 75 m Sieve in Mineral Aggregates by Washing, provides a test procedure. "

2.2.1.3. Hardness

The Los Angeles Abrasion Test, ASTM C131 or C595, is used to determine coarse-aggregate hardness. In these tests, steel balls in a steel tumbler are used to impact the aggregate and break it down. Although the breakdown that results isn't directly related to the wear and tear an aggregate experiences during use, it can be empirically linked to concretes that have long service lives.

2.2.1.4. Soundness

Aggregate soundness is measured by ASTM C88, "Test Method for Soundness of Aggregates by Use of Sodium Sulfate or Magnesium Sulfate." This test measures the amount of aggregate degradation when exposed to alternating cycles of wetting and drying in a sulfate solution.

2.2.1.5. Particle Shape

The particles in natural sand and gravel are smooth and rounded. Coarse and finely crushed aggregate can be shaped like a disk, angular, flat and elongated, or rod-like. The aggregate mineralogy and the crushing machinery used produce these shapes. The amount of cement needed to give strength, the difficulty of finishing, and the effort required to pump the concrete all increase with extreme angularity and elongation. The quantity of mixing water needed is also increased by flat and elongated particles. Compared to smooth particles, angular particles have a stronger bond. By utilizing this feature, properly graded angular particles can reduce the amount of water needed to produce concrete that has the same cement content and strength as a smooth-stone mix.

2.2.1.6. Potential Alkali Reactivity

The alkalies (potassium and sodium oxide) in Portland cement may react with aggregates that contain specific silicas or carbonates. The concrete is cracked by the reaction product, or it might cause pop-outs at the surface. In a warm, humid atmosphere, the reaction is more noticeable with concrete. Because the tests that are currently available do not consistently produce results, testing for potentially reactive aggregates is challenging. There are two types of tests for aggregate potential alkali reactivity: chemical and physical, and pre- or post-concrete. One is chemical, out of the three pre-concrete tests. Only the first aggregate screening should be conducted using the standard chemical test (ASTM C289). The test will produce false-positive results for potentially reactive aggregates, according to experience. The outdated mortar bar test (ASTM C227) moves slowly and might be applied too liberally. Although it is a more rigorous test, the rapid immersion mortar bar test (ASTM C1260) can yield results in two weeks. Concrete's potential alkali reactivity can be assessed visually or by applying an ultraviolet light test method with uranal acetate. Alkali silica reactivity (ASR) can be assessed by petrographic analysis of aggregates and hardened concrete. The best indicator of ASR is extensive field experience with readily available aggregate sources.

2.2.1.7. Resistance to Freezing and Thawing

Aggregate pore structure, absorption, porosity, and permeability are particularly crucial if the concrete they are used in is subjected to frequent freezing and thawing cycles. When aggregates freeze after reaching a critical saturation point, the frozen water cannot expand within them. Based on empirical evidence, the coarse aggregates, not the fine ones, are the ones causing the freeze-thaw deterioration. Concrete performance is measured using a technique outlined in ASTM C666, "Test Method for Resistance of Concrete to Rapid Freezing and Thawing," by varying the weight, decreasing the dynamic modulus of elasticity, and lengthening the sample.

2.2.1.8. Impurities in Aggregates

Organic impurities in the aggregates, particularly the sand, can lead to irregular hardening rates and setting times. The ASTM C40 "Test Method for Organic Impurities in Fine Aggregates for Concrete" provides a method for determining the presence of these impurities. Soft particles, chert, clay lumps and other friable particles, coal, lignite, or other lightweight materials in the aggregates can all result in pop-outs and decreased durability. Concrete surfaces exposed to the elements may also become stained by coal and lignite.

2.2.1.9. Volume Stability

Volume stability refers to the susceptibility of aggregate to expansion when heated or to cyclic expansions and contractions when saturated and dried. Aggregates that are susceptible to volume change due to moisture should be avoided.

2.2.2. Heavyweight and Lightweight Aggregates

Heavyweight aggregates include magnetite, with a specific gravity d of 4.3; barite d =4.2; limonite d = 3.8; ferrophosphorus d = 6.3; and steel shot or punchings d =7.6. Such heavyweight aggregates may be used instead of gravel or crushed stone to produce a dense concrete; for example, for shielding of nuclear reactors as specified in ASTM C637.

2.2.3. Lightweight Aggregates

These fall into two groups: nonstructural and structural. The ASTM C330 and C331 specifications define the structural lightweight aggregates. They can be natural (pumice and scoria) or artificial (expanded clay, shale, slate, or blast furnace slag). Concretes made with these aggregates typically have strengths between 3000 and 4000 psi, though they can be made with stronger strengths, which are covered in Art. 4Point 17. Normal strength lightweight concrete (less than 5000 psi) has an air-dry unit weight between 100 and 115 pcf. The unit weights of high-performance lightweight concrete range from 120 pcf. Vermiculite and perlite are the most widely used nonstructural lightweight aggregates (ASTM C332), though pumice and scoria are also suitable. These substances are utilized in non-structural floor toppings and soundproofing concretes. Concrete made with lightweight aggregates has low thermal conductivities, which means it provides good fire protection. Extreme heat causes the moisture in the concrete to quickly transform from liquid to steam with a volume up to fifteen times greater. Pressure-relieving zones are produced by lightweight aggregates' numerous, large-sized pores.

2.3. ADMIXTURES FOR CONCRETE

Anything added to a concrete mix to change its properties aside from aggregates, water, and Portland cement is called an additive. Chemical admixtures (ASTM C494 and C260), mineral admixtures like fly ash (C618) and silica fume, corrosion inhibitors, colors, fibers, and miscellaneous (pumping aids, damp proofing, gas-forming, permeability-reducing agents) are all included in this definition.

2.3.1. Chemical and Mineral Admixtures

Admixtures of chemicals are typically used in concrete as accelerators, set retarders, water reducers, or a combination of these. The following classification is found in ASTM C494, "Standard Specification for Chemical Admixtures for Concrete":

Type	Property
A	Water reducer
B	Set retarder
C	Set accelerator
D	Water reducer and set retarder
E	Water reducer and set accelerator
F	High-range water reducer
G	High-range water reducer and set retarder

High-range admixtures reduce the amount of water needed to produce concrete of a specific consistency by 12% or more.

2.3.1.1. Water-Reducing Admixtures

By forming a monomolecular layer of admixture at the cement-water interface through a chemical reaction with early hydration products, these reduce the amount of water needed for a concrete mix. By isolating individual cement particles, this layer lowers the energy needed to make the mixture flow. More cement particles are exposed for hydration as a result of the mix being "lubricated." Reducing the amount of mixing water without sacrificing mix slump is made possible by the Type A admixture. This admixture permits the cement content to be reduced without sacrificing strength, or at a consistent water-to-cement ratio. A slump of the mix will increase and strength will also increase if the water content is not decreased because more cement surface area will be exposed for hydration. The effects of Type D and E admixtures are comparable. One should normally anticipate a 5–10% decrease in mixing water. When high-workability concrete is required, type F and G admixtures are used. An admixture-free concrete usually has a 2 to 3-inch slump. Without segregating the mix components, the slump following the addition of the admixture may be between 8 and 10 inches. Particularly helpful for mixes with a low water-to-cement ratio are these admixtures. Their reduction of 12 to 30% in water permits a reduction of cementitious material in proportion.

Typically, hydroxylated carboxylic acids and their salts, lignosulfonic acids, and their salts, or polymers of derivatives of melamines, naphthalenes, or sulfonated hydrocarbons are used to make the water-reducing admixtures. It is important to thoroughly assess and test the admixture combinations used in concrete mixes to make sure the intended qualities are realized. For instance, a retarding admixture might speed up the set, depending on the cement's chemistry and admixture dosage. Also take note that if the dosage is too high, all normal-set admixtures will cause the set to be delayed. Furthermore, there isn't always a direct dosage correspondence between admixtures of the same class due to variations in the percentage of solids across products from different companies. Therefore, when comparing competing admixtures, it's critical to pay close attention to the chemical composition. Superplasticizers are water-reducing admixtures with a high range that satisfy ASTM C494 Type F or G requirements. They are frequently used in conjunction with low water-to-cement ratios, good workability, and low segregation to produce high-strength concrete. They can also be used to create concrete with a consistent water-to-cement ratio and strengths using less cement. Additionally, they can be used to create self-leveling, self-compacting flowing concretes for uses like pumping concrete over long distances from a mixer to formwork or pouring concrete into forms that are clogged with reinforcing steel. These concretes have a significant increase in slump without segregation, but neither the cement content nor the water-cement ratio is decreased. For instance, without adding water and losing strength, a typical concrete mix's initial slump of 3 to 4 inches may be increased to 7 to 8 inches. Superplasticizers can be categorized as modified lignosulfonates, synthetic polymers, sulfonated naphthaline-formaldehyde condensates, or sulfonated melamine-formaldehyde condensates.

2.3.1.2. Air-Entraining Admixtures

To shield the concrete from deterioration brought on by frequent freezing and thawing or exposure to harsh chemicals, these generate a large number of microscopic air spaces within the material. The air gaps allow internal and external water to expand in concrete that is subjected to frequent cycles of freezing and thawing, preventing damage to the material. There are fewer capillaries that extend from the concrete matrix to the surface of air-entrained concrete because it bleeds less than non-air-entrained concrete. As a result, there are fewer opportunities for harmful chemicals to seep into the concrete. Materials for use as air-entraining admixtures to be added to concrete on-site are covered by ASTM C260, "Standard Specification for Air-Entraining Admixtures for Concrete." Types IIA and IIIA Portland cement can also be used to achieve air entrainment.

2.3.1.3. Set-Accelerating Admixtures

These are used to speed up the rate at which concrete gains strength and shorten the time between the start of the water addition process and the initial set. Calcium chloride is the admixture that is most frequently used to accelerate the set. However, there is debate regarding its application when prestressing or reinforcing steel is present. The accelerator may introduce free chloride ions into the concrete, which could exacerbate the steel's corrosion. This is the rationale behind the accelerator. Alternatively, you could use one of the many available admixtures that don't contain chloride.

2.3.1.4. Retarding Admixtures

All common water-reducing admixtures slow down the initial set of concrete to some degree. Concrete can be transported for a longer period of time before the initial set occurs when an admixture of Type B or D is used. The final set is also not ready. Therefore, if walls are to be constructed using retarded concrete, care must be taken. The first set can be delayed for a few hours to a few days, depending on the admixture's base chemical type and dosage. The concrete's compressive strength increases as a positive byproduct of delaying the initial and final sets. When utilized in the same mix design, a Type D admixture—which is frequently used—offers greater 7- and 28-day strengths than a Type A.

2.3.1.5. Mineral Admixtures

Fly ashes, pozzolans, and microsilicates are included in the mineral admixture classification. Natural cement is sometimes used as an admixture.

2.3.1.6. Corrosion Inhibitors

The high alkalinity of the concrete typically forms a passivating layer at the steel surface, protecting the reinforcing steel from corrosion. Iron oxide, a stable substance, makes up this layer. Ferrous-oxide compounds, on the other hand, are more reactive and are found both inside and outside of ferric oxide. Iron-oxide corrosion products are solidified by the reaction of ferrous oxide compounds with oxygen when they come into contact with aggressive substances like chloride ions. They result in a volume gain of four times that of the concrete's tensile strength and an expansion force. The end effect is concrete degradation. Chloride levels between 1.0 and 0.5 lb/yd3 are necessary for corrosion to happen. When chlorides can potentially enter the concrete matrix from outside sources, like deicing salts, the concrete can be protected by adding an internal barrier admixture, calcium nitrate, cathodic protection, increasing the amount of cover over the reinforcing steel, entraining air in the concrete, lowering the water-cement ratio, or a combination of these techniques.

At the time of batching, calcium-nitrate admixtures are added to the concrete to prevent corrosion. They don't erect a physical defense against the intrusion of chloride ions. Instead, they alter the chemistry of the concrete close to the steel surface. Ferrous oxide is oxidized by the nitrite ions, which turns it into ferric oxide. Furthermore, the nitrite strengthens the ferric-oxide passivating layer at the steel surface where it is absorbed. In order for a calcium-nitrite admixture to be effective, the dosage needs to be changed based on how the concrete is exposed to corrosive agents. The dosage should be higher the greater the exposure. The right dosage can only be ascertained using data specific to the proposed admixture on a project-by-project basis. Two types of internal-barrier admixtures exist. One consists of compounds for damp proofing and waterproofing. In addition to the passivating layer, the second group of agents forms an organic film around the reinforcing steel. Regardless of anticipated chloride exposure, the addition of this kind of admixture is encouraged at a set rate.

2.3.1.7. Coloring Admixtures

Concrete is colored for architectural purposes. These could be artificial pigments or mineral oxides. A common material for black color, raw carbon black, significantly lowers the amount of entrained air in a mix. Therefore, either the carbon black should be modified to entrain air, or an additional air-entraining agent may be added to the mix if black concrete is desired for concrete that requires air-entrainment (for freeze-thaw or aggressive chemical exposure). Before being used in construction, the mix design needs to be tested in real-world settings. Maintaining a consistent color at the job site through careful control over materials, batching, and water addition is necessary when using color.

2.3.2. Fibers for Concrete Mixes

Fibers are discrete, discontinuous units found in concrete. Their aspect ratio, or the length to equivalent diameter ratio, can be used to characterize them. Fibers are most useful for controlling cracks in concrete flatwork, particularly in slabs on grade. Synthetic fibers, such as those made of polypropylene, nylon, polyester, and polyethylene, are the most often used kinds in concrete. The fibers aramid, carbon, and acrylic are examples of specialty synthetics. The materials used to make glass-fiber-reinforced concrete are alkali-resistant (AR) glass fibers and E-glass. High-tensile or stainless steel is chopped to create steel fibers. Fibers ought to be evenly distributed throughout a mixture. In concrete, fiber orientation is typically random. In contrast, conventional reinforcement is usually oriented in one or both directions, usually in planes that are parallel to the surface. Additionally, while concrete is being poured, reinforcing steel bars or welded-wire cloth must be kept in place. Whatever their kind, fibers work well to control cracks in concrete because they give the matrix omnidirectional reinforcement. Concrete's impact resistance, toughness, and flexural and fatigue strengths can all be significantly increased with the addition of steel fibers. In concrete flatwork, synthetic fibers are commonly utilized in lieu of welded-wire fabric as a secondary reinforcement to mitigate cracks. The fiber can restrict the size and propagation of drying shrinkage cracks or both plastic and drying shrinkage cracks, depending on the length of the fiber. The ASTM E72, "Standard Methods of Conducting Strength Tests of Panels for Building Construction," test slabs demonstrated that test slabs reinforced with synthetic fibers carried greater uniform loads than slabs containing welded wire fabric, despite the fact that synthetic fibers are not intended to provide structural properties. Although more than 2% of reinforcement ratios have been used in most synthetic fiber research, 0.1% (1.5 lb/yd^3) is the typical field practice. The more cross-sectional area is provided by this dosage than by 10-gage welded wire fabric. The empirical findings show that cracking is both controlled and greatly decreased. Fibers also have the advantage of helping to keep the concrete together after the first cracks.

Fibers made of acrylic, carbon, and aramid have all been researched for use in structural applications, such as adding strength to concrete columns by wrapping them in fiber. Structures resistant to corrosion could be used in additional ways. Specialty synthetics are more expensive, which restricts their application in conventional buildings. Several different kinds of building components, such as architectural wall panels, roofing tiles, and water tanks, are made of glass-fiber-reinforced concrete (GFRC). Because the AR-glass fibers are susceptible to embrittlement, possibly as a result of calcium hydroxide particle infiltration, and the E-glass fibers are alkali reactive, the full potential of GFRC has not been realized. Steel fibers can take the place of traditional reinforcing steel in structural applications. A mix's steel fiber content can range from 0.5 to 2%. A lot of effort has gone into creating quick repair techniques that use cement paste that is squeezed into the steel matrix and thin panels of tightly packed steel fibers. In "Guide for Specifying, Mixing, Placing, and Finishing Steel Fiber Reinforced Concrete," or ACI 544.3R, American Concrete Institute Committee 544 specifies that reinforcing steel must be provided to support the entire tensile load in structural members like beams, columns, and floors that are not on grade. Alternatively, fibers can be used to increase performance or decrease section thickness.

2.3.3. Miscellaneous Admixtures

There are many miscellaneous concrete additives for use as pumping aids and as damp proofing, permeability-reducing, gas-forming agents.

The viscosity of harsh or barely pumpable mixtures is reduced with the use of pumping aids. Fly ash, bentonite, hydrated lime, and synthetic and organic polymers can all be used for this. Results vary depending on the type of concrete used, taking into account factors like higher water requirements and the possibility of reduced strength due to higher water-to-cement ratios. Fly ash is the recommended pumping additive if sand increases the mix's pumpability just a little. It usually doesn't raise the need for water, and when it interacts with the cement's calcium hydroxide, it gives some strength. Stearates, soaps, and other petroleum-based products are examples of damp proofing admixtures. Their aim is to decrease the amount of water and water vapor that can pass through concrete. Use caution when working with these materials as they may raise the mix's water requirement, which will increase the concrete's permeability. The concrete should be thoroughly vibrated and damp cured if dense, low-permeable concrete is desired. The water-cement ratio should be kept to a maximum of 0.50. Fly ash and silica fume can be added to concrete as admixtures to reduce its permeability. Additionally, permeability will be significantly reduced by using a water-cement ratio of less than 0.5 and a high-range water-reducing admixture. Lightweight concrete is created by adding gas-forming admixtures. They are also utilized in grout for masonry where it is intended for the grout to swell and adhere to the concrete masonry unit. They are usually made of powdered aluminum.

2.4. MORTARS AND CONCRETES

Sand, water, fine aggregate, and cementitious material make up mortars. They are utilized for concrete with coarse aggregate added, bedding unit masonry, plasters, and stuccoes. In this case, the mortars used for plasters and unit masonry are the main ones taken into consideration. The characteristics and grading of the sand, the ratio of water to solids, the properties of the cementitious material used, and other factors all affect the properties of mortars, which can vary significantly.

- **Packaging and Proportioning of Mortar:** Typically, mortars are measured in volume. A typical specification calls for using no more than 3 feet of sand and 1 foot of cementitious material. However, it can be challenging to precisely measure the amount of material that makes up a cubic foot. For example, while a 94-pound bag of cement is commonly referred to as a cubic foot when proportioning mortars or concretes, a cubic foot of lime putty can also be used in the same manner. Weights of 40, 42, and 45 lb of hydrated lime have been used as a cubic foot in laboratory studies; however, on the job, a bag is frequently used as a cubic foot. Hydrated limes are sold in 50-lb bags, each of which makes somewhat more than a cubic foot of putty. Masonry cements are offered for sale in 70- to 80-pound bags; one bag is equivalent to one cubic foot.
- **Properties of Mortars:** Types of mortars as a guide in selection for unit masonry. Workability is a crucial characteristic of mortars, especially when they are used with highly absorbent unit masonry. Workability is determined by the type of cement and sand content. A mortar consisting of three parts sand and one part slaked lime putty, for instance, will work better than one consisting of two parts sand and one part Portland cement. However, the 3:1 mortar is not as strong. A suitable compromise can typically be reached by carefully choosing or combining cementitious materials, resulting in a mortar with sufficient strength and workability. An indicator of mortar workability is water retention, which is calculated as the ratio of the flow after a 1-min standard suction to the flow prior to suction. For most uses, a high water retention value is regarded as ideal. Nonetheless, the amount of water retained by mortars made with different limes and cement ratios varies greatly. According to the "Standard Specification for Mortar for Unit Masonry," ASTM C270, mortar must be mixed to an initial flow of 100 to 115, as established by the ASTM C109 test method, and must have a minimum of 75% flow after suction. Although mortar strength has little bearing on masonry strength, it is often used as a specification requirement.

The amount of cement in the matrix has the biggest impact on mortar strength. The age of the material when tested, the curing conditions and the ratio of sand to cementing material are also significant factors. Another crucial feature of mortars is their change in volume. The shrinkage that occurs during early hardening, the shrinkage that occurs on drying, the expansion that occurs on wetting, and temperature changes can all be categorized as normal volume change (as opposed to unsoundness). Mortars swell up again when they get wet after drying. With Portland-cement mortars, alternate wetting and drying results in alternate expansion and contraction, which seems to go on forever.

- **High-Bond Mortars:** Shear, compressive, and bonding strengths are significantly enhanced when polymeric materials like polyvinylidene chloride and styrene-butadiene are mixed with mortar. The quality of the other components, which include sand, water, Type I or III Portland cement, and a workability additive like ground limestone or marble dust, must match that of the regular mortar's ingredients in order to achieve high strength. Masonry is capable of withstanding significant bending and tensile stresses due to the high strength of the mortar. This allows for the pre-laying of single-wythe panels that can be lifted into position and thinner walls.

2.4.1. Portland-Cement Concrete

The ingredients of Portland cement concrete are water, coarse and fine aggregates, and admixtures in the right amounts to create a plastic mass that can be shaped, cast, or molded into shapes that will solidify into a solid mass. Plastic concrete should have the following qualities: it should be workable, placeable, non-segregating, and set within the anticipated time frame. The intended service properties should be offered by the hardened concrete:

- Strength (compressive and flexural)
- Durability (lack of cracks, resistance to freezing and thawing and to chemical attacks, abrasion resistance, and air content)
- Appearance (color, lack of surface imperfections)

Each of these properties affects the final cost of the mix design and the cost of the in-place concrete. These properties are available from normal-weight, lightweight, and heavyweight concretes.

2.4.1.1. Normal-Weight Concrete

In the case of non-air-entrained concrete, the nominal weight of normal concrete is 144 lb/ft^3, whereas for air-entrained concrete it is less. It's common to assume that concrete plus steel reinforcement weighs 150 pounds per foot. The range of strengths for concrete with a normal weight is 2,000–20,000 psi. A common test cylinder with a diameter of 6 inches and a height of 12 inches is used to measure it. When two cylinders with the same load and age are tested together, the average strength of the two cylinders is used to determine the strength of the concrete. Concrete paving mixes can be supported by 6 x 6 x 20-inch flexural beams. The strength increases of concrete with and without air particles are shown graphically. A given mix's strength is defined by its water-to-cement ratio (W/C) and air-entraining effectiveness. Additionally, the maximum size of the aggregate and the desired fluidity (slump) of the concrete at the placement site must be considered in situations where the aggregates and cement to be used lack historical records. For a given mix fluidity, or slump, a given combination of coarse and fine aggregates has a specific water demand.

Two general guidelines are:

1. For a constant slump, the water demand increases with an increase in maximum size aggregate.
2. For a constant maximum-size aggregate, as the slump increases, the water demand increases.

There are many different methods for designing a normal-weight concrete mix. A standard method is given in ACI 211, "Standard Practice for Selecting Proportions for Normal, Heavyweight, and Mass Concrete. For contractors who have to finish and place the concrete into forms, workability is the most crucial attribute. Workability encompasses non-segregation, cohesiveness, and plasticity. Aggregate gradation and shape have a big impact on it. Mixes with flat and elongated aggregates mix with an excessive amount of fines (sand and cement) and mixes deficient in fines are all difficult to pump, place, and finish. You can add 30 to 50 lb/yd3 of fly ash to the sand to increase workability if it lacks fines. The best way to get workable concrete is to use a combined aggregate gradation that is well-graded. Normal-weight concrete has an elastic modulus of 2,000,000–6,000,000 psi. For normal-weight concrete with compressive strengths ranging from 3000 to 5000 psi, the modulus of elasticity can be estimated by multiplying the square root by 57,000. The ASTM C469 procedure should be used to determine the modulus above 5000 psi. Changes in volume can arise from drying shrinkage, creep, or expansion brought on by outside heat sources. The most common cause of issues is drying shrinkage, which results in surface cracks in the concrete. When there is too much water in the mixture, drying shrinkage cracks are the main result. The water's effects are twofold. It first weakens the concrete by raising the water-to-cement ratio (W/C). Second, more water than is required for the cement to hydrate results in an excessive amount of bleed channels to exposed surfaces. These channels are unable to withstand the penetration of water or harsh chemicals when the cement paste experiences its typical drying shrinkage.

When an external load is applied to concrete, it will begin to deform over time in a pattern known as creep. It is a crucial factor to take into account when designing prestressed concrete. When heated, concrete expands, and when cooled, it contracts. Thermal expansion coefficients vary from 3.2 to 7.0 millionths per °F. The movement of external walls, which may bow due to temperature variations, is the most noticeable effect of concrete's reaction to thermal changes. When heated, normal-weight concrete that isn't meant to be exposed to fire expands. A decrease in the modulus of elasticity and some strength loss are side effects. Using a mix with low water content, properly curing the concrete, and adequate air entrainment in the concrete can all help to achieve resistance to freezing and thawing. Chemical attacks can come from the outside (sulfate attack) or the inside (alkali-aggregate reaction). The qualities of the materials that are available and the environment in which the concrete will be used are the primary concerns in either scenario. Although alkali-reactive aggregates ought to be avoided, low-alkali cement that complies with ASTM C150 Type II Modified ought to be chosen if they must be used. A low W/C (0.45 maximum) and air entrainment should be used with either a C150 Type V cement or a C150 Type II cement with C618 Type F fly ash if sulfate attack is a concern. A low W/C (0.45 maximum), more concrete cover over the reinforcing steel, a corrosion-protection additive, or latex-modified concrete should be used to protect against chemical attack. Special exposure conditions are outlined in the "Building Code Requirements for Reinforced Concrete," or ACI 318, published by the American Concrete Institute.

Pavements and hydraulic structures have concerns regarding abrasion resistance. Both call for the use of low W/C, well-cured concrete, and reliable, long-lasting hard-rock aggregates. Reputable craftsmanship and a steady supply of materials are necessary for an acceptable appearance. Before pouring concrete, the formwork has to be properly oiled and sealed. It is not advisable to use wood for forms as it can release sugars into the concrete and cause a delayed surface finish. The concrete should always be workable when it is being placed. To consolidate the concrete, the forms need to vibrate consistently and uniformly.

2.4.1.2. Lightweight Concrete

Using lightweight aggregates or expanding or foaming the concrete can result in concrete that weighs significantly less than the 144 lb/ft^3 of normal-weight concrete. The main purposes of lightweight concrete are to lessen a structure's dead load and to cut foundation costs. The cellular structure of the particles in the aggregates used to make this kind of concrete accounts for their low weight. Because the aggregates or the concrete itself have internal voids, lightweight aggregate concrete, foamed concrete, and expanded concrete all have excellent fire-protective qualities. When lightweight aggregates are utilized, they can be either lightweight coarse and normal-weight fine (sand), lightweight coarse and normal-weight fine, or both fine and coarse. The final combination is the one that is used the least. The unit weights vary from 90 lb/ft^3 (lightweight aggregates) to 115 lb/ft^3 (lightweight sand). Compressive strengths typically fall between 2,500 and 4,000 psi. On the other hand, high-strength lightweight concretes have been created with strengths ranging from 6,000 to 9,000 psi and maximum unit weights of 125 lb/ft^3. The ACI defines structural lightweight concretes as those that have an air-dry unit weight of 115 lb/ft^3 or less and a 28-day compressive strength greater than 2,500 psi.

Using W/C in the design of a lightweight aggregate mix is challenging due to the variable amount of water absorbed in the voids of lightweight aggregates. To avoid segregation, 4 to 6% air entrainment is ideal. The coarse aggregate's maximum size shouldn't be greater than half the depth of the cover over the reinforcing steel. The recommended range for compressive strength for lightweight-aggregate concrete (Etxeberria et al., 2006) exposed to sulfates is 3,750–4,750 psi (refer to ACI 318). A minimum of seven bags of cement should be used per cubic yard of concrete for marine structures, and the W/C should not be greater than 0.40.

The modulus of elasticity E_c of lightweight concrete generally ranges from 1,500,000 to 3,000,000 psi. It may be estimated from

$$E_c = w^{1.5}\sqrt{f_c'}$$

where w unit weight of concrete, lb/ft^3; f'_c= 28-day compressive strength of concrete, psi.

Similar to regular-weight concrete, lightweight concrete experiences volume changes, but it is more stable in a heated environment. The most undesired volume changes result from drying shrinkage because it fractures the concrete's surface. Cracks caused by drying shrinkage are primarily caused by too much water in the mixture. The water's effects are twofold. It first weakens the concrete and raises the W/C. Second, too many bleed channels are created to the exposed surfaces by the extra water added above what is required for the cement to hydrate. These channels are unable to withstand the infiltration of harsh chemicals when the cement paste experiences normal drying shrinkage. Similar to normal-weight concrete, especially prestressed concrete, creep is a major concern for lightweight concrete.

2.4.1.3. Heavyweight Concrete

For shielding and structural reasons, heavyweight aggregate concrete is used in the construction of nuclear reactors and other structures exposed to high-intensity radiation. Because the absorption of such radiation is proportionate to density, heavyweight aggregates are used where heavyweight is required, such as ship ballast and the encasing of underwater pipes. As a result, these aggregates have a greater capacity for absorption than those typically used for regular concrete. They are also used to create shielding concretes. Concrete weighing up to 385 lb/ft^3 can be made using these aggregates. Compressive strengths of 3,200 to 5,700 psi and densities of 210 to 224 lb/ft^3 can be achieved in concrete composed of limonite or magnetite. Concrete that contains barite may weigh 230 lb/ft^3 and have a 6,000 psi strength. Strengths of approximately 5600 psi and densities of 250 to 288 lb/ft^3 can be achieved by using steel shot as fine aggregate and steel punchings and sheared bars as coarse aggregate. The proportions and grading of the mix are typically the same as for regular concrete. Similar to normal-weight concrete, heavyweight concrete has comparable properties. However, because of segregation, mixing and placing procedures are more challenging than for normal-weight concrete. To avoid segregation, use air entrainment, low W/C, high cement content, and good grading. In order to prevent segregation, heavyweight aggregates are occasionally grouted in place. Generally speaking, lightweight concretes are not very resistant to abrasion or weathering.

2.4.1.4. Nonstructural or Foamed Cellular Concretes

These are created by adding admixtures to concrete during the plastic stage that cause or release gas bubbles. For this purpose, aluminum powder is typically used, though hydrogen peroxide, which produces oxygen, or activated carbon, which releases absorbed air, can also be used. Aluminum powder reacts with the alkalies in cement to release hydrogen. As the concrete sets, these foaming agents produce stable, evenly distributed air spaces within it. The two most common aggregates are vermiculite and perlite. The final concrete may have a compressive strength of up to 2,500 psi and weighs 50 lb/ft^3 or less. These lightweight concretes can be used as soundproofing barriers and toppings over structural concrete slabs. The length of time spent handling, mixing, and placing the mixture in relation to the rate of gas generation determines how effective the admixture is. Under normal circumstances, the amount of unpolished aluminum powder to be added to a mix can vary from 0.005 to 0.02% by weight of cement. Lower-strength concretes, however, can be made with larger amounts. For example, twice as much aluminum may be required at 400°F as at 700°F to produce the same amount of concrete expansion. Moreover, alkalies like sodium hydroxide, hydrated lime, or trisodium phosphate may need to be added to the mixture at low temperatures in order to accelerate the production of gas. Aluminum can also be mixed with sand or other admixtures to stop the powder from floating on the water's surface when mixing.

2.4.2. Polymer Concretes

Polymers, which are plastics with long-chain molecules, can be used to replace Portland cement in concrete, add them as an admixture to a mix, or impregnate hardened concrete to improve its properties. Concretes made of polymers instead of Portland cement, like methyl methacrylate and unsaturated polyester, can have a modulus of elasticity and strength that is more than twice that of Portland cement concrete. The polymer concretes have a higher resistance to freezing and thawing cycles and less creep. Polymer concrete can be used right away after curing for a very short period of time—overnight at room temperature, for example—while regular concrete may need to cure for a week or more before being exposed to service loads. Admixtures consisting of monomers and polymers can be utilized to restore and resurface concrete surfaces that have deteriorated. For these uses, latexes consisting of methyl methacrylate, polyester, furans, styrene-butadiene, epoxy-styrene, and vinylidene chloride have been used. Compared to regular concrete, the resultant concrete sets more quickly. It is also possible to enhance Portland cement's bonding qualities by adding a polymer admixture. For this reason, the admixture is added to a mix as an emulsion and provides a large amount of water to the mix, which the cement can use to hydrate. The emulsion is also set by the water's release. Because the emulsion needs to dry in order to gain the appropriate strength, moist curing is therefore not recommended. When applied thinly rather than thickly, grout or mortar containing the bonding admixture develops a higher bond strength and may even outbound the materials being joined.

Concrete can occasionally be impregnated with polymers to strengthen surfaces that are subjected to high traffic. The impregnated concrete's strength and other characteristics are comparable to those of concrete made with polymers in place of Portland cement. To achieve impregnation, the concrete surface must first be heated to a dry temperature before being soaked in a monomer, such as methyl methacrylate, styrene, acrylonitrile, or tert-butyl styrene. After that, heat is used to cure it. Slab Toppers. Epoxies are mixed with sand and other fillers to provide surfacing materials for concrete, at least in part due to their superior adhesion. Epoxy-based surfacing materials have the potential to be thinner than regular concrete topping. They work particularly well for bringing uneven, irregular surfaces into smoothness. Because epoxy cures quickly, the surface can be used right away. Grout. Epoxy grout can be used to fix concrete cracks. For deep penetration, the grout is pushed under pressure into cracks. The epoxy grout can effectively restore strength and seal the crack from liquid infiltration thanks to its strong bonding power.

2.5. BRICKS

Bricks are tiny pieces of construction material that are typically composed of fired clay and bound together with mortar, a mixture of cement, sand, and water. Brick, a material that has long been in use, is heat-resistant, corrosion-resistant, and fire-resistant. Brick is a great material for curved designs and structures in tight spaces because each unit is tiny—typically four inches wide and twice as long.

- **Raw Materials:** Clay minerals found in nature, such as shale and kaolin, comprise the bulk of brick. To create various shades, small amounts of barium, manganese, and other additives are mixed with the clay; barium carbonate is used to increase the chemical resistance of brick against the elements. A wide range of additional additives, such as papermaking byproducts, ammonium compounds, wetting agents, flocculants (which cause particles to form loose clusters), and deflocculates (which disperse such clusters), have also been used in brick production. Certain clays necessitate the inclusion of sand or grog, which are pre-ground and pre-fired materials like leftover brick. Brick with a specific color or surface texture can be created using a wide range of coating materials and techniques. The primary ingredient, sand, is mechanically combined with a colorant to form a typical coating. In order to create surface textures, flux or frit—a glass that contains colorants—may occasionally be added. In order for the sand to adhere to the brick surface, the flux lowers the sand's melting temperature. Nepheline syenite, graded aggregate, and fired and unfired brick are among the additional materials that can be utilized.
- **Why Is Brick Construction so Popular?:** There are many advantages when bricks are used as part of the construction. The following list presents some of the most common advantages when using bricks instead of other construction materials.
    - **Aesthetic:** Bricks offer natural and a variety of colors, including various textures.
    - **Strength:** Bricks offer excellent high compressive strength.
    - **Porosity:** The ability to release and absorb moisture is one of the most important and useful properties of bricks, regulating temperatures and humidity inside structures.
    - **Fire Protection**: When prepared properly a brick structure can give a fire protection maximum rating of 6 hours.
    - **Sound Attenuation**: The brick sound insulation is normally 45 decibels for a 4.5 inches brick thickness and 50 decibels for a nine-inch-thick brick.
    - **Insulation:** Bricks can exhibit above-normal thermal insulation when compared to other building materials. Bricks can help regulate and maintain constant interior temperatures of a structure due to their ability to absorb and slowly release heat. This way bricks can produce significant energy savings, more than 30% of energy savings when compared to wood.

- **Wear Resistant**: A brick is so strong, that its composition provides excellent wear resistance.
- **Efflorescence**: Efflorescence forms on concrete structures and surfaces when soluble salts dissolved in water are deposited and accumulated on surfaces forming a visible scum.
- **Durability:** Brick is extremely durable and perhaps is the most durable man-made structural building material so far.

- **Advantages of Bricks**
  - Economical (Raw material is easily available)
  - Hard and durable
  - Compressive strength is good enough for ordinary construction
  - Different orientations and sizes give different surface textures
  - Very low maintenance cost is required
  - Demolishing of brick structures is very easy, less time-consuming, and hence economic
  - Reusable and Recyclable
  - Highly fire resistant
  - Produces less environmental pollution during the manufacturing process
- **Disadvantages of Bricks**
  - Time-consuming construction
  - Cannot be used in high seismic zones
  - Since bricks absorb water easily, therefore, it cause fluorescence when not exposed to air
  - Very Little tensile strength
  - Rough surfaces of bricks may cause mold growth if not properly cleaned
  - Cleaning brick surfaces is a hard job
  - The color of low-quality brick changes when exposed to the sun for a long period of time

2.5.1. Manufacturing Process

The raw materials are first crushed and ground in a separator and jaw crusher to make brick. Extrusion, molding, or pressing are the three methods used to shape bricks; the first method is the most versatile and therefore the most used. After that, the mixture of ingredients needed for each specific batch is chosen and filtered. After the bricks are formed and any further steps are taken, they are dried to get rid of any extra moisture that could cause them to crack when they are fired. After that, they are cooled and burned in ovens. When all is said and done, they are dehacked, automatically stacked, steel banded, and cushioned with plastic corner protectors.

2.5.1.1. Grinding, Sizing, and Combining Raw Materials

All of the ingredients are first sent to a separator, which takes out any material that is too big. The particles are then squeezed by a jaw crusher with horizontal steel plates, which further reduces their size. Following the selection of the raw materials for each batch of bricks, the various material sizes are frequently separated using a scalping screen. Oversized material is sent to a hammer mill, where steel hammers move quickly to pulverize it, while material that is the right size is sent to storage silos. The discharge from the hammer mill is directed to several vibrating screens that filter out material of the wrong size before it is forwarded to the following stage of production. Another screen is used to regulate the maximum size of particles exiting the mill.

2.5.1.2. Extrusion

The most popular method of forming bricks is extrusion. In this process, water and ground material are fed into one end of a pug mill, which uses rotating shaft knives to cut through and fold the material into a shallow chamber. After that, the mixture is fed into an extruder at the mill's far end. There are typically two chambers in an extruder. The first uses a vacuum to remove air from the ground clay, preventing cracking and other flaws. In order for the auger to extrude the material through the die, it must first be compacted in the second chamber, a high-pressure cylinder. The compressed plastic material is then driven out of the chamber through a die orifice that has been specially shaped. The extruded column's cross-section, known as the pug, takes on the shape of the die. Knives that rotate or stiff wires are used to cut sections to the desired length. Soft, wet clay is molded using a mold, typically a wooden box, in the molding process. Sand is frequently applied inside the box to give it the appropriate texture and make it easier to remove the brick once it has been formed from the mold. Another tool for aiding release is water. A material with a low water content is needed for pressing, the third type of brick forming. A steel plunger with a preset pressure is used to compact the material after it has been put into a die. Pressed bricks are also shaped like frogs and have a sharper outline and a more regular shape than bricks made using the other two methods.

2.5.1.3. Chamfering the Brick

For uses like paving, chamfering machines were created to create a groove in brick. As the brick is extruded, these machines indent it using rollers. Sometimes they have wire cutters attached so that the chamfering and cutting can be completed in a single step. These devices have an hourly capacity of up to 20,000 units.

2.5.1.4. Coating

The type of sand coating that is applied during the extrusion of the brick is determined by the hardness or softness of the extruded material. The soft material is coated using a continuous vibrating feeder; textured material may require brushing or rolling of the coating. Sandblasting is necessary for extremely hard materials, and a pressure roller or compressed air is used for harder materials.

2.5.1.5. Drying

The brick needs to be dried to get rid of extra moisture before it is fired. If this moisture is not eliminated, the water will burn off during firing too quickly, leading to fractures. A pair of dryer types are employed. Tunnel dryers prevent brick from cracking by moving it through zones with controlled humidity using cars. They are made up of a lengthy chamber that the ware is gradually pushed through. To expedite the process, hot air circulated by a fan is forced into the dryer from external sources. Additionally, automatic chamber dryers are in use, particularly in Europe. On two parallel bars, the extruded bricks are automatically arranged in rows. After that, the bricks are fed onto unique racks that have finger-like devices holding multiple pairs of bars in layers. After that, lift trucks or transfer cars with rails are used to move these racks into the dryers.

2.5.1.6. Firing

The brick is dried, then loaded onto cars (often automatically), and heated to a high temperature in kilns. Usually, the cars used to transport the bricks through the tunnel kiln are also the ones that moved them during the drying process. These vehicles are driven through the kiln's constantly regulated temperature zones at a material-specific pace. Though a third of the brick produced today is fired using solid fuels like sawdust and coal, most kilns in the United States run on gas. The high-load, narrow-width tunnel kilns have been replaced by shorter, lower-set, wider kilns that can fire more brick. Additionally, this kind of design has produced low-temperature, long-flame, and high-velocity flame burners that have reduced fuel consumption and increased temperature uniformity.

2.5.1.7. Setting and Packaging

The dehacking process, which has been automated to the extent that nearly all manual brick handling has been eliminated, is used to remove the brick from the kiln car after it has been fired and cooled. Brick can now be set at speeds exceeding 18,000 per hour and rotated 180° by automated setting machines. Typically arranged in eleven-brick-wide rows, a stack is equipped with plastic strips that function as corner guards and are encircled by steel bands. After being packaged, the brick is delivered to the job site and usually unloaded using boom trucks.

2.5.2. Types of Brick

All across the world, bricks are used for pavement and building materials. Bricks were formerly used in American pavement, but they are now more frequently used as decorative surfaces than as building materials. In order to increase the stability and strength of a structure, bricks are typically laid flat and bonded together. Several different kinds of bricks are used, most of them measuring eight inches in length and four inches in thickness.

There are various types of bricks used in masonry.

1. Common Burnt Clay Bricks;
2. Sand Lime Bricks (Calcium Silicate Bricks);
3. Engineering Bricks;
4. Concrete Bricks; and
5. Fly ash Clay Bricks.

In the building industry, bricks are used as siding partly because of their significant qualities and partly because they can be an excellent, reasonably-priced alternative. The advantages and uses of the most widely used kind of bricks are outlined below.

2.5.2.1. Common Burnt Clay Bricks

Using molds, common burnt clay bricks are created. After that, a kiln is used to dry and fire the bricks. Common burnt clay bricks have no particular aesthetic appeal and are used in general construction. These bricks need to be rendered or plastered when they are used to build walls. Bricks made of burned clay are used in construction. Usually, clay is poured into molds, allowed to dry, and then baked or fired in a kiln to create these bricks. Burned bricks typically require plastering to make them aesthetically pleasing. Because of their ability to tolerate high temperatures, they are used in construction projects where mechanical, chemical, and thermal stress are present.

2.5.2.2. Sand Lime Bricks

Sand lime bricks are made by mixing sand, fly ash, and lime followed by a chemical process during wet mixing. The mix is then molded under pressure forming the brick. These bricks can offer advantages over clay bricks such as:

- Their color appearance is gray instead of the regular reddish color.
- Their shape is uniform and presents a smoother finish that doesn't require plastering.
- These bricks offer excellent strength as a load-bearing member.

2.5.2.3. Engineering Bricks

Engineered bricks are produced at extraordinarily high temperatures, which results in a dense, robust brick that can limit strength and water absorption. Excellent load-bearing capacity, damp-proof qualities, and chemical resistance are all offered by engineering bricks. These bricks can be more expensive than standard or conventional bricks and are only utilized in particular projects. Bricks made by engineers are stronger and thicker. They are the least likely to absorb water. High temperatures are used in the production of these bricks. They are employed in civil construction projects where damp-proof characteristics, high load-bearing capacities, and chemical resistance are required.

2.5.2.4. Concrete Bricks

Solid concrete is used to make concrete bricks, which are highly popular among home builders. Concrete bricks have a great visual presence and are typically used in facades and fences. These bricks can be produced in a variety of colors by adding pigment during the manufacturing process. Solid aggregate concrete is used to make concrete bricks. They are heated with steam while being compressed in the mold. Lower-pressure steam is used to create these bricks. Concrete bricks are typically utilized for foundation walls, fences, basement walls, and facades (Casuccio et al., 2008).

2.5.2.5. Fly Ash Clay Bricks

At roughly 1,000°C, clay and fly ash are combined to create fly ash clay bricks. According to certain studies, when these bricks come into contact with moisture or water, they expand and fail poorly, producing pop-outs.

2.5.3. Quality Control

Despite the perception that the brick industry lacks sophistication, a large number of manufacturers are engaged in statistical control and total quality management initiatives. In the latter case, control limits are set for a particular process (e.g., temperature during drying or firing) and the parameter is monitored to ensure that the relevant processes remain within the limits. As a result, defects can be avoided and yields can be increased by controlling the process as it proceeds. The American Society of Testing and Materials (ASTM) has established standards that must be followed when measuring a wide range of mechanical and physical attributes. These characteristics include mechanical strength, density, and physical dimensions. Another crucial characteristic is the brick's ability to withstand freeze-thaw cycles, which is tested in environments meant to mimic those found outside. However, the tests that are in use now are insufficient and don't accurately reflect real-world conditions. It's possible that what passes in the lab won't pass in real life. As a result, the brick industry is working to create a test that is more precise.

Efflorescence, a condition where water dissolves certain elements (salt being one of the most common) in external sources, mortar, or brick itself, presents a comparable issue. Inadequate cleaning can exacerbate the surface discoloration caused by the soluble material's residual deposits. The efflorescence gets worse when salt deposits become insoluble, necessitating thorough cleaning. Even though a brick passes a test in the lab, poor construction methods or design could cause it to fail in real life. As a result, brick-and-mortar businesses are creating their own internal testing protocols, and research is still being done to create a more trustworthy standard test.

2.5.4. Properties and Uses of Bricks

One fundamental material that's frequently used to construct a building's walls is brick. Bricks made of clay are the most widely used kind. It is formed of clay and sand, and the temperature range for its uniform burning is 800–1,200°C. On occasion, surface finishes are applied, e.g., bricks with glaze. Large quantities of glazed bricks are still produced and are available from well-known English brick suppliers and manufacturers. Sand-lime, or calcium silicate, is used to make calcium silicate bricks. In an autoclave, the bricks are steam-treated and put under intense pressure. The bricks have a light color, a smooth texture, and a fine texture. The brick's color is determined by the composition, firing temperature, and material source. Brick sizes have varied significantly over time. The dimensions of some early medieval bricks were 13 by 6 by 2 inches. A charter dated 1,571 specified that bricks must measure 9 inches by 4 inches by 2.25 inches. By the late 15th century, bricks measuring 9.5 inches by 4.5 inches by 2 inches had become the standard. Brick measuring 8 × 25 by 4 inches by 2.5 inches was introduced in the 18th century. According to metric measurement, modern bricks measure 8 3/5 inches by 4 3/5 inches by 2 3/5 inches, or 215 mm by 102 mm by 65 mm. The most common type of brick used in buildings today is modern bricks.

- **Defect of Size:** Poor material selection and preparation or under-firing is the main problem of oversized bricks. Undersized bricks are caused mainly by poor material preparation, faulty molds, and over-firing.
- **Defect of Shape:** Shape defects are caused by a variety of factors. These include incorrect molding technique or mold preparation, improper stacking, rough handling, and uneven drying. Burrs, or partially burned bricks, emerge from the kiln or clamp and fuse together if there is significant over burning of the bricks. Burrs are only appropriate for die-hard enthusiasts.
- **Defects of Body:** Cracking, bloating, and lamination defects are examples of defects that can result from flaws in the raw material. Bricks that have been exposed to direct sunlight or rapid drying winds may be the cause of their initial weakness, which manifests as cracking and warping. A similar result will arise from loading the kiln with green bricks too soon. Large cracks could be the result of rain hitting hot bricks.

- **Defect of Appearance:** These defects are caused primarily by faulty wires in the case of wire-cut bricks. Dark spots are caused by the presence of unevenly distributed iron sulfide in the clay.

2.5.4.1. Types of Brickwork Defects

Bricks are used to build most building components. These days, brickwork flaws are frequently discovered in structures. The severity of this issue is increasing daily. Brickwork flaws could give the occupants a lot of trouble. Differences in temperature and moisture-induced movement are the main causes of problems in buildings with traditional outside bearing walls made of brickwork and an inside bearing structure made of concrete. Wall cracks may be found, and remedies must be developed to address this issue. Building brickwork defects can arise for a number of reasons. This may not come as a surprise given the wide variety of bricks and building construction techniques, the variations in construction sites, and the various owner or occupational uses of the finished building. If the brickwork is not properly maintained, it will decay and deteriorate. Such deterioration can occur for a variety of reasons and in different ways. The primary indicators of the condition of the brickwork include surface growth and discoloration, efflorescence, loose brick coming loose, fractures showing through the mortar or bricks, and more. We will identify the various forms of brickwork defects. It will be simple for owners or tenants to determine the causes of the flaws and then implement the appropriate corrective measures to solve the issue if the various kinds of brickwork defects are recognized.

- **Brick Wall Settlement:** A structural brick wall's bond courses run the risk of being broken by any movement in the wall. Any movement in a brick veneer wall that has caused the veneers' connections to the underlying structure to break or loosen could be dangerous and could cause the masonry to collapse.
- **Bulging Brick Wall:** This is likely to be a bond-brick or bond-course failure. This defect is potentially extremely dangerous and immediate action must be taken. If not, it will cause the building to suddenly collapse.
- **Cracks and Bulges in Brick Walls:** Frost and earth loading can push a below-grade brick foundation wall inwards. The wall is often bulged inwards as well as showing horizontal and step cracking and loose bricks will surface over the bulged area. The damage occurs from slightly above ground level to roughly the frost line.
- **Cracks and Loose Bricks:** Normally, cracks and loose bricks are caused by frost, settlement, expansion, diagonal, and stair-stepped. These defects often happen at building corners where the roof spillage is concentrated.
- **Loose Bricks and Missing or Lost Mortar:** This is the movement where water movement against the foundation from the roof or other sources severely washes out the mortar. When wood blocks that were initially nailed into a wall to allow for the attachment of interior components sustain damage from insects or deterioration, loose and lost bricks may also appear. Similarly, if the wood joists collapse due to rot, insects, or fire, or if they are severely bent, the collapsing joist may cause damage to the

building wall or foundation as its in-wall end moves. Wood joists in brick walls with fire cuts on them were designed to reduce this source of damage by angling the end of the joist where it was inserted into the wall pocket.

- **Spalling Bricks:** Spalling is caused by water and frost, such as water leaking into a brick structure at any entry point: a crack, a brick that has lost its hard surface, or openings by window and door penetrations. It is best to also beware of caulking bricks.
- **Exfoliating-Rust Damage to Bricks:** When caulking is applied incorrectly where it shouldn't have been applied, masonry walls develop these flaws. The most typical illustration of this flaw is the harm done to a brick wall by tightly caulking a steel lintel over a window or door between the brick and the steel. Around the steel lintel, moisture that enters the wall through mortar joints or cracks is trapped. The rust that forms on steel lintels has a great lifting force because the expanding rusting exfoliating metal can cause cracks and damage to the bricks surrounding the lintel.
- **Improper Repair Mortar on Brickwork:** Bricks may spill on the surface during tuckpointing or repointing repair work if the mason applies mortar that is too hard and contains a lot of Portland cement, or if the bricks are soft and the area is exposed to freezing temperatures. Because of the high Portland content, the mortar will have greater waterproofing and harderness than the nearby bricks. Water that becomes trapped around the hard mortar may freeze and cause the bricks to spill out on the surface. This is especially likely to occur when a wall that was originally constructed with soft bricks and soft high-lime mortar is now tuck-pointed with hard high-Portland mortar.
- **Salts on Brickwork:** One of the main causes of degradation is salts. Bricks may become contaminated by leaking water. Seawater can provide salt to coastal areas, and nearby road salt applications during the winter pose a constant risk. Because it causes a constant expansion of crystals within the bricks, salt is harmful. This may ultimately cause the brick's structure to break apart. The application of tainted renders or mortars, or contamination within the bricks themselves, can be the source of the salt. Chimney flues lined with brick are particularly susceptible because condensing flue gases can introduce sulfates. When a chimney is sealed without enough ventilation, this is a common flaw.
- **Efflorescence on Brickwork:** Efflorescence is visible on the brick as a white, powdery deposit. A phenomenon known as "brick surface scum" occurs when soluble salt that has been dissolved in water is carried, deposited, and eventually builds up on brick surfaces. Efflorescence is impossible without water. It's possible that the soluble salts came from brick raw materials. However, salts from outside sources, such as contaminated air, groundwater, mortar ingredients, and other materials in contact

with the bricks, are typically the cause of efflorescence. As much as possible, the building detail should be designed to minimize the risk of saturation—which is unavoidable in brickwork exposed to weathering. Put another way, thoughtful overhangs, copings, and sills with drips must be taken into account. Efflorescence can be "aided on its way" by using gentle brushes to brush the wall dry or sponging clean water down it without getting too wet. Avoid using any other treatments as they could cause the bricks to become disfigured.

- **Moisture Movement on Brickwork:** Moisture has a very different effect on all porous masonry materials, such as mortars, bricks, and concrete masonry units. When combining these materials, such as when brick veneer units are used with block backup, when brick and architectural concrete products are used in the same Wythe bands of precast concrete, or when brick veneer is applied over architectural concrete block, these effects need to be taken into account. Concrete masonry units, mortar, and poured-in-place concrete all shrink as the Portland cement cures after they are first mixed or cast. This is incorporated into the design as an inevitable byproduct of concrete product curing.
- **Sulphate Attack on Brickwork:** The majority of regular clay bricks have calcium, magnesium, or sodium sulfates in them. With calcium sulfate being less soluble than the other two, these salts are soluble in water. Usually, these sulfates are thought of as innocuous efflorescences that only have an esthetic effect and can be easily brushed away. The majority of brick walls that use mortars based on ordinary Portland cement, or OPC, are theoretically vulnerable to sulfate attack. Earth-retaining and parapet walls are the most susceptible, but rendered and facing brickwork can also be affected by sulfate attack. Sulfurate attack on rendered brickwork is characterized by rendering cracking, which is primarily horizontal and aligned with the mortar joints underneath. Early in the attack, the rendering might stick to the bricks fairly well, but as the underlying brickwork expands and the bond between the two materials severs, some areas are likely to become detached.
- **Porosity on Brick:** One crucial aspect of bricks is their porosity. Bricks have more porosity than other molded or pre-cast building materials because of their fine capillaries. The rate of moisture transport in brick is ten times faster than in other building materials because of the capillary effect. During the day, moisture is released, and at night, it is reabsorbed. Brick's capacity to release and absorb moisture through the capillary effect is one of its most advantageous characteristics for controlling the atmosphere's temperature and humidity within a building. Bricks are a great building material because of their unique quality, which makes them especially appropriate for structures in tropical climates. On the other hand, weathering agents like rain, flowing water, and contaminated air can contaminate porous materials

and expose them to chemical attacks.

- **Crack in Brick Faces/Firecracks:** Because clay and manufacturing techniques play a major role, the final look of clay brick can differ significantly. There are certain products that come with naturally occurring cracks of different sizes, also known as fire-cracks. When a product is delivered, fire-cracks are typically visible. Yet, the surface sands and textures used during the manufacturing process can conceal fire cracks; these become apparent once the bricks are laid as a result of exposed brickwork naturally weathering.

2.5.5. Causes of the Brickwork Defects

Wall construction is the main use for brickworks. It took a variety of abilities, including design, experience, material selection, and technique, to build the brick wall. A lack of knowledge could result in building failure or flaws in the brickwork. Brickwork defects can be caused by natural flaws in the material or workmanship, production flaws, design flaws, or environmental agents attacking the structure. Following construction, the brickwork may satisfy a variety of requirements related to climate, loading, and occupational use. The frequency with which defects occur is not surprising. There will be a list of the reasons behind the brickwork defects. It will be simple for the occupational to determine the true causes of the brickwork defects so that the appropriate actions can be taken right away to lessen the likelihood that brickwork defects will occur.

- **Selection of Bricks:** Although cost is a major factor when selecting bricks, the necessary quality must also be carefully taken into account to prevent flaws from arising throughout the material's service life. As a result, choosing the right bricks for a given project frequently requires making trade-offs between requirements for durability and aesthetics and price. It is not always possible to judge the quality of bricks solely by their color; in fact, variations in color among bricks from the same source and production process can frequently be interpreted as a sign of differences in strength and durability.
- **Storage of Bricks:** It is rare to find bricks on site kept under sufficient cover. It is rare for brick defects brought on by inadequate site protection to be noticeable before use. Bad weather, however, can be harmful to bricks because it can cause saturation from rain, which can lead to subsequent efflorescence and decoration defects.
- **Soluble Salts in Bricks:** Burned bricks have a lot of salt underneath. The salts are either formed during the burning process or are derived from the raw clay used to make bricks. Soluble salts are present in most clay bricks, although the kinds and concentrations differ significantly. The strength and pore structure of the bricks have a major influence on how the soluble salts affect efflorescence as well as the more serious issue of brick disintegration.
- **Sulphate Attack:** Salts called sulfates are found naturally in some

ground waters, gypsum products, clay bricks, flue condensates, and industrial waste. When there is continuous moisture present, sulfate, and tricalcium aluminate will slowly react to form calcium sulphoaluminate. The brickwork's render or cement mortar expands as a result of this reaction and finally disintegrates.

- **Solar Radiation:** Although the most frequent cause of brickwork cracking is soil drying out from trees, some clay materials are also highly vulnerable to drying out from direct sunlight. The most common symptom, stepped diagonal cracking, usually appears at the building's northeast and northwest corners, and is most likely to affect the wall facing north.
- **Migration of Moisture:** The long-term dome and saucer effect is a phenomenon caused by moisture moving beneath a building. Moisture moving from the house's perimeter to its center causes the dome effect, which is a slow heaving of the soil; moisture moving in the opposite direction, from the center to the perimeter, is known as the saucer effect. The walls tilt outward due to the dome effect, and the outside wall develops horizontal cracks as a result of the roof's restraint. On the exterior surface as opposed to the interior, these will be wider.
- **Planting of Tree:** Some brickwork flaws could be brought on by large tree roots growing close to the building site. The strength of the footings beneath the house should be taken into consideration when planting trees. Insufficient strength could lead to the tree roots penetrating the ground beneath the structure, causing the brick wall to settle and crack.
- **Uneven Settlement of Foundations:** Movement may happen when a heavy load, like a large column, is placed on the foundation and the soil readjusts or moisture is removed from the soil. When the soil has finally compacted sufficiently to support the load, this consolidation will come to an end. Masonry or brick walls may split as a result of movement during compaction.
- **Excessive Vibration:** Damage caused by vibration from earth tremors, heavy traffic, or pile driving is fairly rare, however, if the vibration is great enough to actually cause the foundation to move, brickwork cracking or defect can occur. The crack will show up irregularly if the sources of the vibration cannot be removed.
- **Additional Building:** Building an additional building onto a house or adding more story of houses can impose a load intensity on the soil different from that which is there already and so cause differential settlement. Even if the loading intensities are similar, the difference in time between when the two settlements occurred can be enough to create the brick wall cracks.
- **Approach of Design:** The design of brick walls is very important. The designer must have knowledge about the brick experience in choosing the types of brick to be used. The designer must be fully aware of the client's needs. Defects often occur because of a lack of understanding of the different

types of bricks and the method of constructing the brick wall.

- **Lack of Experience of the Worker:** When masonry is constructed, workers often do not follow strict construction and standard procedures. This results in incomplete mortar, particularly in vertical mortar joints, and can even lead to dense and permeability seams. Furthermore, the mortar in the brick water was absorbed by the dry brick on the wall, resulting in low mortar strength, brick and mortar separation, and the overall stiffness of the masonry crumbling and gray mortar joints cracking.
- **Overloading:** Brickwork cracks may be caused by the building itself, certain sections of the building, or the ground the structure is sitting on being overloaded. While the overloading may have been partially accommodated by permanent compaction of the bricks, releasing the load does not always allow the masonry wall to return to its initial state. The factors that caused the overloading could be internal—like forces resulting from the installation of very heavy equipment for which the masonry wall was not intended—or external—like an excessive amount of wind. A shift in the profession could yield similar outcomes.
- **Lack of Maintenance:** Maintenance must be taken for all the buildings to minimize the cost of repair work for serious defects happen. Lack of maintenance will make small cracks or defects on brickwork change to more serious and serious.
- **Improper Soil Analysis and Preparation:** It is crucial for a builder to perform some kind of soil preparation and analysis when selecting a plot of land for construction. The analysis will give the builder or developer enough knowledge to select the best construction plan and appropriate soil preparation methods for developing the property. An improperly analyzed and prepared soil can seriously harm a housing development. For instance, a builder may find that the soil on the property is composed of expansive soil. If this is the case, homes must have a foundation that can withstand variations in soil pressure and consistency. This is due to the fact that expansive soil will expand while wet and contract when it dries. Unaware of this type of soil, problems will surely arise when building takes place on it. The masonry wall, or brick wall, will crack. This will require expensive and invasive repair. Regretfully, incorrect preparations and analyzes of the soil occur frequently.
- **Negligence of Construction:** Unfortunately, careless building is a frequent occurrence. A builder must exercise extreme caution when selecting a subcontractor or contractor. Sometimes a builder will use the same contractor for multiple projects. A contractor might know how to lay floor tiles, but he or she may not know how to construct a brick wall. When a defect occurs, multiple parties may bear some of the blame; your attorney will be able to ascertain this.
- **Movement of the Ground:** The earth may move as a result of landslips, mining subsidence, earthquakes, or

moisture changes in shrinkable clay soil. The building's brick wall will fracture because it has moved apart from the other walls without the materials' actual sizes changing. The building can theoretically be restored to its original state by reconnecting the dislodged parts, but in reality, this is rarely possible.

- **Acid Rain:** Rainfall that is significantly more acidic than typical rainfall is referred to as acid rain, and it is typically brought on by pollution from emissions of nitrogen and sulfur. The action of acid rain causes brick masonry to deteriorate through a number of different mechanisms. The bricks' glassy phase dissolves selectively, making them vulnerable to acid rain. The calcareous components' reaction is what primarily affects the mortar. When these reactions' soluble salts combine with rainwater or condensed moisture, they migrate through the masonry's porous matrix. Salts will accumulate in the areas where water evaporates. The masonry structure is mechanically disrupted when these salts dissolve and re-crystallize repeatedly. Depending on their respective porosities, the brick or the mortar will be more severely impacted by the salts because they will concentrate on the more porous material.
- **Environment:** The temperature and wind have both significantly changed in the environment. This is a result of them applying loading to the structure. Strong winds have the potential to shift the building as a whole, including the brick wall, which could cause cracks in certain areas. Applying a high temperature to the brickwork could result in spelling errors, and heating the bricks to a high temperature can cause surface verification.

CASE STUDY

ENERGETICALLY MODIFIED CEMENT

Energetically modified cement (EMCs) are a class of cement made from pozzolans (e.g., fly ash, volcanic ash, pozzolana), silica sand, blast furnace slag, or Portland cement (or blends of these ingredients). The term "energetically modified" arises by virtue of the mechanochemistry process applied to the raw material, more accurately classified as "high energy ball milling" (HEBM). At its simplest, this means a milling method that invokes high kinetics by subjecting "powders to the repeated action of hitting balls" as compared to (say) the low kinetics of rotating ball mills. This causes, amongst others, a thermodynamic transformation in the material to increase its chemical reactivity. For EMCs, the HEBM process used is a unique form of specialized vibratory milling discovered in Sweden and applied only to cementitious materials, here called "EMC Activation."

By improving the reactivity of pozzolans, their strength-development rate is increased. This allows for compliance with modern product-performance requirements ("technical standards") for concretes and mortars. In turn, this allows for the replacement of Portland cement in the concrete and mortar mixes. This has a number of benefits to their long-term qualities. Energetically modified cements have a wide range of uses. For example, EMCs have been used in concretes for large infrastructure projects in the United States, meeting U.S. concrete standards.

Justification

The term "energetically modified cement" incorporates a simple thermodynamic descriptor to refer to a class of cements produced using a specialized highly intensive milling process first discovered in 1993 at Luleå University of Technology (LTU) in Sweden. The transformatory process is initiated entirely mechanically as opposed to heating the materials directly. The mechanisms of mechanochemical transformations are often complex and different from "traditional" thermal or photochemical mechanisms. HEBM can transform both the physical and thermodynamic properties that for example, "can lead to glass formation from elemental powder mixtures as well as by amorphization of intermetallic compound powders." The effects of HEBM-transformation cause a thermodynamic change that resides ultimately in a modified Gibbs Energy. The process increases the binding capacity and chemical reactivity rates of the materials transformed.

Continuing academic work and research regarding "self-healing" properties of energetically modified cements is ongoing at LTU. For example, EMCs has received awards from the *Elsa ⊠ Sven Thysells stiftelse för konstruktionsteknisk forskning* (Elsa & Sven Thysell Foundation for Construction Engineering Research) of Sweden. The contribution of EMCs to the domain of mechanochemistry itself has also been recognized.

Etymology

The term "energetically modified cement" was first used in 1992 by Vladimir Ronin, introduced in a paper by Ronin et al. dated 1993 and presented at a formal meeting of the academic Nordic Concrete Research group. The process was refined by Ronin and others, including Lennart Elfgren (now Professor Emeritus of LTU, Department of Civil, Environmental and Natural Resources Engineering). In 2023, LTU awarded Elfgren the "Vice-Chancellor's Medal for Merit for outstanding and meritorious work" by virtue of his work *"for the spread of new knowledge and understanding of, in particular, the concrete construction field."*

At the 45th World Exhibition of Invention, Research and Innovation, held in 1996 in Brussels, Belgium, EMC Activation was awarded a gold medal with mention by EUREKA, the European inter-governmental (research and development) organization, for *"modification énergique de ciments."*

The term "energetically modified" has been used elsewhere—for example as recently as 2017—although such usage does not denote the method used was EMC Activation as defined here.

Overview

The claims made include:

- An EMC is a fine powder (typical of all cement) whose color depends on the material processed.
- EMCs are produced using only a "fraction" of the energy used in Portland cement production (claimed ~100 KWh/ton, <8% of Portland cement).
- No CO_2 is released by the process. It is "zero emissions."
- The purpose of an EMC is to replace the Portland cement requirement in the mortar or concrete being used. More than 70% replacement is claimed.
- EMC Activation is a dry process.
- No noxious fumes are released.
- EMC Activation is a low-temperature process, even though temperatures can be "momentarily extreme" at "sub-micron" scales.
- EMCs require no chemicals for their thermodynamic transformation.
- There are several types of EMCs, depending on the raw materials transformed.
- Depending on user requirements, delivered dry products may comprise also a minority proportion of "high clinker" Portland cement.
- Each type of EMC has its own performance characteristics, including mechanical load and strength development. Concrete cast from EMCs may yield significant "self-healing" capabilities.

- The most frequently used EMCs are made from fly ash and natural pozzolans. These are relatively abundant materials, and the performance characteristics can exceed those of Portland cement.
- In 2009, fly ash EMCs were demonstrated to exceed the 'Grade 120 Slag' benchmark per ASTM C989 — the most reactive form of cementitious blast furnace slag.
- Silica sand and granite can also be treated by the process to replace Portland cement.
- EMC products have been extensively tested by independent labs and certified for use by several US DOTs including in Federal Highway Administration projects.
- EMCs comply with respective technical standards, such as ASTM C618–19 (U.S.); EN-197, EN-206 and EN 450–1:2012 (CEN territories, including EEA); BS 86151:2019 (U.K.).
- Compared to using Portland cement, the resulting concrete-mix using EMC does not require a higher "total cementitious content" to meet strength-development requirements.
- In testing by BASF, the 28-day strength-development for 55% replacement of Portland cement by a natural pozzolanic EMC was 14,000 psi/96.5 MPa (i.e., > C95). This comprised a "total cementitious content" of 335 kg/m^3 (564 lbs/CY) concrete mix.

EMCs as "Low Carbon" Cements

Unlike Portland Cement, an EMC's production releases no carbon dioxide whatsoever. This makes EMCs "low carbon cements."

The first cited claims for EMC's CO_2-reduction capabilities were made in 1999, when worldwide Portland cement production stood at 1.6 billion tons per year. From 2011 to 2019, worldwide Portland cement production increased from 3.6 to 4.1 billion tons per year. Energetically modified cement's potential for contributing to a worldwide reduction of CO_2 has been *externally* recognized since 2002 and has been ongoing. Recent recognition has included the 2019 Energy Transitions Commission (Lord Adair Turner and Lord Stern) report *Mission Possible sectoral focus: cement* (2019). Recognition of the "Zero-Carbon" potential was set out by McKinsey & Co in its 2020 report *Laying the foundation for zero-carbon cement*. In 2023, the contribution offered by EMCs in achieving "low carbon" materials was further acknowledged within the academic domain of mechanochemistry.

CLASS ACTIVITY

CONCRETE MIX DESIGN AND TESTING

- **Objective:** To understand the properties and importance of different construction materials, especially concrete, by designing, mixing, and testing concrete samples.
- **Materials Needed:** Portland cement, Sand, Gravel, Water, Measuring tools, and Mixing containers, Concrete molds etc.
- **Learning Outcomes:** Students will understand the importance of mix design and how different ratios affect concrete strength. They will gain hands-on experience in mixing and testing concrete.
- **Assessment:**
  - Participation and teamwork during the activity.
  - Accuracy and completeness of the recorded observations and predictions.
  - Quality of the analysis and conclusions drawn from the testing results

SUMMARY

- The engineering structures are composed of materials. These materials are known as engineering materials or building materials or materials of construction. It is necessary for the civil engineer to become conversant with the properties of such materials.
- Cementitious materials include the many products that are mixed with either water or some other liquid or both to form a cementing paste that may be formed or molded while plastic but will set into a rigid shape.
- Masonry cements or mortar cements are widely used because of their convenience. While they are, in general, mixtures of one or more of the above-mentioned cements with some admixtures, they deserve special consideration because of their economies.
- Aluminous cement is much more resistant to the action of sulfate waters than Portland cement. They also appear to be much more resistant to attack by water containing aggressive carbon dioxide or weak mineral acids than silicate cement.
- Natural cement is formed by calcining a naturally occurring mixture of calcareous and argillaceous substances at a temperature below that at which sintering takes place.
- Hydrated limes are prepared from quicklimes with the addition of a limited amount of water. After hydration ceases to evolve heat, the resulting product is a fine, dry powder. It is then classified by air-classification methods to remove undesirable oversized particles and packaged in 50-lb sacks.
- Lightly calcined magnesium oxide mixed with a solution of magnesium chloride forms a cement known as magnesium oxychloride cement or Sorel cement. It is particularly useful in making flooring compositions in which it is mixed with colored aggregates.
- Aggregate is a broad encompassing boulder, cobbles, crushed stone, gravel, aircooled blast furnace slag, native and manufactured sands, and manufactured and natural lightweight aggregates. Aggregates may be further described by their respective sizes.
- The distribution of aggregate sizes in a concrete mix is important because it directly influences the amount of cement required for a given strength, workability of the mix (and amount of effort to place the mix in the forms), in-place durability, and overall economy.

REVIEW QUESTIONS

1. What are the various types of cementitious materials and their applications in modern construction?
2. How do low-temperature gypsum derivatives influence the properties and performance of cementitious composites?

3. What roles do heavyweight and lightweight aggregates play in the engineering and structural characteristics of concrete?
4. How do chemical and mineral admixtures modify the behavior and durability of cement-based materials?
5. What are the fundamental components and characteristics of Portland cement concrete, and how do they affect its usage in construction projects?

REFERENCES

1. Abbas, A., Fathifazl, G., Isgor, O. B., Razaqpur, A. G., Fournier, B., & Foo, S. (2009). Durability of recycled aggregate concrete designed with equivalent mortar volume method. *Cement and Concrete Composites, 31*, 555–563.
2. Acker, A. V. (1998). Recycling of concrete at a precast concrete plant. In *Sustainable Construction Use of Recycled Concrete Aggregate* (pp. 321–332). Proceedings of the international symposium, London, UK: Thomas Telford.
3. Ahmed, A., & Struble, L. (1995). Effects of microstructure on fracture behavior of hardened cement paste. In *Microstructure of Cement-Based Systems/Bonding and Interfaces in Cementitious Materials: Symposium* (pp. 99–108).
4. Aitcin, P. C., & Neville, A. M. (1993). High performance concrete demystified. *Concrete International, 15*, 21–26.
5. Ajdukiewicz, A., & Kliszczewicz, A. (2002). Influence of recycled aggregates on mechanical properties of high strength/high performance concrete. *Cement and Concrete Composites, 24*, 269–279.
6. Alexander, M. G. (1996). The effects of ageing on the interfacial zone in concrete. In *Interfacial Transition Zone in Concrete: State-of-the-Art Report* (pp. 150–174).
7. Ali Elhakam, A., Mohamed, A. E., & Eslam, A. (2012). Influence of self-healing mixing method and adding silica fume on mechanical properties of recycled aggregate concrete. *Construction and Building Materials, 35*, 421–427.
8. American Association of State Highway and Transportation Officials. (1983). *Rapid Determination of the Chloride Permeability of Concrete* (AASHTO T277). American Association of State Highway and Transportation Officials.
9. Ann, K. Y., Moon, H. Y., Kim, Y. B., & Ryou, J. (2008). Durability of recycled aggregate concrete using pozzolanic materials. *Manual of Waste Management, 28*(6), 993–999.
10. Casuccio, M., Torrijos, M. C., Giaccio, G., & Zerbino, R. (2008). Failure mechanism of recycled aggregate concrete. *Construction and Building Materials, 22*, 1500–1506.
11. Chakradhara, R. M., Bhattacharyya, S., & Barai, S. (2011). Influence of field recycled coarse aggregate on properties of concrete. *Materials and Structures*, 205–220.
12. Edwards, B. (1999). *Sustainable Architecture: European Directives and Building Design* (2nd ed.). Oxford Architectural Press.
13. Eguchi, K., Teranishi, K., Nakagome, A., Kishimoto, H., Shinozaki, K., & Narikawa, M. (2007). Application of recycled coarse aggregate by mixture to concrete construction. *Construction and Building Materials, 21*, 1542–1551.

14. Erdogdu, S. (2005). Compatibility of superplasticizers with cements different in composition. *Cement and Concrete Research, 30*, 767–773.

15. Etxeberria, M., Vázquez, E., & Mari, A. R. (2006). Microstructure analysis of hardened recycled aggregate concrete. *Magazine of Concrete Research, 58*, 683–690.

CHAPTER

3

Structural Mechanics

LEARNING OBJECTIVES

After studying this chapter, you will be able to:

- Discuss the basic principles of statics and dynamics
- Explain the introduction to structural elements
- Focus on simple structural analysis techniques

KEY TERMS FROM THIS CHAPTER

Civil engineering
Dynamic forces
Finite element analysis
Static analysis
Dynamic analysis
Equilibrium
Frequency
Trusses

INTRODUCTION

Engineering's field of structural mechanics studies how structures behave under load and how to design them for performance and safety. It uses mathematical and physical concepts to examine and forecast how various forces will affect structures like buildings, bridges, dams, and mechanical systems. Understanding the ideas of forces, stresses, and strains is fundamental to structural mechanics. Structures are subject to forces from the sun, wind, earthquakes, and dynamic loads like machinery or people moving through them. The materials of the structure experience internal stresses as a result of these forces, which need to be controlled to avoid failure. Whereas strain gauges a material's deformation under stress is the amount of internal force per unit area. These factors are used by engineers to determine whether a structure can support anticipated loads without going over its material limitations.

Structural mechanics design involves a number of important steps. Initially, engineers examine the anticipated loads and environmental circumstances that a structure will face throughout its lifespan. This includes accounting for the worst-case situations, like hurricanes or earthquakes. They forecast how the structure will react to these loads using computer simulations and mathematical models, making sure safety margins are included to account for unforeseen circumstances and uncertainties.

Structural mechanics relies heavily on materials. Engineers choose materials according to their mechanical attributes, like cost-effectiveness, strength, stiffness, and durability. Common materials include steel, wood, concrete, and composites. These materials are selected for particular applications based on how well they function in various environments. Technological developments have improved the field of structural mechanics significantly. Before construction starts, engineers can create detailed models of structures and simulate their behavior under different conditions using computer-aided design (CAD) and finite element analysis (FEA) software. This makes more efficient designs possible and lowers the chance of structural failure.

Environmental impact and sustainability are also taken into account in structural mechanics. When designing a structure, engineers try to use as little material as possible, optimize it for energy efficiency, and use recycled materials whenever possible. By taking a comprehensive approach, buildings are guaranteed to function well and enhance their surroundings. In order to design safe, effective, and long-lasting structures that can endure the forces placed upon them over time, structural mechanics is a prerequisite. Structural mechanics continues to advance through the application of physics, mathematics, and engineering principles, embracing new materials and technologies to address the demands of contemporary infrastructure development and construction.

3.1. BASIC PRINCIPLES OF STATICS AND DYNAMICS

In the field of civil engineering, the safety, stability, and effectiveness of infrastructure and structures are largely dependent on both statics and dynamics.

Let's delve into how these principles are applied in this specific field:

3.1.1. Principles of Statics in Civil Engineering

1. **Force and Moment Equilibrium:** To analyze and design structures that are in static equilibrium, one must have a solid understanding of statics principles. By using these principles, civil engineers can make sure that structures like buildings, bridges, dams, and other infrastructure hold up against different types of loads, including dead loads (e.g., the structure's weight), live loads (e.g., people), wind loads, and seismic pressures.
2. **Structural Analysis:** Statics is a tool used by engineers to compute internal forces in structural elements like trusses, columns, and beams. Engineers make sure that these elements don't exceed the strength and capacity of the material by analyzing the forces (such as tension and compression) and moments acting on them. This helps to prevent structural failure.
3. **Support Reactions:** Determining support reactions at the points where a structure connects to its foundation or other supporting elements requires an understanding of statics principles. This entails figuring out the forces and moments at the supports to make sure the structure is sufficiently stable and that they can safely bear the applied loads.
4. **Design of Structural Elements:** Statics principles govern the sizing and arrangement of structural elements in civil engineering design. Engineers use the expected forces and moments ascertained by static analysis to compute cross-sectional dimensions, reinforcement needs, and connection details.

3.1.2. Principles of Dynamics in Civil Engineering

1. **Dynamic Loads Analysis:** The effects of dynamic loads, such as wind gusts, vehicle-induced vibrations, and earthquake ground motions, on structures are analyzed by civil engineers using principles of dynamics. Dynamics aids in forecasting how these fluctuating loads may cause a structure to accelerate, vibrate, or deform, affecting the resilience and safety of the structure's design.
2. **Response Spectrum Analysis:** Engineers use response spectrum analysis, which is based on dynamics principles, to assess how a structure will react to ground motions during an earthquake for seismic design. This technique estimates the maximum forces and deformations likely to occur during an earthquake by taking

into account the dynamic properties of the structure and the frequency content of seismic waves.

3. **Vibration Analysis:** In civil engineering, dynamics principles are used to analyze and reduce vibrations in structures that are caused by wind, traffic, machinery, or pedestrian activity. To make sure that vibrations do not jeopardize comfort, safety, or structural integrity, engineers examine the inherent frequencies and mode shapes of structures.
4. **Dynamic Stability Considerations:** Civil engineers evaluate a structure's dynamic stability in addition to its static stability when it comes to wind-induced oscillations or fluid-structure interactions (e.g., bridges spanning waterways). Dynamics aids in the design of structures to reduce the possibility of instability and resonance under dynamic loading circumstances.

3.1.2.1. Integration and Application

To guarantee complete safety and functionality, both statics and dynamics are incorporated into the design and analysis processes of civil engineering projects. Engineers can accurately simulate and predict the behavior of structures under a variety of static and dynamic conditions by using sophisticated computational tools like computer-aided design (CAD) and finite element analysis (FEA). The fundamental tenets of civil engineering are statics and dynamics, which direct the analysis, design, and evaluation of structures to effectively withstand both static and dynamic loads. Through the application of these principles, engineers can maximize the resilience, longevity, and performance of infrastructure, satisfying both societal demands for sustainable development and safety requirements.

3.1.3. Dynamic Analysis of Structures

When it comes to comprehending how structures behave when exposed to dynamic loads like wind, vibrations, and earthquakes, dynamic analysis is indispensable. The structural response is better-understood thanks to this analysis, which also ensures the dependability, effectiveness, and safety of different engineering systems. Because structures are subjected to a variety of dynamic forces on a constant basis, it is critical to assess how they react to these loads. A subfield of structural engineering known as "dynamic analysis" is devoted to analyzing how structures behave in dynamic situations. Dynamic analysis takes into account the time-dependent effects on structures, in contrast to static analysis, which considers the equilibrium of forces (Wodek Gawronski, 2004).

The process of researching and assessing how a structure behaves under dynamic loads, such as wind, vibrations, earthquakes, and other external forces, is known as "dynamic analysis of structures." It entails examining how a structure responds to loads that change over time in order to comprehend how it acts and responds to these dynamic forces.

The dynamic features of a structure, such as its inherent frequencies, mode shapes, damping qualities, and reaction to different kinds of dynamic loads, are the main focus of structural dynamics. The goal of the analysis is to ascertain the structural response in terms of displacements, velocities, accelerations, and internal forces while taking into account the interactions between the structure and the applied forces. In order to make sure that a structure can withstand the dynamic forces it may encounter throughout its lifetime, engineers and designers rely heavily on dynamic analysis when designing and evaluating structures. Additionally, it helps optimize the design to reduce oscillations, vibrations, and possible failure modes.

Structural dynamics is studied using a variety of techniques, including finite element analysis, response spectrum analysis, modal analysis, and time history analysis. To forecast the behavior and response of the structure under dynamic loading conditions, these approaches combine computational techniques, numerical simulations, and mathematical modeling. Many different types of structures, such as buildings, bridges, dams, offshore platforms, aerospace structures, and mechanical systems, are subject to the application of dynamic analysis. Engineers can guarantee the safety, stability, and performance of these structures under various dynamic loading scenarios by comprehending their dynamic behavior.

3.1.4. Types of Dynamic Analysis

- **Modal Analysis:** A basic method for determining a structure's inherent frequencies and mode shapes is modal analysis. To ascertain the dynamic properties of the structural system, entails solving the eigenvalue problem. In order to support the further stages of analysis, modal analysis offers vital information regarding the predominant modes of vibration and their associated frequencies.
- **Response Spectrum Analysis:** Response spectrum analysis is a commonly used method for assessing structures that are subjected to earthquake loads. It makes use of a response spectrum, which shows a structure's maximum response at various frequencies. Engineers can evaluate a structure's performance under seismic motions and design it to withstand anticipated ground motions by using the response spectrum on the structure.
- **Time History Analysis:** The process of time history analysis entails modeling the real-world time-varying loads or ground movements that a structure might encounter. It records the structure's dynamic response over time while taking into account the particulars of the applied forces. When dealing with intricate loading patterns or non-linear behavior of the structure, time history analysis comes in handy (Figure 3.1).

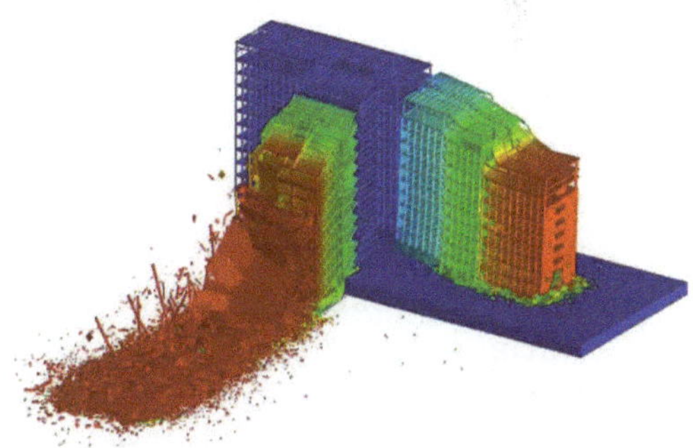

Figure 3.1. *It captures the dynamic response of the structure over time.*

Source: https://www.structuralguide.com/wp-content/uploads/2023/05/Dynamic-Analysis-of-Structures.jpg.webp.

3.1.5. Steps Involved in Dynamic Analysis

Dynamic analysis follows a systematic approach to evaluate the response of structures to dynamic loads. The key steps involved are:

- **Structural Modeling:** The creation of a precise and realistic model of the structure is the first stage in dynamic analysis. Determining the geometry, material characteristics, and boundary conditions are part of this. To guarantee accurate results, the structural model should include all of the important characteristics of the actual structure.
- **Load Definition:** Determining the dynamic loads operating on the structure is the next step. These loads may differ based on the particular application and analysis type. For instance, response spectra or ground motion records are used as input in seismic analysis. The distribution of wind pressure is taken into account in wind analysis. For consistent results, it's critical to carefully choose and apply the right loads.
- **Selection of Analysis Method:** The selection of an appropriate analysis method is contingent upon the nature of the problem and the desired level of accuracy. Initial evaluations and mode shape extraction are frequent uses of modal analysis. Analysis of the response spectrum is appropriate for seismic design. When a thorough understanding of dynamic behavior is needed, time history analysis is used. The particular goals of the study will determine which analysis method is used.
- **Calculation of Natural Frequencies and Mode Shapes:** The natural frequencies and matching mode shapes of the structure are computed in modal analysis. These are the structure's innate qualities that govern its dynamic behavior. The mode shapes show the patterns of deformation connected to each frequency, while the natural frequencies show the frequencies at which the structure tends to vibrate.
- **Evaluation of Dynamic Response:** The dynamic response of the structure to the applied loads can be assessed once the natural frequencies and mode shapes are established. This entails figuring out the internal forces, displacements, and accelerations at different points throughout the structure. The response analysis assists in evaluating the structural performance, pinpointing areas of importance, and guaranteeing that the design satisfies the intended standards.

3.1.6. Applications of Dynamic Analysis

Structural dynamics finds wide applications in various fields of engineering. Some notable applications include:

- **Seismic Design of Buildings:** When it comes to seismic design, dynamic analysis is essential because it allows engineers to assess how structures will react to an earthquake. Engineers can design buildings that can withstand ground motions and minimize damage during seismic events by taking into account the dynamic characteristics of the soil-structure system and the expected seismic forces.
- **Bridge Design and Evaluation:**

Dynamic forces like wind, earthquakes, and traffic volumes can affect bridges. By taking the dynamic behavior, natural frequencies, and mode shapes into account, dynamic analysis helps to design bridge structures that are both safe and effective. It guarantees that bridges are stable and serviceable, and able to bear the dynamic loads.

- **Vibrations in Mechanical Systems:** For mechanical systems to be understood and vibrations to be mitigated, dynamic analysis is necessary. It aids in the analysis of the dynamic response of other mechanical systems, automobiles, aerospace structures, and rotating machinery. Engineers can reduce fatigue failures, vibrations, and noise by maximizing design and identifying potential resonance conditions.

  An essential tool for engineering design is the dynamic analysis of structures, which sheds light on how these structures respond to dynamic loads. Engineers can optimize designs, guarantee safety, and enhance performance by knowing how structures react to wind, vibrations, earthquakes, and other dynamic forces.

For many applications, such as seismic design, bridge evaluation, and vibration control in mechanical systems, dynamic analysis is still essential, even though it can be difficult to capture complex behavior and meet computational demands. It enables engineers to design structures that can withstand dynamic forces while maintaining dependability and safety.

3.1.7. Analysis of Forces and Equilibrium

Engineers utilize forces and equilibrium as fundamental principles in structural analysis to guarantee the stability and safety of civil engineering structures. Designing structures that can withstand the different loads they may encounter over their lifespan requires an understanding of how forces interact and equilibrium conditions are maintained.

3.1.7.1. Forces

In structural analysis, forces are the interactions between bodies that cause them to accelerate or deform. These forces can be broadly categorized into external forces and internal forces:

1. **External Forces:** These are forces applied to a structure from outside, such as:
   - **Dead Loads:** Permanent, static forces due to the weight of the structure itself and any permanent attachments.
   - **Live Loads:** Variable forces caused by occupants, furniture, vehicles, or other moving objects.
   - **Environmental Loads:** Forces from natural elements like wind, snow, earthquakes, and thermal expansion.
2. **Internal Forces:** These are forces that develop within the structure in response to external forces and are necessary to maintain equilibrium:
   - **Axial Forces:** Forces along the axis of a structural member, either tension (pulling apart) or compression (pushing together).

- **Shear Forces:** Forces parallel to the surface of a structural member that causes one part of the material to slide relative to an adjacent part.
- **Bending Moments:** Forces that cause a structural element to bend, resulting in tension on one side and compression on the other.
- **Torsional Forces:** Forces that cause twisting in a structural member.

3.1.7.2. Equilibrium

The state in which the total of all forces and moments acting on a structure is zero is known as equilibrium, and it is a fundamental idea in structural mechanics. This requirement guarantees that there is no net acceleration and that the structure stays balanced and stable.

Equilibrium is typically analyzed using two primary conditions:

1. **Translational Equilibrium:** This condition states that the vector sum of all external forces acting on a structure must equal zero.
2. **Rotational Equilibrium:** This condition states that the sum of all external moments (torques) acting on a structure about any point must equal zero.

Engineers can forecast and design structures to withstand the forces to which they will be subjected throughout their service life by analyzing forces and ensuring equilibrium. This entails figuring out the internal forces in structural members, making sure they don't go over the material limits, and creating the right supports and connections to efficiently distribute loads. In civil engineering, structural analysis is based on forces and equilibrium.

These ideas are used by engineers to create effective and safe structures that can withstand a variety of loads and environmental factors. The precision and effectiveness of force analysis are continually being improved by developments in computational tools and simulation techniques, giving engineers the ability to optimize designs and guarantee the resilience of civil infrastructure.

3.2. INTRODUCTION TO STRUCTURAL ELEMENTS

In order to form a larger structure, indivisible pieces come together to form structural elements. Since they form the fundamental basis for all structures and have no parts or characteristics that could be made any simpler, they are regarded as indivisible. To facilitate analysis, structural elements are used to divide a building or construction into several smaller pieces. The structural elements consist of five main parts: diaphragms, columns, beams, struts, and ties. Buildings, bridges, and other architectural constructions can all be described by structural systems that are made up of many more intricate structural elements.

3.2.1. Structural Engineering

The structural design of man-made structures is the focus of the civil engineering specialty known as structural engineering. These engineers, who are frequently referred to as creating the "bones and muscles" of structures, must comprehend the stiffness, stability, and strength of both building and non-building structures. Within the field of civil engineering, structural engineering applies the principles of mathematics, physics, and empirical knowledge to the safe design of the load-bearing components and "bones" of man-made structures. The vast and intricate body of knowledge available to structural engineers today allows them to precisely forecast how various forms and materials will behave in relation to loads and stresses placed on their structures.

The field of structural engineering is a subset of civil engineering, wherein structural engineers are skilled in designing the "bones and joints" that give structures constructed by humans their form and appearance. The stability, strength, rigidity, and earthquake susceptibility of constructed structures, both for buildings and nonbuilding structures, must also be understood and calculated by structural engineers. The structural designs are combined with those of other designers, including building services engineers and architects, and they frequently oversee the on-site construction of projects by contractors. Additionally, they can be involved in the design of vehicles, machinery, and medical equipment when structural integrity has an impact on safety and functionality.

Refer to the structural engineering glossary. The foundation of structural engineering theory is empirical knowledge of the structural performance of various materials and geometries, as well as applied physical laws. Complex structural systems are constructed through structural engineering design using a variety of comparatively straightforward structural concepts. To accomplish these objectives, structural engineers must find innovative and effective ways to use resources, structural components, and materials.

1. Engineering Design Process

i. Conceptual Stage:

- Needs are identified then objectives are expressed to meet these needs
- Input from:
- Clients
- Governmental regulatory agencies
- Architects, planners, and engineers

ii. Preliminary Design Stage:

- Creative ideas by the engineers
- Keep in mind construction aspects
- Thorough consideration of expected loads on the structure at all construction stages and during occupancy of the finished structures.
- Sizing of structural elements for safety and serviceability
- Architectural Constraints
- Simplicity and duplication
- Fabrication and construction procedures
- Preliminary design approximate theories of structural analysis are used to minimize the time during this phase

iii. Selection Stage:

- At this stage, all alternatives are presented and all parties involved participate in the selection stage so the final design stage can begin

iv. Final Design Stage:

- Loads are determined with greater accuracy than in the preliminary stage.
- All loading combinations are examined in this stage.
- Structural analysis is carried out with greater accuracy than the preliminary stage with the elimination of all approximations
- The results are presented in sets of drawings and specifications showing:
    - o Sizing of Members
    - o Detailing
    - o Quality of workmanship
    - o Design/building codes used
    - o Bill of materials
    - o Total cost

2. Structural Analysis

- *Structural Analysis*: the determination of the structural response to specific loads and actions.
- *Response*: measured by establishing the forces and deformations throughout the structure.
- *Analysis*: based on engineering mechanics theory, laboratory research, experience, and engineering judgment.

3. Structural Form

The form of the structure depends on many considerations as:

- Functional requirements
- Aesthetic (Beauty) requirement

- Surface and subsurface conditions
- Material availability
- Construction Expertise
- Economical limitations
- Environmental impact
- Safety

3.2.2. Structural Elements

Any structure is essentially made up of only a small number of different types of elements:

- Columns
- Tie Rods
- Beams
- Plates
- Arches
- Shells
- Catenaries

Many of these elements can be classified according to form (straight, plane/curve) and dimensionality (one-dimensional/two-dimensional) (Figure 3.2):

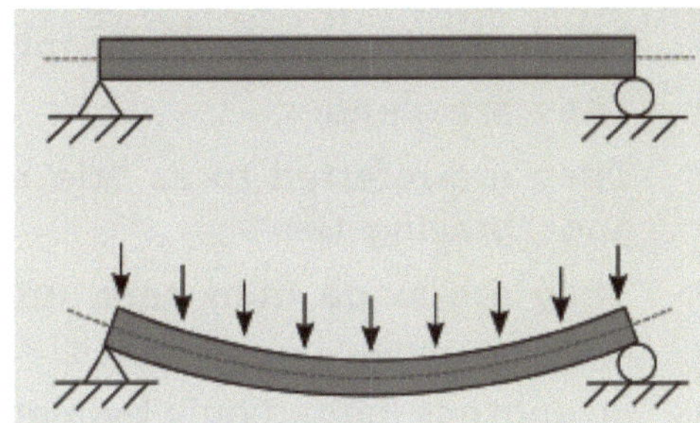

Figure 3.2. *A statically determinate simply supported beam, bending under an evenly distributed load.*

Source: https://upload.wikimedia.org/wikipedia/commons/thumb/c/c9/Bending.svg/320px-Bending.svg.png.

	One-Dimensional		Two-Dimensional	
	Straight	Curve	Plane	Curve
(predominantly) bending	beam	continuous arch	plate, concrete slab	lamina, dome
(predominant) tensile stress	rope, tie	Catenary	shell	
(predominant) compression	pier, column		Load-bearing wall	

3.2.2.1. Columns

Columns are elements that can support bending in addition to axial force (compression) or both (a technically named beam column is actually just a column). The axial and buckling capacities of an element must be verified in the design of a column.

The element's ability to resist the tendency to buckle is known as its buckling capacity. Its capacity is determined by the column's geometry, material, and effective length, which is determined by the constraints at the column's top and bottom. The effective length is k* l where l is the real length of the column and K is the factor dependent on the restraint conditions. The capacity of a column to carry axial load depends on the degree of bending it is subjected to, and vice versa. This is represented on an interaction chart and is a complex non-linear relationship.

3.2.2.2. Tie Rods

A tie rod or tie bar (also known as a hanger rod if vertical) is a slender structural unit used as a tie and (in most applications) capable of carrying tensile loads only.

- Subjected to tensile force only.
- They are slender.
- They are referred to as "tie rods" and "bracing ties"
- They are made from bars, angles, and channels.
- Strength is limited only by material strength

Subtypes and Examples of Applications

- In airplane structures, tie rods are sometimes used in the fuselage or wings.
- Tie rods are often used in steel structures, such as bridges, industrial buildings, tanks, towers, and cranes.
- Sometimes tie rods are retrofitted to bowing or subsiding masonry walls (brick, block, stone, etc.) to keep them from succumbing to lateral forces.
- The rebar used in reinforced concrete is not referred to as a "tie rod," but it essentially performs some of the same tension-force-counteracting purposes that tie rods perform.
- In automobiles, the *tie rods* are part of the steering mechanism. They differ from the archetypal tie rod by both pushing and pulling (operating in both tension and compression). In the UK, these items are generally referred to as *track rods*.
- In steam locomotives, a *tie rod* is a rod that connects several driving wheels to transmit the power from the connecting rod.
- Tie rods known as *sag rods* are sometimes used in connection with purlins to take the component of the loads which is parallel to the roof.
- The spokes of bicycle wheels are tie rods.
- In ships, tie rods are bolts that keep the whole engine structure under compression. They provide for fatigue strength. They also provide for proper running gear alignment which prevents fretting. They help to reduce the bending stress being transmitted to the transverse girder (Figure 3.3).

Figure 3.3. *Tie rods and anchor plates in the ruins of Coventry cathedral.*

Source: https://upload.wikimedia.org/wikipedia/commons/thumb/1/10/Coventry_Cathedral_ruins_tie_rods.JPG/450px-Coventry_Cathedral_ruins_tie_rods.JPG?20090312150008.

3.2.2.3. Beams

A beam may be defined as an element in which one dimension is much greater than the other two and the applied loads are usually normal to the main axis of the element. Beams and columns are called line elements and are often represented by simple lines in structural modeling.

- cantilevered (supported at one end only with a fixed connection)
- simply supported (fixed against vertical translation at each end and horizontal translation at one end only, and able to rotate at the supports)
- fixed (supported in all directions for translation and rotation at each end)
- continuous (supported by three or more supports)
- a combination of the above (ex. supported at one end and in the middle)

Elements that only carry pure bending are called beams. When a beam is bent, its length is divided into two sections, one of which is put into compression and the other into tension. While the tension part needs to be able to sufficiently withstand the tension, the compression part needs to be built to resist buckling and crushing (Figure 3.4).

Figure 3.4. *Beams are elements that carry pure bending only.*

Source: https://upload.wikimedia.org/wikipedia/commons/thumb/b/b1/The_Little_Belt_Bridge_%281935%29.jpeg/800px-The_Little_Belt_Bridge_%281935%29.jpeg?20180202005203.

3.2.2.4. Trusses

A truss is a construction made up of nodes, connection points, and members. Members can act in tension or compression when forces are applied at nodes where they are connected. Members that act in tension are known as tension members or ties, and members that act in compression are known as compression members or struts. Gusset plates are typically used in trusses to join intersecting elements. Because of their relative flexibility, gusset plates cannot transfer bending moments. Truss members can act in pure tension or compression because the connection is often set up so that the lines of force in the members coincide at the joint. When using solid beams would not be cost-effective for large-span structures, trusses are typically employed (Figure 3.5).

Figure 3.5. *A truss is a structure comprising members and connection points or nodes.*

Source: https://upload.wikimedia.org/wikipedia/commons/f/ff/McDonnell-Planetarium.jpg.

3.2.2.5. Plates

Bending can occur in two directions on plates. One example of a plate is a level slab of concrete. Although continuum mechanics can be used to understand plates, computer analysis or a codified empirical approach are typically used in their design due to their complexity. They can also be designed using yield line theory, which provides an upper bound on the collapse load by analyzing an assumed collapse mechanism.

In practice, this method is employed, but because it offers an upper-bound (i.e., It is crucial to take great care to make sure that the assumed collapse mechanism is realistic in order to avoid making an unsafe prediction of the collapse load for poorly thought-out collapse mechanisms.

3.2.2.6. Shells

Shells derive their strength from their form and carry forces in compression in two directions. A dome is an example of a shell. They can be designed by making a hanging-chain model, which will act as a catenary in pure tension and inverting the form to achieve pure compression.

3.2.2.7. Arches

Arches carry forces in compression in one direction only, which is why it is appropriate to build arches out of masonry. They are designed to ensure that the line of thrust of the force remains within the depth of the arch. It is mainly used to increase the bountifulness of any structure.

3.2.2.8. Catenaries

Catenaries derive their strength from their form and carry transverse forces in pure tension by deflecting (just as a tightrope will sag when someone walks on it). They are almost always cable or fabric structures. A fabric structure acts as a catenary in two directions.

3.2.3. What Do Structural Engineers Do?

The trained experts known as structural engineers are in charge of ensuring that the tall buildings and bridges we use on a daily basis are sturdy, safe, and won't collapse when loads are applied. They accomplish this by using their technical expertise to specify various building materials in a range of forms and geometries and to design structures that are resilient to environmental stresses and pressures like earthquakes, storms, and gravity loads. In my capacity as a structural engineer, I collaborate closely with my clients to comprehend the purpose of the project and then offer a safe, practical, and affordable solution to successfully finish it. In general, there are three steps involved in the structural design of a building: load calculation, structural analysis, and design.

Standards and building codes can be used to estimate the weight of the building and the loads that snow, wind, and earthquakes will place on the structures, depending on the function of the building. After the loads operating on the structure are identified, the structure is examined to ascertain how each load affects each of the component parts. Structural engineers can analyze various types of structures using engineering software. It is usually a good idea to use fundamental structural analysis formulas and first principle knowledge to verify the accuracy of the software results. Building components such as the floor system, vertical supports, foundations, and other structural elements can be designed using the forces and stresses obtained from the structural analysis software.

My primary concern as a structural engineer is public safety. The effects of a collapsed structure can be disastrous when it comes to the infrastructure and buildings that are used in our society. For this reason, factors of safety are used by structural engineers in their designs to take into consideration unforeseen deficiencies in construction materials or workmanship as well as uncertainty in load assumptions.

3.2.4. Where Do Structural Engineers Work?

When a building owner plans to add a significant expansion, add more floors, or change the use of a building, structural engineers are brought into the project. It's critical to realize that making changes to any structural component without first seeking advice from a qualified engineer could cause major structural damage and, in extreme circumstances, cause the building to collapse partially or completely.

When a structure sustains damage from fire, corrosion, environmental deterioration, impact, or wear and tear that could cause a loss of capacity and endanger public safety, structural engineers are also called upon. When a structural engineer is called in to evaluate an existing building, they visually inspect the building to ascertain the load-bearing elements' structural integrity, identify any potential safety concerns for the occupants, and recommend repair methods and structural details to return the building to its original state so that the applied loads can be withstood.

3.2.4.1. Are Architects & Engineers the Same Thing?

Engineers and architects play crucial but distinct roles in a building's design.

Figure 3.6. *Architects and engineers the same thing.*

Source: https://static.vecteezy.com/system/resources/previews/007/679/668/non_2x/architect-or-engineer-cartoon-illustration-using-a-multipurpose-board-table-to-sketch-building-constructions-and-project-miniatures-concept-vector.jpg.

If the building's interior spaces are cozy and welcoming in addition to having a pleasing aesthetic, the architect did a good job creating the structure. However, the stability of the structure under various loading scenarios clearly demonstrates the structural engineer's work. Regardless of the amount of snow on the roof or the wind speed outside, the building stays intact and provides its occupants with a safe and comfortable environment (Figure 3.6).

With the help of this explanation, hopefully, you will have a clear idea of structural engineering's definition, applications, and significance in a society where structures are all around us and we rely on them for daily needs. Like other engineers, structural engineers bear a heavy burden of making sure the public is protected, and cared for, and that their best interests are taken into account.

3.3. SIMPLE STRUCTURAL ANALYSIS TECHNIQUES

Basic structural analysis methods are essential tools in civil engineering that are used to assess how different structural elements behave and perform under various loads and conditions. These methods are essential for guaranteeing the effectiveness, dependability, and safety of infrastructure, including buildings, bridges, dams, and other structures.

The joints in truss analysis method is among the most widely used and fundamental approaches. Trusses are structural elements made up of straight members joined at joints. The internal forces (tensions and compressions) in each member are found by engineers analyzing each joint and applying equilibrium equations (sum of forces and sum of moments). This method makes member force calculations faster by breaking down complicated structures into smaller, more manageable components.

The method of sections is another crucial method that is frequently applied to the analysis of beams and frames. Using this technique, a section of the structure is cut through and isolated for close inspection. In order to ascertain the internal forces and moments at particular points along the beam or frame, engineers apply equilibrium equations to the isolated segment. This facilitates the assessment of crucial segments where the highest stresses or deformations could transpire, directing the structural element design and reinforcement.

When evaluating the impact of moving loads, such as vehicles, on structural systems exposed to external loads, like buildings and bridges, engineers use influence lines. Influence lines show how a particular reaction (e.g., shear force, and bending moment) at a structural point changes when a load passes over it. Plotting these lines helps engineers design supportive elements and load distribution strategies by pointing out critical locations where loads induce maximum forces or moments.

For more intricate structural analysis, engineers use matrix methods in addition to these basic approaches. In matrix structural analysis, the behavior of interconnected structural elements is modeled and solved through the use of computational techniques and linear algebra. When analyzing large and irregular structures, where traditional hand calculations become impractical, this approach is especially helpful.

Furthermore, structural analysis has been transformed by developments in finite element analysis (FEA) and computer-aided design (CAD). Engineers can produce intricate 3D models of structures with CAD software, which makes it easier to see and assess intricate geometries and component interactions. By breaking down structures into smaller components and using numerical techniques to simulate real-world conditions, FEA software further improves analysis capabilities by accurately predicting stresses, deformations, and failure modes. In civil engineering, basic structural analysis methods serve as the basis for evaluating how structures behave and function under different loads. With the aid of these methods, engineers can create structures that are economical, safe, and effective by comprehending internal forces, stresses, and deformations.

As technology develops, the incorporation of sophisticated analysis techniques and computational tools enhances the precision and effectiveness of structural assessment and design, guaranteeing the resilience and durability of civil infrastructure against changing demands and challenges.

3.3.1. Types of Structural Analysis: Concept, Importance, and Methods

An essential component of construction is structural analysis, which involves determining the stability and strength of a building. Determining how a structure behaves under various loads and conditions is a methodical process that uses computational and mathematical models. When designing and constructing buildings, bridges, and other structures, structural analysis is essential. It aids engineers in spotting possible issues and adjusting the design as needed to guarantee the structure will function as intended (Figure 3.7).

Building structures merely on the basis of educated guesses and gut feeling was a thing of the past. The development of cutting-edge technologies has allowed engineers to analyze structures in unprecedented detail and sophistication. Examples of these technologies are computer-aided design and advanced material sciences. This helps them to decide on the components, layout, and building techniques that will guarantee the structure is both efficient and safe (Figure 3.8).

Figure 3.7. *Performing structural analysis is a crucial process during construction.*

Source: https://housing.com/news/wp-content/uploads/2023/03/structural-compressed.jpg.

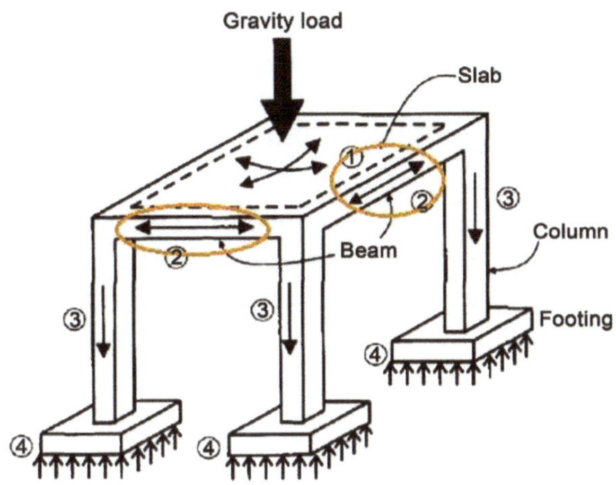

Figure 3.8. *This enables them to make informed decisions about the materials.*

Source: https://housing.com/news/wp-content/uploads/2023/03/What-is-Slab-Beam-Column-and-Footing-Construction_-The-Constructor.jpg.

3.3.2. Structural Analysis: Concept

A subfield of solid mechanics called structural analysis makes use of condensed models of solids like bars, beams, and shells to aid in engineering decision-making. Determining the impact of loads on the actual structures and their constituent parts is its primary goal. The models used in structure analysis are frequently differential equations in a single spatial variable, in contrast to the theory of elasticity. Any structure that has to support weights, such as ships, airplanes, bridges, and buildings, is subject to this kind of analysis. Structural analysis computes a structure's deformations, internal forces, stresses, support reactions, velocity, accelerations, and stability using concepts from applied mechanics, materials science, and applied mathematics. Physical testing is frequently avoided when a structure's suitability for use is confirmed by the analysis's findings. Thus, a crucial component of the engineering design of structures is structural analysis (Amin Ghali et al., 2016).

The scientific field of structural analysis studies the stability and strength of structures like towers, bridges, and buildings. It involves analyzing how a structure behaves under various loads and circumstances using computational and mathematical models. Verifying a structure's safety, stability, and ability to function as intended for the duration of its design is the aim of structural analysis.

The first step in the structural analysis process is to have a thorough understanding of all the loads that will be placed on the structure, including internal and external loads like wind, earthquakes, and temperature changes. After that, a mathematical model of the structure is created using this information, and it is solved to find the distribution of the forces and moments inside the structure. Subsequently, the outcomes of the structural analysis are employed to make well-informed decisions regarding the material selection, the dimensions and configuration of the structural components, and the construction techniques to be employed. Structural analysis is an iterative process; until the structure satisfies the design requirements, the design may be improved in response to the analysis's findings.

3.3.3. Structural Analysis: Importance

An essential component of building design that contributes to the longevity, stability, and safety of structures is structural analysis. It is a methodical procedure that includes utilizing computational and mathematical models to assess a building's stability and strength (Figure 3.9).

The importance of structural analysis in building design can be seen in several ways:

1. **Safety:** Structural analysis helps identify potential problems and weaknesses in the design that could lead to structural failure. This allows engineers to make necessary changes to the design to ensure the building will be safe and secure for its occupants.
2. **Optimization:** By performing structural analysis, engineers can optimize the design of the structure, making it more efficient and reducing the amount of material needed. This leads to cost savings and a more sustainable structure.
3. **Performance:** Structural analysis helps engineers understand how a structure will perform under various loads and conditions. This enables them to make informed decisions about the materials and construction methods that will ensure the structure will perform as intended.
4. **Future Proofing:** Structural analysis also helps engineers anticipate how a building will perform over time. This includes how it will respond to changes in the environment, such as wind, earthquakes, and temperature fluctuations. This information can be used to design the building to withstand these conditions, ensuring its longevity.

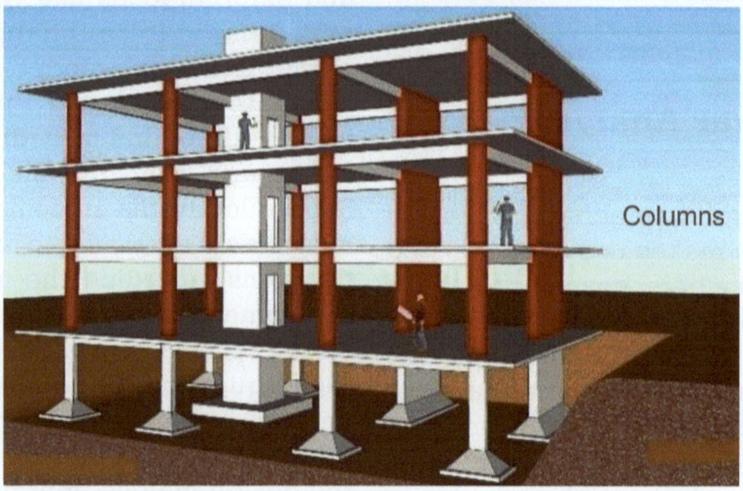

Figure 3.9. *Structural analysis is a crucial aspect of building design.*

Source: https://housing.com/news/wp-content/uploads/2023/03/Major-Parts-of-Reinforced-Concrete-Buildings-_-Framed-Structures-Components-CivilDigital--637x400.jpg.

3.3.4. Structural Analysis: Types

There are several types of structural analysis that are commonly used in construction, each with its own set of advantages and disadvantages. Some of the most common types of structural analysis include:

1. **Statics:** Statics is a type of structural analysis that focuses on the equilibrium of structures under static loads, such as gravity. It is used to determine the

forces and moments in a structure and how they are distributed throughout the structure.

2. **Dynamics:** Dynamics is a type of structural analysis that focuses on the response of structures to dynamic loads, such as earthquakes or wind. It is used to determine the natural frequencies and mode shapes of a structure and how it will respond to dynamic loads.
3. **Stability:** Stability analysis focuses on the ability of a structure to resist collapse or lateral displacement. This includes the evaluation of buckling, lateral drift, and torsional stability.
4. **Finite Element Analysis (FEA):** FEA is a type of computational analysis that involves dividing a structure into a large number of smaller elements and using computer algorithms to solve for the behavior of each element. This type of analysis is used to determine the distribution of stresses and strains in a structure and how they change over time.
5. **Linear and Nonlinear Analysis:** Linear analysis assumes that the relationship between loads and deflections is linear, while nonlinear analysis accounts for nonlinear behavior such as material yielding or plastic behavior.

These are just a few of the most common types of structural analysis used in construction. By understanding the different types of structural analysis and their strengths and weaknesses, engineers can make informed decisions about which type of analysis to use for a particular project.

3.3.4.1. Structures and Loads

A structure is defined as a body or a system of connected parts that is used to support a load in the context of structural analysis. Buildings, bridges, and towers are significant examples of civil engineering; in other engineering disciplines, ship and aircraft frames, tanks, pressure vessels, mechanical systems, and electrical supporting structures are significant examples. An engineer's job is to design a structure with economic and environmental constraints taken into consideration, as well as safety, aesthetics, and serviceability. Numerous different types of non-building structures are the focus of other engineering specialties.

3.3.4.2. Classification of Structures

The combination of structural elements and their materials is known as a structural system. It is crucial for a structural engineer to be able to identify the different components of a structure and categorize it according to its form or function. In addition to connecting rods, trusses, beams, and columns, other structural elements that direct the systemic forces through the materials include cables, arches, cavities, and channels; they can even be angles, surface structures, or frames.

3.3.4.3. Loads

The loads that a structure must support must be identified once the dimensional requirements have been established. Therefore, defining the loads acting on the structure is the first step in structural design. Building codes often specify the design loading for a structure. Engineers must adhere to all requirements of both general building codes and design codes for the structure to continue to be dependable.

In the design process, structural engineers have to deal with two different kinds of loads.

The weights of the various structural members and any items that are fixedly fastened to the structure make up the first category of loads, known as dead loads. The floor slab, roofing, walls, windows, plumbing, electrical fixtures, and other various attachments are a few examples of what this includes. Live loads are the second kind of load, and they differ in size and location. Live loads come in a variety of forms, including those from buildings, railroad and highway bridges, impact loads, wind, snow, earthquakes, and other natural loads.

3.3.5. Analytical Methods

A structural engineer needs to ascertain details like structural loads, geometry, support conditions, and material properties in order to conduct an accurate analysis. Such an analysis usually yields displacements, stresses, and support reactions. After that, this data is contrasted with standards that denote the circumstances of failure. Stability, non-linear behavior, and dynamic response may all be examined by advanced structural analysis. The elasticity theory approach, which is essentially a special case of the more general field of continuum mechanics, the finite element approach, and the mechanics of materials approach—also known as strength of materials—are the three methods used in the analysis. The first two use analytical formulations that can usually be solved by hand. They apply primarily simple linear elastic models and result in closed-form solutions. In reality, the finite element approach is a numerical technique for resolving differential equations produced by mechanics theories like strength of materials and elasticity theory. The finite-element approach, on the other hand, is more suitable for structures of any size and complexity and is largely dependent on computer processing power.

The formulation relies on three fundamental relations, namely equilibrium, constitutive, and compatibility, irrespective of the approach taken. When any of these relations are only partially satisfied, or only an approximation of reality, the solutions are approximate.

Structural analysis is a critical aspect of construction that involves evaluating a building's strength and stability. There are several methods of structural analysis used in construction, including:

1. **Hand Calculations:** Hand calculations are a simple and straightforward method of structural analysis that involves solving for the forces and moments in a structure using basic mathematical equations. This method is often used for simple structures and is a good starting point for more complex analyzes.
2. **Computer-Aided Analysis:** Computer-aided analysis is a type of structural analysis that involves using computer software to solve the behavior of a structure. This method is faster and more accurate than hand calculations and can be used to analyzes complex structures with many variables.
3. **Experimental Testing:** Experimental testing is a type of structural analysis that involves physically testing a structure to determine its behavior under loads. This method is often used in conjunction with computer-aided analysis and hand calculations to

validate the results of the analysis.

4. **Load Testing:** Load testing is a type of experimental testing that involves applying loads to a structure to determine its behavior under those loads. This method is used to determine the load-bearing capacity of a structure and to validate the results of the analysis.
5. **Monitoring:** Monitoring is a type of structural analysis that involves observing a structure over time to determine its behavior. This method is often used to monitor the performance of structures in real-world conditions and to validate the results of the analysis

3.3.5.1. Limitations

Every technique has some notable drawbacks. The application of material mechanics is restricted to extremely basic structural components with comparatively straightforward loading scenarios. Nonetheless, the permitted loading conditions and structural components are adequate to address a wide range of practical engineering issues. In principle, structural elements with any geometry can be solved under any loading condition thanks to the theory of elasticity. However, analytical solutions are only applicable in mildly complex situations. In contrast to mechanics of materials problems, which only call for the solution of an ordinary differential equation, elasticity problems also involve the solution of a system of partial differential equations, which is a significantly more complex mathematical task. It's possible that the finite element method is both the most practical and the most restrictive. The method itself depends on solving equations based on other structural theories (like the other two covered here). With the caveat that there will always be some numerical error, it does, however, generally make it possible to solve these equations, even with extremely complex geometry and loading conditions. A thorough grasp of this method's limitations is necessary for its effective and dependable application.

3.3.5.2. Strength of Materials Methods (Classical Methods)

The mechanics of materials method, the most straightforward of the three approaches here, can be applied to basic structural elements subject to particular loadings, such as circular shafts subject to torsion, prismatic beams in a state of pure bending, and axially loaded bars. The superposition principle can be used to superimpose the solutions under specific circumstances in order to analyze a member undergoing combined loading. There are solutions for unique situations involving common structures like pressure vessels with thin walls.

The method of sections and method of joints for truss analysis, the moment distribution method for small rigid frames, and the portal frame and cantilever method for large rigid frames are all products of using this approach in combination with statics for the analysis of entire systems.

These techniques were developed in the second half of the nineteenth century, with the exception of moment distribution, which was used in the 1930s. They are still utilized for both small-scale constructions and the initial design of larger ones.

Euler-Bernoulli beam theory and linear isotropic infinitesimal elasticity serve as the foundation for the solutions. Stated differently, these presumptions include but are not limited to, the following: the materials in question are elastic; stress and strain are linearly related; the material (but not the structure) behaves the same regardless of the direction of the applied load; all deformations are small; and beams are long in relation to their depth. The more an engineering model deviates from reality, the less useful (and riskier) the outcome becomes, as is the case with any simplifying assumption.

3.3.6. Methods Using Numerical Approximation

It is customary to base structural analysis on approximations of differential equation solutions. Usually, numerical approximation techniques are used for this. The Finite Element Method is the most widely used numerical approximation in structural analysis.

A structure is approximated by the finite element method as an assembly of elements or components with different types of connections between them, each of which has a stiffness assigned to it. Since a continuous system, like a plate or shell, has a finite number of elements connected at a finite number of nodes, it can be modeled as a discrete system in which the total stiffness is the sum of the stiffnesses of the individual elements. The stiffness (or flexibility) relation of an element determines how that element behaves. The system's stiffness or flexibility relation is obtained by assembling the different stiffnesses into a master stiffness matrix that depicts the entire structure. For basic one-dimensional bar elements, we can use the mechanics of materials approach to determine the stiffness (or flexibility) of a given element; for more complex two- and three-dimensional elements, we can use the elasticity approach. The most effective way to progress analytically and computationally is to solve partial differential equations using matrix algebra.

Later and more sophisticated matrix methods, known as finite element analysis, model an entire structure with one-, two-, and three-dimensional elements and can be used for articulated systems along with continuous systems like a pressure vessel, plates, shells, and three-dimensional solids. Earlier applications of matrix methods were used to model articulated frameworks with truss, beam, and column elements. Matrix finite-element analysis, which can be further divided into two main approaches: the displacement or stiffness method and the force or flexibility method, is commonly used in commercial computer software for structural analysis. Due to its simplicity in formulation for sophisticated applications and ease of implementation, the stiffness method is by far the most widely used. These days, finite-element technology is sophisticated enough to handle almost any system, given enough processing power. It can be applied to a variety of situations, such as static and dynamic external effects, solid and fluid interactions, anisotropic, orthotropic, and isotropic materials, and linear and non-linear analysis. However, since a lot depends on the model and the accuracy of the data input, this does not imply that the computed solution will always be reliable.

CASE STUDY

STRUCTURAL MECHANIC

Background

The original Tacoma Narrows Bridge was completed in July 1940 and spanned the Tacoma Narrows strait in Washington State, USA. Designed by Leon Moisseiff, the bridge quickly gained attention for its unique aerodynamic design, featuring a slim and flexible deck.

- **Failure:** Just months after its completion, on November 7, 1940, the bridge collapsed due to strong winds. The collapse was primarily attributed to aeroelastic flutter, a phenomenon where the bridge deck oscillated vertically in response to wind, leading to destructive vibrations. The bridge's failure occurred in dramatic fashion, captured on film, as the main span twisted and collapsed into the water below.
- **Lessons Learned:** The collapse of the Tacoma Narrows Bridge highlighted several crucial lessons in structural mechanics and bridge engineering:
  1. Dynamic Forces: Engineers learned that dynamic forces, such as wind-induced vibrations and aeroelastic effects, must be carefully considered in bridge design. The natural frequency of the bridge structure, combined with the wind speed and direction, can lead to resonance and catastrophic failure if not adequately addressed.
  2. Importance of Aerodynamics: The collapse underscored the importance of aerodynamic stability in bridge design. Subsequent bridge designs incorporated lessons from the Tacoma Narrows Bridge failure to improve aerodynamic profiles and reduce susceptibility to wind-induced oscillations.
  3. Advancements in Structural Analysis: The failure prompted advancements in structural analysis techniques, including an improved understanding of aeroelasticity and the development of computational tools to simulate and predict bridge behavior under dynamic loads.
  4. Engineering Ethics and Public Safety: The incident emphasized the ethical responsibility of engineers to prioritize public safety and consider all potential failure modes during the design and construction phases of infrastructure projects.
- **Legacy and Impact:** Despite its tragic collapse, the Tacoma Narrows Bridge failure led to significant advancements in bridge engineering. The lessons learned from this case study have influenced modern bridge design practices worldwide, ensuring that bridges are not only structurally sound but also resilient to dynamic forces and environmental conditions.

CLASS ACTIVITY

In a typical class activity on structural mechanics in civil engineering, students often engage in hands-on experiments or simulations to understand fundamental principles. They learn to analyze forces, stresses, and deformations in structural elements like beams and trusses using theoretical calculations and software tools. Practical sessions may include building and testing simple models to observe how different materials and configurations respond to loads. This interactive approach helps students grasp essential concepts of equilibrium, stability, and structural behavior, preparing them for real-world applications in designing safe and efficient civil engineering structures.

SUMMARY

- Structural mechanics is a branch of engineering that deals with the behavior of structures under loads and their design to ensure safety and performance.
- Dynamic analysis plays a crucial role in understanding the behavior of structures subjected to dynamic loads such as earthquakes, wind, and vibrations.
- Forces and equilibrium play a critical role in structural analysis, providing the foundational principles that engineers use to ensure the stability and safety of civil engineering structures.
- Structural engineers are trained professionals who are responsible for making sure that the structures we use in our daily lives, like bridges and tall buildings, are safe, and stable and don't collapse under applied loads.
- Structural analysis is a branch of solid mechanics that uses simplified models for solids like bars, beams, and shells for engineering decision-making.
- It is common practice to use approximate solutions of differential equations as the basis for structural analysis. This is usually done using numerical approximation techniques. The most commonly used numerical approximation in structural analysis is the Finite Element Method.

REVIEW QUESTIONS

1. Explain the significance of equilibrium in the analysis of structures. How does it relate to the principles of statics?
2. How do the principles of dynamics differ from those of statics in civil engineering? Provide an example of a scenario where dynamic analysis is crucial.
3. Discuss the role of natural frequency in the dynamic analysis of structures. Why is it important to determine the natural frequency of a building?

4. Describe the process and purpose of response spectrum analysis in civil engineering. How does it help in designing structures to withstand seismic events?
5. Explain how the method of joints and the method of sections are used to analyze forces in trusses. What are the advantages of each method?
6. Identify and describe the primary structural elements in a typical building. How do these elements contribute to the overall stability and load distribution of the structure?

REFERENCES

1. Bažant, Z. P., & Le, J.-L. (2003). *Computational Mechanics of Reinforced Concrete Structures*. CRC Press.
2. Blockley, D. (2014). *A Very Short Introduction to Structural Engineering*. Oxford University Press.
3. Boresi, A. P., & Schmidt, R. J. (2003). *Advanced Mechanics of Materials* (6th ed.). Wiley.
4. Chen, W.-F., & Duan, L. (2014). *Bridge Engineering Handbook* (2nd ed.). CRC Press.
5. Gawronski, W. (Ed.). (2004). *Advanced Structural Dynamics and Active Control of Structures*. Springer.
6. Ghali, A., Neville, A., & Brown, T. G. (2016). *Structural Analysis: A Unified Classical and Matrix Approach*. CRC Press.
7. Hewson, N. R. (2003). *Prestressed Concrete Bridges: Design and Construction*. Thomas Telford.
8. Hibbeler, R. C. (2010). *Structural Analysis*. Prentice Hall.
9. Hosford, W. F. (2005). *Mechanical Behavior of Materials*. Cambridge University Press.
10. Kurowski, P. M. (2008). *Finite Element Analysis for Design Engineers*. SAE International.
11. Paz, M., & Leigh, W. (2019). *Structural Dynamics: Theory and Computation* (5th ed.). Springer.
12. Sezen, H., & Kunnath, S. K. (2007). *Earthquake Engineering: Theory and Implementation*. ASCE Press.
13. Simiu, E., & Scanlan, R. H. (2015). *Wind Effects on Structures: Modern Structural Design for Wind* (3rd ed.). Wiley.

CHAPTER 4

Surveying and Leveling

LEARNING OBJECTIVES

After studying this chapter, you will be able to:

- Understand the surveying and leveling; and
- Define leveling methods.

KEY TERMS FROM THIS CHAPTER

Alidade	Auto-level
Benchmark	Contour lines
Differential leveling	Electronic distance measurement
Geodetic surveying	Geographic information system
GPS	Leveling staff
Photogrammetry	Remote sensing
Theodolite	Topographic surveying
Total station	Traverse
Triangulation	Trigonometric leveling

INTRODUCTION

Surveying and leveling play indispensable roles in civil construction, forming the backbone of any successful project. Surveying, the process of measuring and mapping the Earth's surface, is essential for determining the precise positions of points and the distances and angles between them. This information is crucial for designing and planning construction projects with accuracy and efficiency.

In civil construction, surveying begins with establishing control points, typically using GPS technology or traditional methods like total stations. These control points serve as reference marks for subsequent measurements, ensuring consistency and reliability throughout the project. Surveyors then collect data on the terrain, existing structures, and other relevant features using a variety of instruments and techniques. Leveling is a critical component of surveying, focusing on determining and establishing the elevations of points on the Earth's surface relative to a reference datum. This process is essential for ensuring proper grading, drainage, and alignment of structures, roads, and utilities. Leveling involves using optical or digital levels to measure vertical height differences between points, allowing engineers to create precise elevation profiles and contour maps.

Together, surveying and leveling provide the essential groundwork for civil construction projects, enabling engineers to accurately plan, design, and execute infrastructure developments. By meticulously mapping the terrain and establishing precise elevations, surveyors and leveling professionals ensure that construction projects proceed smoothly, efficiently, and safely, ultimately contributing to the successful completion of vital infrastructure that serves communities for years to come.

4.1. SURVEYING

Surveying is an essential aspect of civil engineering, serving as the foundation for planning, designing, and construction projects. It involves the measurement and mapping of various features of the Earth's surface to determine precise positions, distances, and elevations (Hong-Sen Yan & Marco Ceccarelli, 2009).

4.1.1. Importance of Surveying in Civil Engineering

1. **Pre-Construction Planning**: Surveying provides vital data for pre-construction planning, including site selection, land-use planning, and environmental assessments. It helps engineers understand the terrain, existing infrastructure, and potential challenges before initiating a project.
2. **Design and Layout**: Surveying aids in the design and layout of infrastructure projects such as roads, bridges, buildings, and pipelines. Engineers rely on accurate measurements and mapping to create detailed designs that adhere to safety standards and regulatory requirements.
3. **Construction Control**: During construction, surveying ensures that structures are built according to design specifications. It involves setting out reference points, establishing alignments, and monitoring progress to maintain accuracy and integrity throughout the construction process.
4. **Property Boundary Determination**: Surveying plays a crucial role in defining property boundaries and resolving disputes related to land ownership. By precisely measuring and mapping land parcels, surveyors help establish legal boundaries and property lines.
5. **Infrastructure Maintenance**: Surveying is essential for the maintenance and management of existing infrastructure. It allows engineers to assess the condition of roads, bridges, dams, and other structures, facilitating timely repairs and improvements.

4.1.2. Methods of Surveying

Surveying techniques can be broadly categorized into two main methods:

1. **Traditional Surveying Methods**: Traditional methods involve the use of basic instruments such as chains, tapes, and compasses for linear and angular measurements. These methods are time-consuming and labor-intensive but can still be useful in certain situations, especially in areas with limited access to advanced technology.
2. **Modern Surveying Methods**: Modern methods leverage advanced technologies such as Global Navigation Satellite Systems (GNSS), total stations, LiDAR (Light Detection and Ranging), and drones (Unmanned Aerial Vehicles, UAVs). These technologies enable faster, more accurate data collection and processing, enhancing the efficiency and precision of surveying operations.

4.1.3. Types of Surveying

There are two types of surveying: these are:

1. **Plane Surveying:** The majority of surveying measurements are made as though the earth's surface were perfectly flat. Plane surveying is the name of the surveying technique that is predicated on this notion. When conducting plane surveys, the earth's curvature is ignored, and the results are calculated using the concepts of plane geometry and plane trigonometry.

   A surveyor's work is made simpler by the use of plane surveying techniques. The effect of the earth's curvature on our measurement is negligible within 20 km, making it difficult to detect. Stated differently, the practical equivalent of the horizontal distance measured between two points along a truly level line is the distance measured along the straight chord that connects the two points (Figure 4.1).

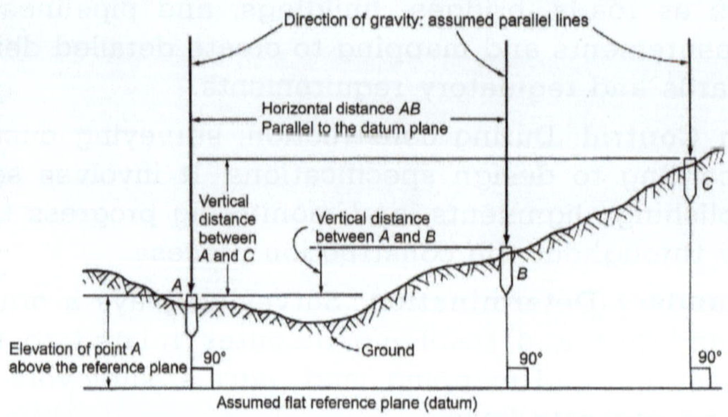

Figure 4.1. *In plane surveying, the curvature of the earth is neglected, and vertical distances are measured with reference of a flat plane.*

Source: https://www.cartercenter.org/resources/pdfs/health/ephti/library/lecture_notes/env_health_science_students/Surveying.pdf.

2. **Geodetic Surveying:** Geodetic surveys are those that account for the curvature of the earth. Organizations such as the Geological Survey typically take these kinds of surveys into consideration. Large-scale mapping projects and the creation of extensive point networks for horizontal and vertical control on Earth are typically accomplished through the use of geodetic surveying techniques.

4.1.4. Surveying Instruments

Surveying instruments are devices used to measure and collect data about the physical characteristics of land, buildings, and other objects. They are employed in diverse fields such as civil engineering, construction, mapping, mining, archaeology, and environmental studies. These instruments enable accurate measurement of distances, angles, elevations, and positions on the Earth's surface.

Some of the major Surveying Instruments are explained below:

1. **Theodolites (Figure 4.2):**
   - Optical instruments for measuring horizontal and vertical angles.
   - Commonly used in construction, mapping, and geodetic surveys.
   - Types include transit theodolites, digital theodolites, and electronic theodolites.

Figure 4.2. *Theodolites.*

Source: https://www.topconpositioning.com/us/en/solutions/technology/infrastructure-products/theodolites.

2. **Total Stations:**
   - Advanced electronic instruments combining theodolite functionality with distance measurement.
   - Measure angles and distances simultaneously, offering high accuracy.
   - Equipped with electronic data collection capabilities and often integrated with GPS technology (Figure 4.3).

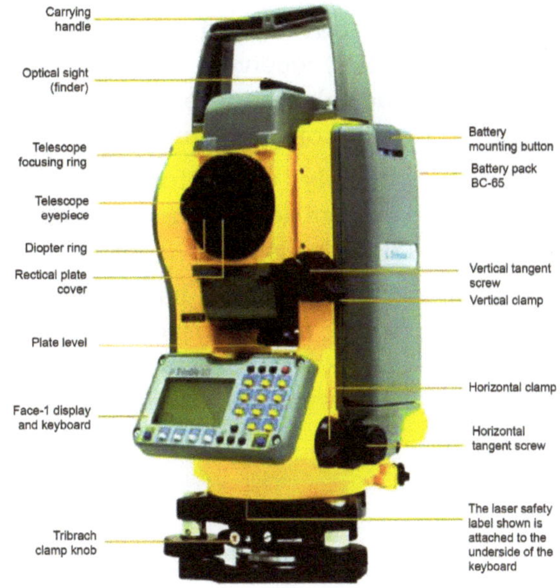

Figure 4.3. *Total stations.*

Source: http://gyaninformationpedia.blogspot.com/2015/12/fundamentals-of-surveying-total-station.html.

3. **Levels:**
   - Instruments for measuring height differentials or elevations.
   - Common types include dumpy levels, automatic levels, and digital levels.
   - Used in construction, roadwork, and land surveying (Figure 4.4).

Figure 4.4. *Automatic level.*

Source: https://www.maramani.com/blogs/home-design-ideas/equipment-used-surveying.

4. **GPS Receivers:**
   - Global Positioning System (GPS) receivers utilize satellite signals to determine precise positions on Earth.
   - Widely used for mapping, navigation, and geospatial data collection.
   - Types include handheld GPS units, survey-grade receivers, and Real-Time Kinematic (RTK) GPS systems (Figure 4.5).

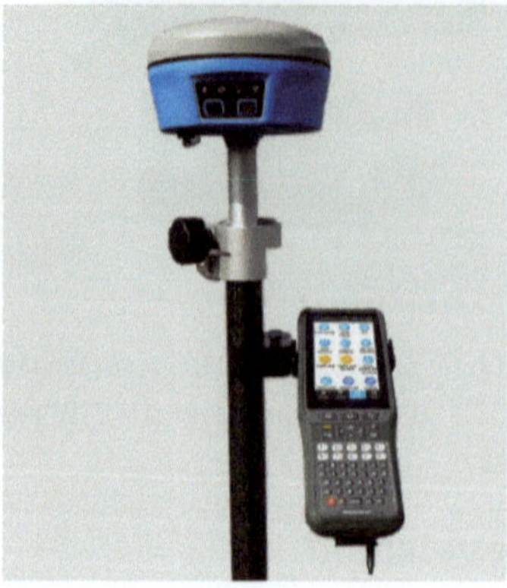

Figure 4.5. *GPS receivers.*

Source: https://5.imimg.com/data5/SELLER/Default/2022/7/VP/OE/DW/78820955/dgps-500x500.png.

5. **Laser Scanners:**
   - High-tech instruments for capturing 3D data of surfaces and objects.
   - Emit laser beams to measure distances and create detailed point clouds.
   - Applications include building information modeling (BIM), archaeological surveys, and topographic mapping.
6. **GNSS Receivers:**
   - Global Navigation Satellite System (GNSS) receivers are similar to GPS receivers but can utilize signals from multiple satellite constellations.
   - Offer enhanced accuracy and reliability compared to traditional GPS receivers.
   - Used in land surveying, precision agriculture, and geospatial applications.
7. **Digital Photogrammetry Systems:**
   - Systems that utilize aerial or terrestrial photographs to create accurate maps and 3D models.
   - Combine principles of photography, geometry, and computer vision.
   - Applications include mapping, land-use planning, and disaster assessment.
8. **Magnetic Locators:**
   - Instruments used to detect underground metal objects, such as utility pipes and property markers.
   - Utilize magnetic fields to locate ferrous materials.
   - Commonly employed in utility surveys and archaeological investigations.
9. **Ground Penetrating Radar (GPR):**
   - Non-destructive geophysical method for detecting subsurface features.
   - Emit electromagnetic pulses into the ground and record reflections to identify buried objects and geological structures.
   - Used in archaeological surveys, utility mapping, and geotechnical investigations (Figure 4.6).

Figure 4.6. *Ground penetrating radar (GPR).*

Source: https://www.linkedin.com/pulse/ground-penetrating-radar-gpr-most-advance-way-akshay-bhosale.

10. Compasses:

- Traditional instruments for measuring direction or bearing.
- Types include magnetic compasses, gyrocompasses, and digital compasses.
- Widely used in orienteering, navigation, and geological surveys.

11. Measuring Wheels:

- Simple devices for measuring distances on the ground.
- Consist of a wheel attached to a handle with an odometer-like mechanism.
- Used for quick distance measurements in construction, land surveying, and outdoor activities (Figure 4.7).

Figure 4.7. *Measuring wheels (model 1).*

Source: https://5.imimg.com/data5/SELLER/Default/2022/9/GP/IP/XK/138336373/road-measuring-wheels-500x500.jpg.

12. Inclinometers:

- Instruments for measuring angles of slope or inclination.
- Types include clinometers, digital inclinometers, and tilt sensors.
- Used in geotechnical surveys, slope stability analysis, and structural monitoring (Figure 4.8).

Figure 4.8. *Measuring wheels (model 2).*

Source: https://ptoflosgatos.com/wp-content/uploads/2009/12/inclinometer1.jpg.

13. Altimeters:

- Devices for measuring altitude or vertical elevation above a reference point.
- Types include barometric altimeters, GPS altimeters, and laser altimeters.
- Utilized in aviation, hiking, and environmental monitoring.

14. Spirit Levels:

- Tools for determining horizontal or vertical alignment.
- Consist of a glass tube filled with liquid (usually alcohol) and an air bubble.

- Used in construction, carpentry, and surveying for precise leveling.

15. **Protractors:**
    - Instruments for measuring angles or determining angular relationships.
    - Commonly used in map drafting, engineering design, and trigonometric calculations.
    - Types include circular protractors, digital protractors, and bevel protractors.
16. **Rangefinders:**
    - Instruments for measuring distances to a target.
    - Utilize various methods such as laser, ultrasonic, or optical principles.
    - Used in hunting, golfing, and surveying for accurate distance measurements.
17. **Tape Measures:**
    - Flexible rulers for measuring distances.
    - Commonly used in construction, interior design, and land surveying.
    - Available in various lengths and materials, including steel, fiberglass, and cloth.
18. **Handheld Computers and Data Collectors:**
    - Portable devices for collecting, storing, and managing survey data.
    - Allow surveyors to input field measurements, notes, and observations directly.
    - Integrated with surveying software for data processing and analysis.
19. **Environmental Monitoring Instruments:**
    - Instruments for measuring environmental parameters such as temperature, humidity, and air quality.
    - Used in environmental surveys, pollution monitoring, and climate research.
    - Examples include weather stations, water quality meters, and air pollution monitors.
20. **Automatic Monitoring Stations:**
    - Automated systems for continuous monitoring of geophysical or environmental parameters.
    - Include sensors, data loggers, and telemetry systems for remote data transmission.
    - Used in earthquake monitoring, weather forecasting, and hydrological studies.

4.1.5. Classification of Surveying

Surveying can be classified into different categories based on various factors. Here are the main classifications: (Lewis, 2001).

1. Based on Purpose
    - **Archaeological Survey:** Archaeological survey is the systematic examination and documentation of archaeological sites and landscapes to gather information about past human activities and cultural remains.
    - **Geological Survey:** It involves the mapping of Earth's geology including rocks, minerals, etc.

- **Mine Survey:** It involves the mapping of underground mines so as to ensure the safe extraction of minerals.
- **Military Survey:** It collects geospatial data for military operations.

2. Based on the Nature of the Surveyed Area
   - **Land Surveying:** This type focuses on measuring and mapping land surfaces for various purposes. It includes Cadastral surveying, Engineering surveying, and Astronomical surveying.
   - Cadastral Surveying: This type of land survey establishes and defines land boundaries for legal and administrative purposes.
   - **Engineering Surveying:** This type of land survey is conducted for construction and engineering projects. It ensures accurate positioning and measurements.
   - **Astronomical Surveying:** This surveying involves the observation and mapping of celestial objects of the universe.
   - **Marine Surveying:** It involves surveys conducted in bodies of water, such as rivers, lakes, and oceans.
   - **Hydrographic Surveying:** This survey is done in rivers, lakes, and oceans. It measures the levels of the water, the tides, the currents, and the features under water. It is used for navigation, building ports, etc.
3. Based on Instruments Used
   - **Optical Surveying:** This surveying uses instruments like theodolites and levels to make angular and distance measurements.
   - **Remote Sensing:** It uses aerial or satellite imagery to collect data about an area. No direct measurement is involved.
   - **Compass Surveying:** This method uses a compass to determine directions and angles. Notably, there are two types of compass used – Prismatic Compass and the Surveyors Compass. The principle of Compass Surveying is Traversing.
   - **Theodolite Surveying:** It uses a theodolite. Theodolite is a precise instrument used to measure angles and distances.
   - **Chain Surveying:** It uses chains or tapes to measure distances and angles.
   - **GPS Surveying:** This method utilizes Global Positioning System (GPS) technology to determine accurate positions using satellites.
4. Based on Method Employed
   - **Triangulation Surveying**
     - o Triangulation is a surveying method that measures the angles within a triangle. This triangle is formed by three survey control points.
     - o By using the trigonometric formulas and measured length of one side of the triangle, surveyors can

calculate the distances of the remaining sides.

- o This method uses the principles of trigonometry. This principle is used to determine accurate measurements. It also allows accurate mapping and surveying of land.
- o This method is used to create detailed representations of the surveyed terrain. It is used in various fields such as engineering, architecture, and land management.

- **Traverse Surveying**
  - o Traverse survey is a method used to establish a series of interconnected lines.
  - o The lengths of the lines are measured using a tape or chain. The directions are determined using the angle-measuring instrument.
  - o It is used to depict the relative positions and distances between various points on the land.
  - o Traverses are commonly used in surveying projects to create a framework for mapping and determining boundaries. These are also used to calculate the areas and volumes of specific regions.
  - o This method is used to capture geometric data. This data enables accurate representations and analyzes.

4.1.6. Advancements in Surveying Technology

1. **Integration of GNSS and INS**: The integration of GNSS (Global Navigation Satellite Systems) with Inertial Navigation Systems (INS) enables precise positioning and orientation determination, even in challenging environments such as urban canyons or dense forests.
2. **3D Laser Scanning**: Advancements in 3D laser scanning technology allow for rapid and accurate capture of detailed surface topography and object geometry. It is widely used in architectural design, heritage preservation, and forensic investigations.
3. **Mobile Mapping Systems**: Mobile mapping systems combine LiDAR, cameras, and GNSS receivers mounted on vehicles to capture geospatial data along roadways and infrastructure corridors. They offer efficient data collection for transportation planning, asset management, and urban modeling.
4. **UAV Lidar Systems**: Miniaturization and cost reduction of LiDAR sensors have led to the widespread adoption of UAV-based LiDAR systems for aerial surveying applications. These systems offer high-resolution, point cloud data acquisition capabilities with increased flexibility and accessibility.
5. **Real-time Data Processing**: Advances in real-time data processing algorithms enable on-the-fly analysis and visualization of survey data in the field. This allows surveyors to make informed decisions quickly, improving efficiency and productivity.

4.2. LEVELLING

In surveying, the main purpose of leveling is to ascertain the relative heights of different points on, above, and below the surface of the ground. It is a technique for figuring out how much two points on the surface of the earth differ in terms of elevation or level. Every measurement of distance starts from a reference point that has known parameters (height above mean sea level, elevation, etc.). In surveying, this part of measuring vertical distance is called leveling (Guy Bomford, 1980).

4.2.1. Basic Terms Used in Leveling in Surveying

It is necessary to be familiar with all of the fundamental terms used in surveying when leveling. See Figure 4.9 below for a more lucid visual representation.

- **Level Surface:** A curved surface that has all of its points at the same distance from the surface of the earth is called a level surface. A level surface is a continuous surface that is perpendicular to the mean spheroid of the earth.
- **Level Line:** The line that depicts a level surface is referred to as a "level line." The plumb or vertical lines and the level line always make a right angle at each point. It shows that every point on a straight line has the same distance to the planet's center. Despite being a curved surface, smaller area works are handled like planar surfaces.
- **Horizontal Plane: It** is a plane tangential to the level surface at any point under consideration.
- **Horizontal Line:** It refers to any line that lies in the horizontal plane.

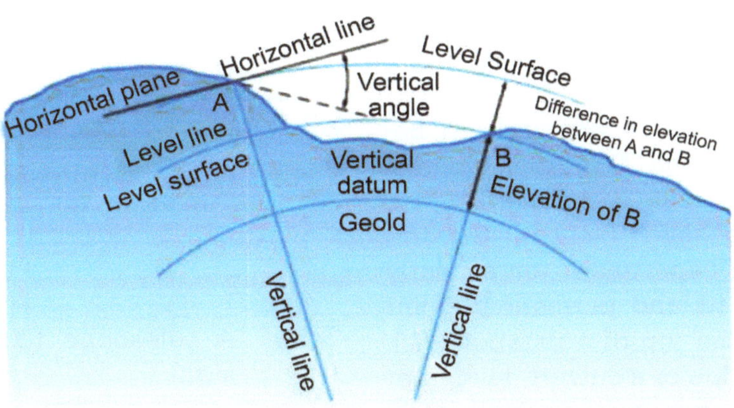

Figure 4.9. *Basic terms used in leveling in surveying.*

Source: https://testbook.com/civil-engineering/leveling-in-surveying-definition-and-types.

- *Vertical Line:* It is a line from any point of the Earth's surface to the center of the Earth. A plumb line can be regarded as a vertical line.
- *Datum:* Datum is the reference with respect to which other stations are fixed.
- *Vertical Angle:* The angle between two connected lines in a vertical plane is called a vertical angle.
- *Mean Sea Level:* It is a reference point and is obtained by averaging the elevations of high and low tides, at several points, for a long period of time, about 19 years.
- *Benchmark:* It is the fixed reference point of known elevation. Depending on the permanency and the precision with which the benchmark is located, benchmarks are classified as:
    - *Great Trigonometric Survey (GTS) Benchmark:* These benchmarks are established by the Survey of India at an interval of about 100 km all over the country, with the datum being taken as the mean sea level at Bombay Port. The elevations of the benchmarks are shown in the GTS Maps.
    - *Permanent Benchmark:* In between the GTS benchmarks, permanent benchmarks are established by the Government departments on a clearly defined and permanent point like the top of a parapet wall, a bridge or a culvert, kilometer stone, a railway platform, etc.
    - *Arbitrary Benchmark:* These are the reference points whose elevations are arbitrarily assumed generally for a small leveling purpose. It should be noted that arbitrary benchmarks do not imply any fixed datum.
    - *Temporary Benchmark:* These are the reference points on which a day's work is closed and from where leveling is continued the next day. Examples of such types of benchmarks are the kilometer stones, parapets, etc.

- **Reduced Level in Surveying:** In surveying, the term "reduced level" describes the process of equating survey point elevations in relation to a widely accepted datum. It is the vertical separation between the selected datum plane and the survey point. As a result, it is regarded as the base elevation, which serves as a benchmark for determining the depths or heights of other significant locations. Here, "level" refers to elevations, and "reduced" means equating. A location with an elevation of straight zero that is designated as the datum can be real or imaginary.

4.2.2. Leveling Instruments

A level is an instrument giving horizontal line of sight and magnifying the reading at a far away distance. It consists of the following parts:

i. A telescope to provide a line of sight
ii. A level tube to make the line of sight horizontal
iii. A leveling head to level the instrument.

The following types of levels are available:

i. Dumpy level
ii. Wye (or, Y) level
iii. Cooke's reversible level
iv. Cushing's level
v. Tilting level
vi. Auto level.

i. **Dumpy Level:** This instrument is sturdy and short, with a telescope tube that is fixedly attached to a vertical spindle. The level tube is therefore unable to move in a vertical plane. It is unable to be taken out of its support. That's why it's called the dumpy level. The leveling head's socket allows the telescope to rotate in a horizontal plane. The telescope's top is fitted with a bubble tube. A typical dumpy level is depicted in Figure 4.11. Figure 4.10 displays a picture of it.

Figure 4.10. *Dumpy level.*

Source: S.S. Bhavikatti, Basic civil engineering, ISBN (13): 978-81-224-2853-7.

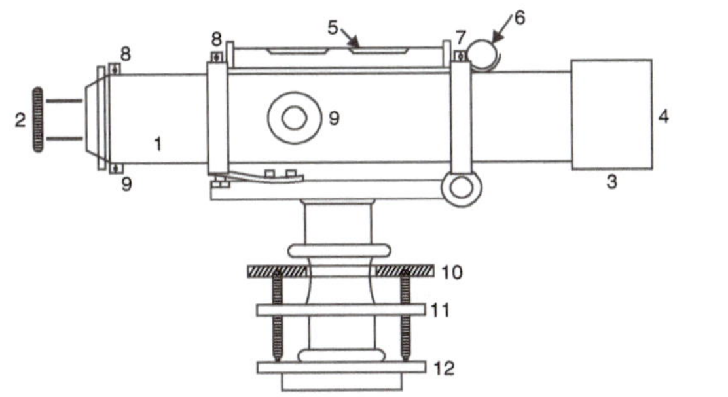

Figure 4.11. *Typical dumpy level.*

Source: S.S. Bhavikatti, Basic civil engineering, ISBN (13): 978-81-224-2853-7.

A telescope is a tube that has an eyepiece and an object glass. Prior to viewing the object through the graduated staff held in place, the object glass can be adjusted using the focusing screw. By turning the eyepiece, one can observe that the crosshairs clearly emerge and the parallel is eliminated. As long as the same person takes the readings, an adjusted eyepiece doesn't need to be changed. Over the level tube is a glass tube that has a slight curvature to it. Ether or alcohol is poured into the tube, leaving a small space that forms a bubble in the air. Every time, the air bubble reaches its maximum height. The telescope is horizontal when the bubble is centered because the level tube is fixed with its axis parallel to the telescope tube. To determine how far the bubble is out of center, the tube is graduated on both sides of its center. The glass tube is inserted into an open brass tube, and capstan-headed nuts are used to secure it to the telescope tube from below. If the bubble tube is out of order, these nuts are used to adjust it (John S. Scott, 1992).

The leveling head is made up of three-foot screws holding two parallel plates together. The term "trivet" refers to the lower plate, and "tribratch plate" to the upper one. The tripod stand can be attached to the lower plate using screws. The instrument can be leveled to achieve a perfect horizontal line of sight by adjusting the screws. To use a dumpy level in the field, it must be mounted on a tripod stand. The tripod stand is made up of three legs that are attached to a head that can accommodate the level's lower plate. To ensure a firm grip on the ground, metal shoes are fitted to the lower part of the legs.

ii. **Wye or Y-Level:** This kind of level has two Y-shaped supports for the telescope, which can be fixed with the aid of curved clips. It is possible to open the clips and fit the telescope by reversing it end to end. If readings are taken in both telescope directions, this level has the advantage of having some errors eliminated.

iii. **Cooke's Reversible Level:** This device has two stiff sockets that allow the telescope to be inserted from either end and screwed into place. The telescope is removed and reversed end to end after the screw has been loosened in order to take readings in the reversed position. Thus, it combines the reversibility of the Y-level with the rigidity of the dumpy level.

iv. **Cushing's Level:** Since both collars are identical, reversing the telescope end to end in this case involves switching out the objective and eyepiece.

v. **Tilting Level:** This telescope has a tilting screw that allows it to be tilted through approximately four degrees. Bubble can therefore be easily centered. However, the bubble must be centered before taking each reading. Therefore, it is helpful if there are few readings that need to be taken at each instrument setting.

vi. **Auto Level:** The self-aligning level is known as the auto-level or automatic-level. An inclination-compensating device achieves automatic leveling within a specific tilt range. The advantages of this instrument are its high speed, accuracy, and comfort during operation.

4.2.3. Dumpy Leveling in Surveying

The Dumpy Level is used to locate the points on the same horizontal plane. It is also referred to as the builder's level or the automatic level. The surveyor can determine the elevations of various points at the same elevation with the use of dumpy levels. Since the telescope is fixed to its supports, it is not possible to rotate the dumpy level in the vertical direction. William Gravatt created the Dumpy Level in 1832. The vertical spindle and telescope are cast in one piece at a dumpy level. The leveling head is depicted in the figure. The telescope is mounted on the upper plate of Figure 4.12, which is essentially made up of two plates. The lower plate is fixed directly onto a tripod. The instrument can be accurately leveled by adjusting the three foot screws, also known as leveling screws, which hold the two plates apart.

Once this is done, the telescope should stay level no matter which way it is pointed, thanks to the assistance of the bubble that is fixed to it. The telescope must be oriented horizontally in all directions using the foot screws before any readings are taken. A tangent screw and clamp are included with the instrument to enable you to aim the telescope in any direction. Generally speaking, the level tube's sensitivity is 20" per 2 mm. The line of collimation is the segment that connects the optical center of the objective with the point where the crosshairs intersect. The line of collimation is normal to the vertical axis when the instrument is in a state of permanent adjustment. Additionally, the line of collimation is horizontal and parallel to the axis of the bubble tube. Leveling a Dumpy Level follows essentially the same steps as leveling a theodolite. The bubble is said to traverse when it is in the center, regardless of which direction the telescope is pointed.

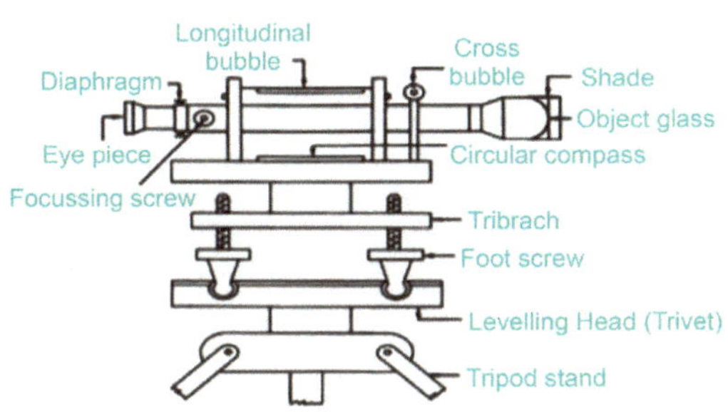

Figure 4.12. *Dumpy level and its component.*

Source: https://testbook.com/civil-engineering/leveling-in-surveying-definition-and-types.

4.2.3.1. Temporary Adjustment of Dumpy Level

- **Selection of Suitable Position:** The level is set by choosing an appropriate position. From this vantage point, more observations ought to be easy to make because the surface ought to be relatively level and stable.
- **Fixing Level of Tripod Stand:** The tripod stand is set up as needed, legs wide apart, and firmly planted in the ground. The level is secured to the tripod stand's top using the fixing mechanism meant for that specific level. It should be kept in mind that no station should have the level set up.
- **Approximate Leveling by a Leg of Tripod Stand:** They bring the foot screw to the middle of their run. The tripod stand's two legs are securely anchored in the earth. After that, the third leg is shifted in or out to the left or right until the bubble is roughly in the middle of its run.
- **Perfect Leveling by Foot Screw:** The telescope is positioned with the longitudinal bubble on top, parallel to any foot screws that may be present. (i.e., first position), and the bubble is centered by evenly rotating the foot screw in both inward and outward directions. Next, the telescope is rotated through 900 to reach a third foot screw. The bubble is then centered by rotating this foot screw in either a clockwise or counterclockwise direction. The bubble is moved to the center and the telescope is returned to its initial position.
- **Focusing the Eyepiece:** A piece of white paper is held in front of the object glass and the eyepiece is moved in or out by turning it clockwise or anticlockwise until the crosshair can be seen clearly.

- **Focusing the Object Glass:** The telescope pointed at the staff members leveling. Turning the focusing screw clockwise or counterclockwise will remove parallax and make the staff's graduation clearly visible through the eyepiece.
- **Taking the Staff Reading:** Finally, the level of the instrument is verified by turning the telescope in any direction.

4.2.3.2. Permanent Adjustment of Dumpy Level

For the dumpy level to be in proper adjustments, the following 3 conditions should be satisfied.

i. The axis of the bubble tube should be perpendicular to the vertical axis of the instrument.

ii. The horizontal cross-hair should lie in the plane perpendicular to the vertical axis.

iii. The line of sight should be parallel to the axis of the bubble tube.

4.3. LEVELING METHODS

4.3.1. Direct Leveling

Direct leveling is a common form of leveling. In this method, the telescope is made horizontal, and the horizontality is checked using a spirit level.

Types of Direct Leveling

- Simple Leveling
- Differential Leveling
- Reciprocal Leveling
- Precise Leveling
- Fly Leveling

1. **Simple Leveling:** When finding the elevation difference between two points that are each visible from a single level position is required, this is the easiest leveling operation to perform. On a staff that is held successively upon these points, the readings can be obtained (Ghilani Charles, 2010).

   The exact state of the level doesn't matter, but the sight distances to the two staff positions should be maintained as close to equal as possible in order to remove the instrumental error. The level is placed anywhere there is solid ground; it need not be in the same vertical plane as the two staff positions.

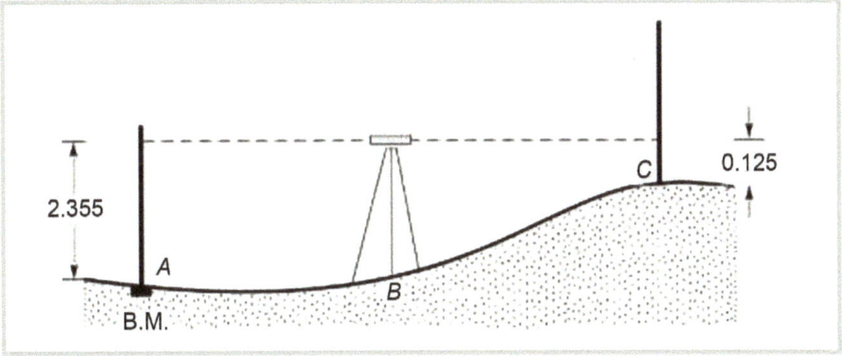

Figure 4.13. *Sampling leveling.*

Source: https://civiljungle.com/leveling-methods/#google_vignette.

2. **Differential Leveling:** Determining this difference in elevation between two or more points without any regard to the alignment of the points is called differential leveling. It is used when:

   i. Two points are a large distance apart (Figure 4.14)

ii. The difference in elevation between the two points is large (Figure 4.15) and

iii. Some obstacle intervenes between the points (Figure 4.16).

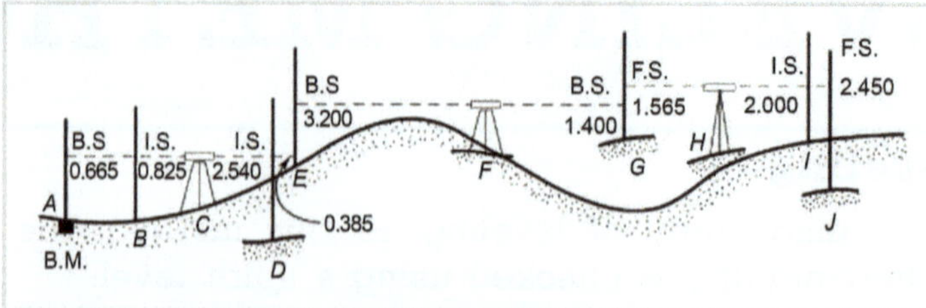

Figure 4.14. *Differential leveling (point far apart).*

Source: https://civiljungle.com/leveling-methods/#google_vignette.

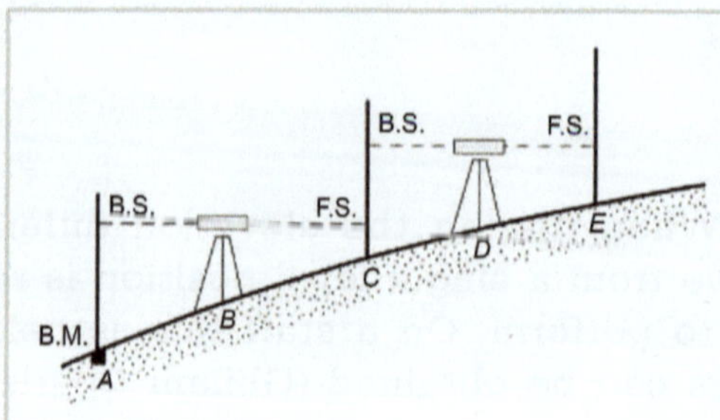

Figure 4.15. *Differential leveling (point with large difference of elevation).*

Source: https://civiljungle.com/leveling-methods/#google_vignette.

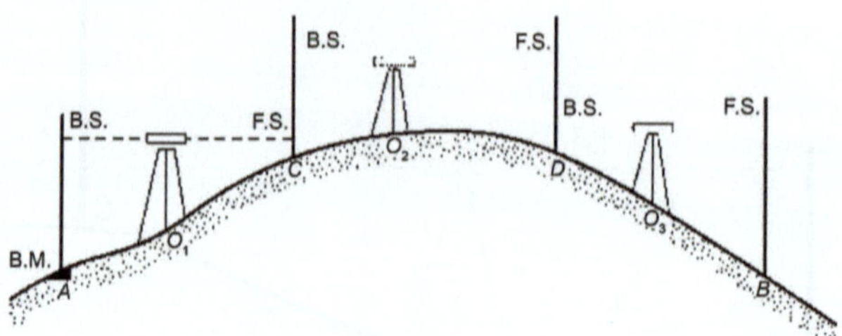

Figure 4.16. *Differential leveling (point having an obstacle in between).*

Source: https://civiljungle.com/leveling-methods/#google_vignette.

3. **Reciprocal Leveling:** The process of leveling occurs when two sets of reciprocal observations accurately determine the elevation difference between two points. When a valley, river, or other obstruction prevents the instrument from being set up between the two points, this method comes in very handy, as well as if the sights are significantly longer than allowed.

   The earth's curvature, inaccurate staff readings, and faulty instrument adjustments become noticeable for such long sights. To reduce these errors, specific techniques like reciprocal leveling should be applied (Figure 4.17).

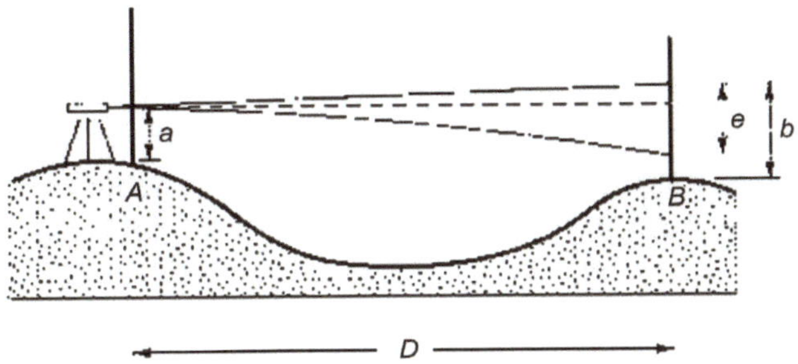

Figure 4.17. *Reciprocal leveling.*

Source: https://civiljungle.com/leveling-methods/#google_vignette.

4. **Precise Leveling:** This is the leveling process where specific tools are employed. Ordinary leveling and precise leveling are equivalent in theory. In the former, the elevations attained are suitable for regular use, and the intervals between checkpoints are reasonably short. On the other hand, the level loop may be quite long for accurate leveling, and every effort is taken to control the sources of errors. The foresight and backsight distance balance is the primary error control in precise leveling.

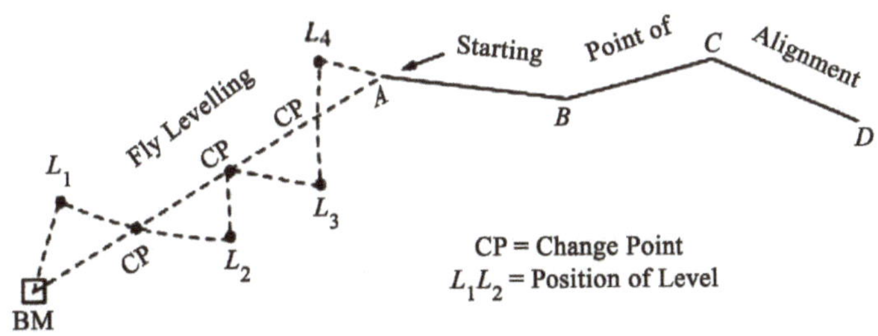

Figure 4.18. *Fly leveling.*

Source: https://civiljungle.com/leveling-methods/#google_vignette.

5. **Fly Leveling:** This leveling operation involves running a line of levels to estimate the elevations along a path. Roads, railroads, tunnels, canals, and other linear structures are the targets of this reconnaissance work (Figure 4.18).

4.3.2. Barometric Leveling

The elevation of a point is inversely proportional to the weight of the air column above the observer, according to the principle used in barometric leveling. Because air is compressible, the relationship between pressure and elevation is not constant. The pressure is also impacted by abrupt changes in humidity, temperature, and meteorological conditions brought on by storms.

The barometric techniques are especially well-suited for work in rugged terrain where extreme precision is not required. When measured electronically, these are also used to bring the slope distances closer to the horizontal.

A barometer is the name of the device used to measure pressure. A modified barometer called an altimeter is used to determine the relative elevations of locations on Earth's surface. Though it operates simply, its sensitivity to variations in atmospheric pressure is high. A single-base method refers to the process of using an altimeter to measure elevation. There must be two altimeters. The readings are taken at regular intervals at the control point, which is a known elevation where one altimeter and one thermometer are situated.

The desired elevations of the points are reached by the other altimeter, known as the roving altimeter. The roving altimeter's readings at the desired locations are subsequently adjusted in accordance with variations in temperature and other factors noted at the control point. The difference in elevation between the two points may be obtained by the following formula:

$$H = 18336.6 \times (\log10\ h1 - \log\ 10\ h2) \times (1 + ((T1 + T2)/500))$$

H = The difference in elevation between the two points. h1. h2 = Thc Barometric (in cm) at the lower and higher point respectively, and T1.T2 =Temperatures pf air (in °C) at the lower and higher points. Respectively. Types of barometric leveling are:

i. **Barometer:** In aerial photogrammetry, barometers are used for leveling to approximate elevations, determine elevation differences, and measure airplane flying height. Additionally, they are employed in the computation of the refraction correction in specific types of astronomical observations. Since leveling with a barometer is not very accurate, it is typically limited to small-scale topographical and reconnaissance surveys where precise elevation determination is not necessary. There are two types of barometers: the mercurial and the aneroid. Although the former is easier to carry and less prone to breakage, it is less accurate (Figure 4.19).

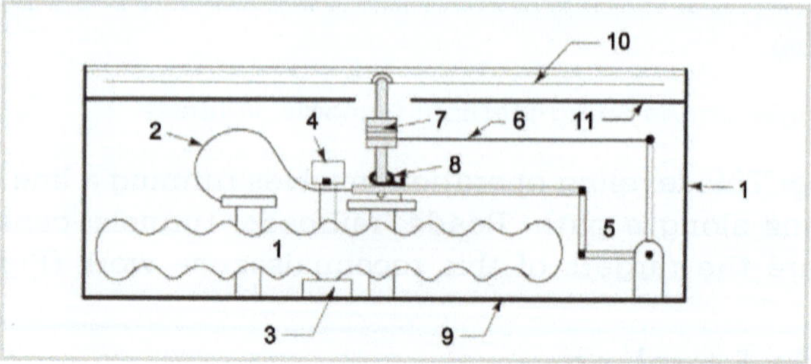

Figure 4.19. *Barometer.*

Source: https://civiljungle.com/leveling-methods/#google_vignette.

ii. **Mercurial Barometric:** The mercury barometer operates on the basis of balancing a mercury column against the atmospheric pressure, where the atmospheric pressure at the observation point is determined by the point's elevation above mean sea level. Cistern and siphon are the two primary varieties of mercurial barometers. Because they require vertical support, mercurial barometers are frequently suspended using a gimbal mounting that is fastened to a specialized tripod.

The lower open end of the glass tube, which is approximately 85 cm long and has a closed upper end, is submerged in a mercury cistern that is exposed to the atmosphere in the cistern type of barometer. The air in the tube is forced out, creating a vacuum in the area above the mercury level.

A mercury column is maintained in the tube; its height is dependent upon the pressure on the mercury's surface in the cistern because the pressure on the mercury in the cistern is atmospheric and there is no pressure on the upper end of the column. The mercury-filled tube in a siphon-type mercury barometer is bent into a U shape at the bottom. The U-tube has one branch that is kept shorter than the other. The long branch is closed at the top with the vacuum at its top, but the short branch has a tiny aperture at the top to allow air to pass through. This kind of barometer is less common and less accurate than the cistern type.

4.3.3. Hypsometric Leveling

One tool that can be used to determine the altitudes of different points is a hypsometer. It operates on the tenet that a liquid boils when the atmospheric pressure and its vapor pressure are equal. It should be noted that when pressure drops, water's boiling point drops as well. e. as one reaches a greater altitude.

The method, therefore, consists of determining the boiling point temperatures at various stations. The corresponding atmospheric pressures may be obtained from the tables. In the absence of tables, the following approximate formula may be used:

$$h = 76.00 \pm 2.679\ t$$

where, t is the difference in boiling point from 100°C, and It is the pressure in cm.

The difference in elevations may be obtained by using the formula given below formula:

$$H = 18336.6 \times (\log 10\ h1 - \log 10\ h2) \times (1 + ((T1 + T2)/500))$$

H = the difference in elevation between the two points; h1. h2 = The Barometric (in cm) at the lower and higher point, respectively; and T1.T2 =Temperatures pf air (in °C) at the lower and higher points, respectively.

The hypsometer consists of a thermometer graduated to 0.1°C. It is fitted inside a telescopic tube and is suspended over a small boiler filled with distilled water. The thermometer is kept in steam and is adjusted so as not to touch the water. The temperature of this air in the shade is also observed simultaneously with a detached thermometer.

4.3.4. Stadia Leveling

Another name for it is Tacheometric Surveying. Differential leveling is a common technique for measuring vertical distances, while chaining is used to measure horizontal distances. These two approaches yield results with the necessary accuracy. However, chaining on uneven terrain does not produce very precise outcomes. The tachometer is the preferred instrument when the ground is uneven and more observations made more quickly with standard precision are acceptable.

Data collection for the purpose of creating contours on a topographic map is one application of a tachometer for the aforementioned circumstances. Tachymetric distances are less accurate than chaining on level terrain, but more accurate on steep and uneven terrain. An optical method of measuring distance is called a tachometer. This method of surveying is very quick and easy, but it is less accurate.

The terms tachymetry and telemetry are also used to refer to the tachometer. The creation of a contoured plan is the main goal of a tachymetric survey. This tool is especially useful for filling in specific details on topographical maps, initial location surveys (e.g., for roads, bridges, reservoirs, railroads, etc.), and surveying areas with steep terrain, fractured borders, waterways, etc.

Additionally, it can be used to quickly verify distances measured with a chain or tape on surveys with greater accuracy. In essence, a tachometer is a transit theodolite with additional stadia wires on its diaphragm for measuring distances in addition to cross-wires. A Stadia rod is used for observations; it is typically a level staff with a larger least count of 1 cm. From these observed readings, both horizontal and vertical distances are calculated (Figure 4.20).

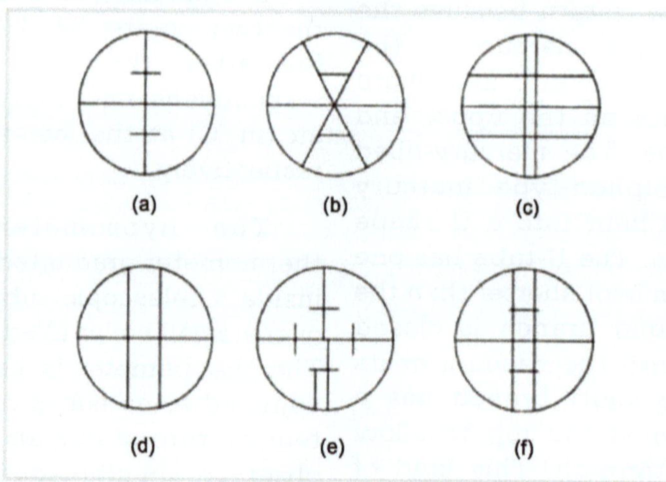

Figure 4.20. *Stadia digraph.*

Source: https://civiljungle.com/leveling-methods/#google_vignette.

4.3.5. Indirect Leveling (Trigonometric Leveling)

This is an indirect leveling technique where the observed distances and vertical angles are used to calculate the points' differences in elevation. The distances can be calculated trigonometrically or directly, and the vertical angles can be measured using transit. In topographical work, trigonometrical leveling is frequently used to determine the elevation of building tops, church spires, chimneys, and other structures.

Additionally, it can be advantageous in challenging environments like mountainous regions. There could be countless instances, depending on the field circumstances and the measurements that could be taken using the available instruments.

- **The Base of the Object Accessible—the Object may be Vertical or Inclined:** As per Figure 4.21, AF is the vertical object, D is the horizontal distance between the object and instrument, S is the reading on the leveling staff held vertical on the B.M., h1 is the height of the instrument, h is the height FE, and ⊠ is the angle of elevation on the top of the object.

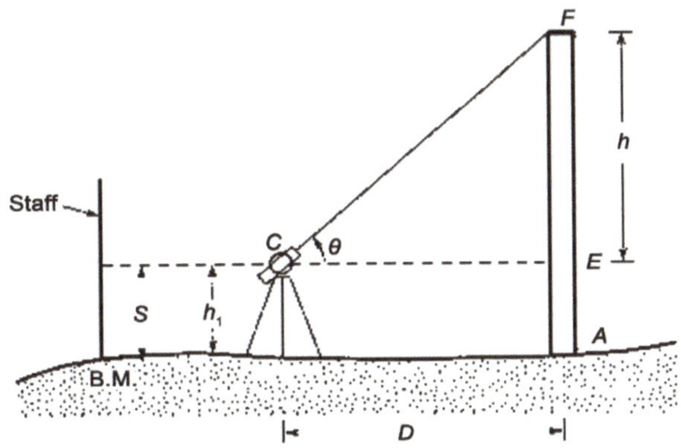

Figure 4.21. *Reduce the level of the top of a vertical object.*

Source: https://civiljungle.com/leveling-methods/#google_vignette.

- From triangle CEF;
  - EF = CE tan ⊠
  - h = D tan ⊠
  - Reduced Level of F = R.L of B.M. + S +h
  - Reduced Level of F = R.L of B.M. + S + D tan ⊠

As per Figure 4.22, AF is the inclined object, x is the distance between the foot of the object and the projection F' of the top, O1, and O2 are the instrument stations such that O1, O2, and A are in the same vertical plane, D1 and D2 are the distances of the foot of the object from the instrument stations O1 and O2, respectively.

S1 and S2 are the staff readings on B.M. from instrument positions at O1 and O2, respectively, and ⊠**1** and ⊠**2** are the angles of elevation from O1 and O2, respectively.

Reduced Level of F = R.L of B.M. + S1 +h1

Reduced Level of F = R.L of B.M. + S + (D1 + x) tan ⊠ 1

Reduced Level of F = R.L of B.M. + S1 +h1

Reduced Level of F = R.L of B.M. + S + (D2 – x) tan ⊠ 2

$$x = \frac{(S_2 - S_1) + (D_2 \tan\theta_2 - D_1 \tan\theta_1)}{\tan\theta_1 + \tan\theta_2}$$

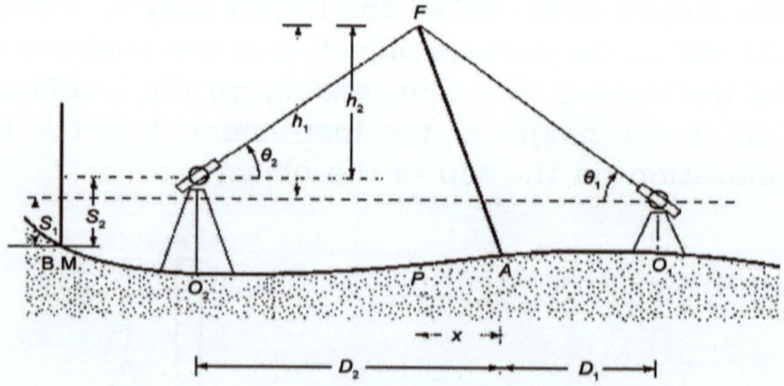

Figure 4.22. *Reduce the level of the top of an inclined object.*

Source: https://civiljungle.com/leveling-methods/#google_vignette.

CASE STUDY

ENHANCING INFRASTRUCTURE DEVELOPMENT THROUGH EFFECTIVE SURVEYING AND LEVELING

Introduction

Surveying and leveling are fundamental processes in civil engineering projects, crucial for accurate mapping, design, and construction. This case study explores the application of surveying and leveling techniques in a large-scale infrastructure project, focusing on their importance, the challenges faced, and the strategies employed to overcome them.

Project Overview

The case study revolves around the development of a new highway network connecting urban centers. The project encompasses various terrains, including plains, hills, and water bodies, making it imperative to employ precise surveying and leveling methodologies.

Importance of Surveying and Leveling

Accurate surveying and leveling lay the foundation for successful infrastructure development. They ensure precise alignment of roads, bridges, and other structures, optimizing safety, efficiency, and cost-effectiveness. Inaccuracies in surveying and leveling can lead to design flaws, construction delays, and compromised structural integrity.

Challenges Faced: The project encountered several challenges, including:

1. **Diverse Terrain:** Surveying and leveling across different terrains required adaptable methodologies to maintain accuracy.
2. **Environmental Factors:** Weather fluctuations, particularly during the rainy season, posed challenges to surveying operations.
3. **Remote Locations:** Some project sites were located in remote areas with limited access, necessitating innovative solutions for data collection and equipment transportation.
4. **Accuracy Requirements:** The project demanded high precision, especially in areas prone to natural hazards like landslides and floods.

Strategies Employed: To address these challenges, the project team implemented various strategies:

1. **Advanced Surveying Technologies:** Utilizing state-of-the-art surveying equipment such as GPS, total stations, and LiDAR technology enhanced data accuracy and efficiency.

2. **Remote Sensing:** Integration of satellite imagery and aerial surveys facilitated comprehensive mapping of large and inaccessible areas.
3. **Real-time Monitoring:** Continuous monitoring of survey data in real-time enabled prompt adjustments and mitigated errors.
4. **Collaborative Approach:** Collaboration between surveyors, engineers, and environmental specialists ensured holistic problem-solving and decision-making.

Results and Outcomes

Through effective surveying and leveling practices, the project achieved several significant outcomes:

1. **Optimized Design:** Accurate topographic data facilitated optimized design solutions, minimizing earthwork requirements and construction costs.
2. **Improved Safety:** Precise alignment and leveling reduced the risk of accidents and structural failures, ensuring the safety of both construction workers and future road users.
3. **Environmental Conservation:** Detailed environmental surveys enabled the identification and mitigation of potential ecological impacts, promoting sustainable development practices.
4. **Timely Completion:** Efficient surveying and leveling methodologies contributed to the project's timely completion, meeting deadlines and budgetary constraints.

Conclusion

This case study highlights the indispensable role of surveying and leveling in infrastructure development projects. By overcoming various challenges through innovative strategies, the project exemplifies how precise surveying techniques can enhance safety, efficiency, and sustainability in large-scale construction endeavors.

CLASS ACTIVITY

"Students will form teams and use basic surveying equipment to measure the height of various objects around the school campus. They will then compare their measurements and discuss the importance of accuracy and precision in surveying and leveling."

SUMMARY

- Surveying is an essential aspect of civil engineering, serving as the foundation for planning, designing, and construction projects. It involves the measurement and mapping of various features of the Earth's surface to determine precise positions, distances, and elevations.
- A surveying, which takes the earth's curvature into account is called a Geodetic survey. These types of surveys are usually considered by agencies like the Geological Survey. Geodetic surveying methods are generally used to map large areas and to establish large-scale networks of points on the earth for horizontal and vertical control.
- Surveying instruments are devices used to measure and collect data about the physical characteristics of land, buildings, and other objects. They are employed in diverse fields such as civil engineering, construction, mapping, mining, archaeology, and environmental studies. These instruments enable accurate measurement of distances, angles, elevations, and positions on the Earth's surface.
- Leveling in surveying is primarily used to determine the relative height of various points on, above, and below the ground's surface. It is a method of calculating the difference in elevations or levels between two points on the earth's surface.
- The auto-level or the automatic-level is a self-aligning level. Within a certain range of tilt automatic leveling is achieved by an inclination-compensating device. The operational comfort, high speed, and precision are the advantages of this instrument.
- Fly Leveling is an operation of leveling in which a line of levels is run to determine the approximate elevations along a route. It is carried out for reconnaissance of linear structures such as roads, railways, tunnels, canals, etc.
- In the Indirect Method of contouring, levels are taken at some selected points and their levels are reduced. Hereby in this method, horizontal control is established first and then the levels of those points are found.

REVIEW QUESTIONS

1. What is the purpose of a survey in civil engineering?
2. Differentiate between plane surveying and geodetic surveying.
3. Explain the principle of leveling and its importance in surveying.
4. What are the main types of leveling instruments used in surveying?
5. What is direct leveling? Explain the types of direct leveling.
6. How is a theodolite different from a total station in surveying applications?

REFERENCES

1. Basak, N. N. (2014). *Surveying and Levelling*. McGraw Hill Education. p. 542. ISBN 9789332901537.
2. Bomford, G. (1980). *Geodesy* (4th ed.). Clarendon Press. p. 204. ISBN 0-19-851946-X.
3. Genovese, I. (2005). *Definitions of Surveying and Associated Terms*. ACSM. 314pp. ISBN 0-9765991-0-4.
4. Ghilani, C. (2010). *Elementary Surveying: An Introduction to Geomatics* (13th ed.). Pearson. pp. 90–91. ISBN 978-0-13-255434-3.
5. Lewis, M. J. T. (2001). *Surveying Instruments of Greece and Rome*. Cambridge University Press. ISBN 9780521792974.
6. Pugh, J. C. (1975). *Surveying for Field Scientists*. Methuen. 230pp. ISBN 0-416-07530-4.
7. Punmia, B. C. (2005). *Surveying by B. C. Punmia*. Firewall Media. p. 2. ISBN 9788170088530.
8. Ronalds, B. F. (2016). *Sir Francis Ronalds: Father of the Electric Telegraph*. Imperial College Press. ISBN 978-1-78326-917-4.
9. Scott, J. S. (1992). *Dictionary of Civil Engineering*. Springer Science+Business Media. p. 252. ISBN 0-412-98421-0.
10. Yan, H. S., & Ceccarelli, M. (2009). *International Symposium on History of Machines and Mechanisms: Proceedings of HMM 2008*. Springer. p. 107. ISBN 978-1-4020-9484-2.

CHAPTER 5

Geotechnical Engineering

LEARNING OBJECTIVES

After studying this chapter, you will be able to:

- Describe the soil properties and classification;
- Know about site investigation techniques;
- Learn about bearing capacity and settlement of foundations; and
- Discuss about slope stability and landslides.

KEY TERMS FROM THIS CHAPTER

Aeolian soils
Alluvial soils
Bearing capacity
Clay
Earthquakes
Glacial soils
Lacustrine soils
Landslides
Loam
Marine soils
Remote sensing
Residual soils
Sand
Shear failure
Silt
Site investigation
Soil mass
Soil permeability

INTRODUCTION

The area of civil engineering that focuses on soils, structures, or structural elements that are located on or within these soils is called geotechnical engineering. The physical characteristics required to comprehend the mechanisms underlying the behavior of soils under external influences, such as those brought on by structural loads or variations in water content, are provided by the mechanics of soils. Therefore, the primary focus of soil mechanics is on how soils deform in order to improve the accuracy of failure load predictions. As a result, thinking of the issue as a soil-structure system—which consists of a foundation or other geotechnical engineering structure—and a porous soil medium—a soil profile with multiple layers—is both more practical and possibly more accurate. When designing these soil-structure systems, the two main requirements that must be met in order to withstand any external static or dynamic load are settlement and ultimate bearing capacity. It is anticipated that any soil-structure system will meet these needs over the course of the project once such stability criteria have been established.

The collection of subfields in a broad range of real-world issues involving soils, rocks, slopes, foundations, walls, etc., is known as geotechnical engineering. Although the fundamentals of soil mechanics serve as the ultimate guide, each problem has a unique recipe, so engineers, who serve as the practice's chefs, must be able to determine what the most appetizing and desired food is (i.e., engineering solution) by utilizing mathematics, their ultimate resource. In order to create the most accurate, practical, and efficient solution possible, geotechnical engineers are increasingly using emerging technologies in their engineering solutions. This is due to the rapid advancement of technology. As geotechnical engineers, we still utilize the following three instruments to accomplish such a complex task: (i) experimental methods, (ii) numerical methods, and (iii) analytical methods.

5.1. SOIL PROPERTIES AND CLASSIFICATION

The meaning of the term "soil" varies based on the field in which it is used. A geologist refers to the substance found in the relatively thin region of the Earth's surface where roots are found, and which was created as a byproduct of previous surface processes. Rock is the term used to describe the remaining crust.

To an engineer, it is a material that can be:

- **Built on:** foundations of buildings, bridges
- **Built in:** basements, culverts, tunnels
- **Built with:** embankments, roads, dams
- **Supported**: retaining walls

5.1.1. Definition of Soil

In a broad sense, soil can be defined as a naturally occurring mixture of loose or moderately cohesive mineral grains that can be separated through basic mechanical processes, whether they are inorganic or organic in nature. It helps to understand the various types of soils.

5.1.2. Soil Mechanics

Soil mechanics is a branch of civil engineering that deals with the behavior and study of soil in the context of engineering applications. Regardless of whether they contain an admixture of organic constituents or not, sediments and other unconsolidated accumulations of solid particles produced by the mechanical and chemical disintegration of rocks are the subject of soil mechanics, which is the application of mechanics and hydraulics laws to engineering problems. A multiphase aggregation of solid particles, water, and air makes up soil. Some of the most fundamental engineering mechanics concepts are needed to describe the mechanical behavior of this fundamental composition, which gives rise to distinctive engineering properties. The mechanical characteristics of soil, such as its strength, stiffness, and permeability, are of interest to engineers. These are mostly determined by the type of soil grains, the amount of water present, the stress level, and the unit weight.

5.1.3. Formation of Soil

On the Earth's surface, rocks extend up to as much as 20 km depth. The major rock types are categorized as igneous, sedimentary, and metamorphic.

- **Igneous Rocks:** formed from crystalline bodies of cooled magma.
- **Sedimentary Rocks:** formed from layers of cemented sediments.
- **Metamorphic Rocks:** formed by the alteration of existing rocks due to heat from igneous intrusions or pressure due to crustal movement.

Materials that have come from the breakdown of rocks through different physical and chemical weathering processes make up soils. The nature and structure of a given soil depends on the processes and conditions that formed it:

- Breakdown of parent rock: weathering, decomposition, erosion.
- Transportation to the site of final deposition: gravity, flowing water, ice, wind.
- Environment of final deposition: flood plain, river terrace, glacial moraine, lacustrine or marine.
- Subsequent conditions of loading and drainage: little or no surcharge, heavy surcharge due to ice or overlying deposits, change from saline to freshwater, leaching, contamination

All soils originate, directly or indirectly, from different rock types.

5.1.4. Soil Types

Rocks that are either igneous, sedimentary, or metamorphic are the source of the soils. Because of the interaction of water, carbon dioxide, and organic matter, new minerals may form in the soils or the mineral composition of the parent rock may remain the same. The various types of soils are as follows (Figure 5.1):

1. Residual soils
2. Transported soils

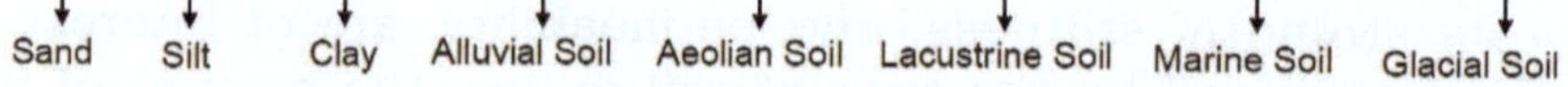

Figure 5.1. *Types of soils.*

Source: https://egyankosh.ac.in/bitstream/123456789/29713/1/Unit-1.pdf.

1. **Residual Soils:** Soils that remain where they were formed due to rock disintegration are known as residual soils. Clays, silts, and sands fall into this group. Sand is loose-textured soil. Its shape can range from angular to sub-angular, rounded to well-rounded, and is typically made up of unaltered mineral

grains. Fine-grained silty soil has little to no malleability. Soil made of clay is cohesive. It is made up of weather rock particles that are microscopic and submicroscopic, and it is flexible in a variety of water contents.

2. **Transported Soils:** Soils that are carried to a different location by the forces of gravity, water, wind, and ice after breaking down in rock are known as transported soils. Transported soils may be classified according to their mode of transportation and deposition:
   i. Alluvial soils are those that have been transported by running water, e.g., river deposits.
   ii. Aeolian soils are those that have been transported by wind, e.g., sand dunes.
   iii. Lacustrine soils are those which are deposited at the bottom of lakes.
   iv. Marine soils are those which are deposited in seawater.
   v. Glacial soils are those which have been deposited as a result of glacial activity.

In addition to the above classification, it is desirable to be familiar with the following types of soils also:

i. Organic silt
ii. Organic clay
iii. Bentonite
iv. Black cotton soils
v. Peat
vi. Varved clay
vii. Hard Pan
viii. Glacial till
ix. Boulder soils
x. Cohesive oils
xi. Cohesionless soils
xii. Bentonite
xiii. Calcareous soils
xiv. Diatamaceous earth
xv. Dune sands
xvi. Latersite
xvii. Loam
xviii Kankar

5.1.5. Physical Properties of Soil

The physical characteristics of soil are texture, structure, bulk density, porosity, consistency, temperature, color, and resistivity, in decreasing order of significance for ecosystem services like crop production (Gardner Catriona et al., 1999).

In the field, characteristics of the soil horizons and profile are frequently explained in terms of the physical characteristics of the soil. Horizons are delineated by variations in their physical characteristics. The physical characteristics of soil influence how a soil looks and feels.

The major soil physical properties are:

- Soil Texture
- Soil Structure
- Soil Consistence/Soil Strength
- Soil Color
- Soil Permeability
- Soil Temperature

5.1.5.1. Soil Texture

A unique physical size group is represented by each soil separation. Minerals with an equivalent diameter of less than two millimeters and falling within a range of size restrictions. The soil separates that are recognized in the United States have the following names and sizes (Table 5.1):

Table 5.1. Names and Sizes of the Soil

Very Coarse Sand	2.0 - 1.0 mm
Coarse Sand	1.0 - 0.5 mm
Medium Sand	0.5 - 0.25 mm
Fine Sand	0.25 - 0.10 mm
Very Fine Sand	0.10 - 0.05 mm
Silt	0.05 - 0.002 mm
Clay	0.002 mm

Source: https://sist.sathyabama.ac.in/sist_coursematerial/uploads/SCI1304.pdf.

General classification is as follows (Table 5.2):

Table 5.2. Classification of the Soil

Sand	2.0 - 0.05 mm
Silt	0.05 - 0.002 mm
Clay	<0.002 mm

Source: https://sist.sathyabama.ac.in/sist_coursematerial/uploads/SCI1304.pdf.

5.1.5.2. Soil Structure

The organization of primary sand, silt, and clay particles into secondary aggregates known as peds, or structural units, which have recognizable shapes, is known as structure. The term "structural type" refers to these aggregates with distinct shapes.

The five basic types of structural units are as follows:

1. **Platy:**
   - i. Plate-like aggregates that form parallel to the horizons like pages in a book.
   - ii. This type of structure may reduce air, water and root movement.
   - iii. Common structure in an E horizon and usually not seen in other horizons.
2. **Blocky:**
   - i. Two types--angular blocky and subangular blocky
   - ii. These types of structures are commonly seen in the B horizon.
   - iii. Angular is cube-like with sharp corners while subangular blocky has rounded corners.
3. **Prismatic:**
   - i. The vertical axis is longer than the horizontal axis. If the top is flat, it is referred to as prismatic.
   - ii. If the top is rounded, it is called columnar.
4. **Granular:**
   - i. Peds are round and porous, spheroidal. This is usually the structure of A horizons.
5. **Structureless:**
   - i. No observable aggregation or structural units.
     - a. Single grain-sand
     - b. Massive-solid mass without aggregates

5.1.5.3. Soil Consistency

Expresses cohesive and adhesive forces holding soil particles together; varies with moisture content. Describes the resistance of soil at various moisture contents to mechanical stress or manipulation. It is described at three moisture levels:

i. Wet
   - Stickiness (non-sticky, slightly sticky, sticky, very sticky)
   - Plasticity (non-plastic, slightly plastic, plastic, very plastic)

ii. Moist
   - Very friable, friable, firm, very firm

iii. Dry
   - Loose. soft. slightly hard, hard, very hard, extremely hard

It shows the kind and quantity of clay material, the state of the tillage, and the possibility of compaction. The soil's resistance to rupturing or deforming is known as its consistency. The cohesive and adhesional forces holding the peds together are known as soil consistency. It speaks to the soil's level of stickiness and flexibility. Clay content in the soil has an impact on the kind and quantity of soil consistency.

Soil consistency indicates:

- Where are zones that may restrict root growth and seedling emergence?
- Whether a soil is likely to develop compacted zones; ruts, crusts, hardpans, etc.

Determining Soil Consistency

- **Fingers:** squeeze aggregates or push fingers into the soil.
- **Penetrometer:** measures how hard it is to push into the soil. This would be the same effect as a plant root.
- **Examine Roots:** J roots or a root mat indicate problems.

Sand has a very weak consistency, there is little force between the particles. This means that a car tire can easily push the sand apart and it is easy to get stuck.

Factors Affecting Soil Consistence

- Water Content
- Soil Texture
- Soil Density

5.1.5.4. Soil Color

It is the most apparent and determinable feature of the soil. It is an indicator of the properties of the soil, but it has little direct impact on the soil. Nonetheless, the color alone can reveal a great deal about the soil.

- **Organic Matter Content:** the more organic content the darker the soil color
- **Soil Color and Soil Temperature:** dark colored soils absorb more heat so they warm up quicker and have higher soil temperatures.
- **Soil Color and Parent Material:** generally dark parent material will develop into dark soils.
- **Soil Color and Drainage:** soil drainage refers to the length of time a soil is waterlogged. Not how fast the soil is drained.

5.1.5.5. Soil Permeability

Soil permeability is a crucial engineering characteristic. The quality of the soil that allows liquids (usually water) to pass through its interconnected voids or pores is known as permeability.

A material having continuous voids is called a permeable material. E.g., Sand and gravel.

An impermeable material is one that has the fewest voids, like stiff clay. Granular materials are regarded as highly permeable in all practical senses.

Necessity of Studying Permeability

The following are the factors which influences the study of the percolation (movement) of water through the soil.

- Determination of the rate of settlement of a saturated compressible soil layer.
- Calculation of seepage through the body of the earthen dam and stability of their slopes.
- Calculation of uplift pressure under hydraulic structure and their safety against piping.
- Ground water flow towards the well and drainage of soils.

5.1.5.6. Soil Temperature

The ratio of energy absorbed to energy lost determines the temperature of the soil. Based on biomes, the mean annual temperature of soil ranges from -10 to 26°C. The temperature in the soil controls a variety of processes, including plant and root growth (McMichael Bobbie L, & Burke John, 1998), nutrient availability, seed germination, and seed dormancy breaking (Herranz José et al., 1998). There are significant seasonal, monthly, and daily variations in soil temperature; these variations are much smaller as soil depth increases. Thick mulching is a type of soil cover that can both lessen surface temperature fluctuations and slow the warming of the soil in the summer.

5.1.6. Constituents of Soil

Both solids and voids make up the mass of soil. Mineral matter makes up the solids. Air and water may be present in the voids. Air or water may fill the spaces entirely or partially. It has been noted that the voids and solids in a soil sample can be thought of as occupying distinct volumes rather than distinct spaces.

5.1.7. Phase Diagram for Soil

The bulk of soil is made up of materials that are typically found in three states: solid, liquid, and gaseous. Air forms the gaseous phase, water forms the liquid phase, and soil particles form the solid phase. As seen in Figure 5.2(a), these phases coexist in real soils in a mixed manner. For the sake of engineering studies, these stages are regarded as independent and distinct. Figure 5.2(b) depicts the conceptual three-phase model used in soil mechanics.

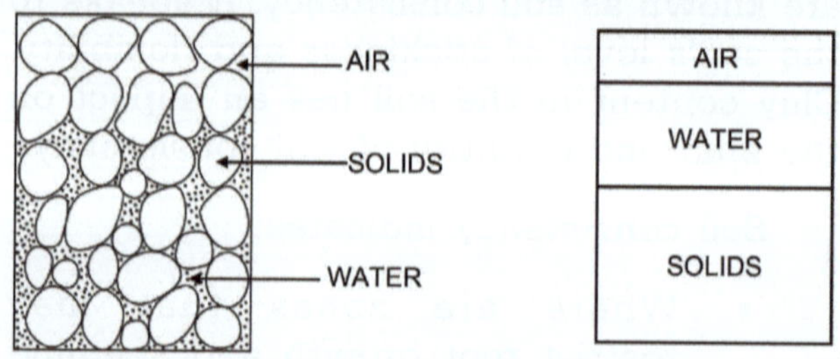

Figure 5.2. *(a) soil three-phase system; (b) conceptual three-phase model.*

Source: https://egyankosh.ac.in/bitstream/123456789/29713/1/Unit-1.pdf.

5.1.8. Phase Relationships

To derive the phase relationships, the weights and the volumes of the three phases are designated as shown in Figure 5.3.

The symbols used in Figure 5.3 have the following meanings:

W = Total weight of given soil sample;

W_a = Weight of air in the voids;

W_w = Weight of water in the voids;

W_s = Weight of solids;

V = Total volume of soil sample;

V_a = Volume of air in the voids;

V_w = Volume of water in the voids;

V_v = Volume of voids;

V_s = Volume of solids.

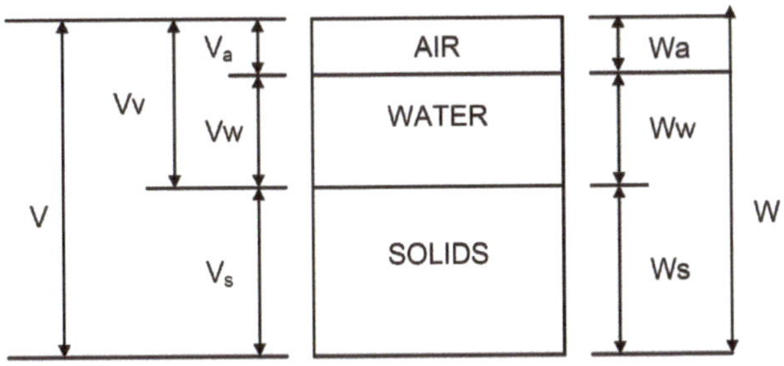

Figure 5.3. *Weight-volume relationships.*

Source: https://egyankosh.ac.in/bitstream/123456789/29713/1/Unit-1.pdf.

The total weight can be written as:

$W = W_a + W_w + W_s$

or

$W = W_w + W_s$ (neglecting the weight of air) (5.1)

The total volume can be written as

$V = V_a + V_w + V_s = V_v + V_s$ (5.2)

5.1.8.1. Volume Relationships

The percentage air voids (n_a), air content (a_c), porosity (n), degree of saturation (s), and voids ratio (e) are the volume relationships that are frequently utilized in a soil element.

- **Void Ratio:** Voids ratio, e, of a soil, is defined as the ratio of the total volume of voids to the volume of solids, or

$$e = \frac{V_v}{V_s} \quad (5.3)$$

Because the volume of voids is smaller than the volume of solids, the value of e is typically less than one. In macroporous soils such as loess, and soil suspensions, the value of e is larger than 1.

- **Porosity:** Porosity, n, is defined as the ratio of the volume of voids to the total volume of the given soil mass, or

$$n = \frac{V_v}{V} \quad (5.4)$$

Porosity is usually expressed as a percentage.

Relationships between void ratio and porosity.

$$e = \frac{V_v}{V_s} = \frac{V_v}{V - V_v} = \frac{\left(\frac{V_v}{V}\right)}{1 - \left(\frac{V_v}{V}\right)} = \frac{n}{1-n} \quad (5.5)$$

We can also write Eqn. (5.5) as:

$$e\,(1-n) = n$$

or

$$e - en = n$$

or

$$e = n + ne$$

$$n = \frac{e}{(1+e)} \quad (5.6)$$

- **Degree of Saturation:** The degree of saturation, S, is defined as the ratio of the volume of water to the total volume of voids present in a given soil mass, or

$$S = \frac{V_w}{V_v} \tag{5.7}$$

A percentage is typically used to express the saturation level. For a completely dry soil sample, its value is zero percent; for a fully saturated soil sample, it is 100%.

- **Percentage Air Voids:** Percentage air voids, n_a, is defined as the ratio of the volume of air voids to the total volume of a given soil mass, or

$$n_a = \frac{V_a}{V} \times 100 \tag{5.8}$$

It is expressed as a percentage.

- Air Content: The air content, a_c, is defined as the ratio of the volume of air voids to the total volume of voids in a given soil mass

$$a_c = \frac{V_a}{V_v} = \frac{V_v - V_w}{V_v} = \frac{V_v}{V_v} - \frac{V_w}{V_v}$$

$$= 1 - \frac{V_w}{V_v} = 1 - S \tag{5.9}$$

5.1.8.2. Weight Relationships

Water content, specific gravity, and unit weight (bulk, moist, dry, saturated, and submerged) are the most often utilized weight relationships.

- **Water Content:** Water content, w, is defined as the ratio of the weight of water to the weight of solids in a given mass of soil, or

$$w = \frac{W_w}{W_s} = \frac{W_w}{W_d} \tag{5.10}$$

- **Unit Weight:** The unit weight of a soil mass, γ, is defined as the ratio of the weight of soil to its volume.
- **Bulk or Moist Unit Weight:** Bulk or moist unit weight, γ_{bulk} is defined as the ratio of total weight of the soil to its total volume.

$$\gamma_{bulk} = \frac{W}{V} \tag{5.11}$$

- **Dry Unit Weight:** Dry unit weight γ_{dry} is defined as the ratio of the weight of soil solids to its total volume, or

$$\gamma_{dry} = \frac{W_s}{V} = \frac{W_{dry}}{V} \tag{5.12}$$

- **Saturated Unit Weight:** Saturated unit weight γ_{sat}, is defined as the ratio of the total weight of a saturated soil sample to its total volume

$$\gamma_{sat} = \frac{W_{sat}}{V} \tag{5.13}$$

- **Submerged Unit Weight:** Submerged unit weight, γ_{sub}, is defined as the ratio of the submerged weight of soil solids to its total volume.

$$\gamma_{sub} = \frac{(W_{dry})_{sub}}{V} = \frac{W_{sat} - W_w}{V}$$

$$= \frac{W_{sat}}{V} - \frac{W_w}{V} = \gamma_{sat} - \gamma_w \tag{5.14}$$

- **Unit Weight of Soil Solids:** The unit weight of soil solids, γ_s, is defined as the ratio of the weight of soil solids to the volume of solids.

$$\gamma_s = \frac{W_s}{V_s} = \frac{W_{dry}}{V_s} \quad (5.15)$$

- **Specific Gravity:** The specific gravity of soil solids, G_s, is defined as the ratio of the weight of a given volume of soil solids at 27°C to the weight of an equal volume of distilled water at 27°C, both weights being taken in air.

$$G_S = \frac{\gamma_S}{\gamma_W} \quad (5.16)$$

5.1.8.3. Interrelationships

From the various parameters defined in the previous sections, we can now derive useful interrelationships. Connection among w, e, s, and G.

From the previous sections, we know that

$$w = \frac{W_w}{W_s},\ e = \frac{V_v}{V_s},\ S = \frac{V_w}{V_v}$$

and

$$G = \frac{\gamma_s}{\gamma_w}, = \frac{W_s}{V_s \gamma_w}, = \frac{W_s}{V_s} \times \frac{V_w}{W_w} \quad (5.17)$$

$$e = \frac{V_v}{V_s} = \frac{V_v}{\left(\frac{W_s V_w}{G\, W_w}\right)} = G \frac{W_w}{W_s} \frac{V_v}{V_w}$$

$$= G\left(\frac{W_v}{W_s}\right) \frac{1}{\left(\frac{V_w}{V_w}\right)}$$

$$= \frac{Gw}{S}$$

or

$$Se = w\,G$$

or

$$w\,G = Se \quad (5.18)$$

For a fully saturated soil, S = 1 and $w = w_{sat}$

$$e = w_{sat}\, G \quad (5.19)$$

Relation between γ_{bulk}, γ_{dry}, γ_{sat}, γ_{sub}, w, G, S, e and γ_w

Now;

$$\gamma_{bulk} = \frac{W}{V} = \frac{W_w + W_s}{V_v + V_s}$$

$$= \frac{W_s\left(\frac{W_w}{W_s} + 1\right)}{V_s\left(\frac{V_v}{V_s} + 1\right)} \quad (5.20)$$

We know that

$$\frac{W_w}{W_s} = w$$

$$\frac{V_v}{V_s} = e$$

and

$$\frac{W_s}{V_s} = \gamma_s = G\gamma_w$$

Putting the values, we get

$$\gamma_{bulk} = G\gamma_w \frac{(1+w)}{(1+e)} \quad (5.21)$$

For a dry soil mass, w = 0.

Putting w = 0 in Eqn., (5.21), we get a dry unit weight of soil (γ_{dry})

$$\gamma_{dry} = G\gamma_w \frac{(1+0)}{(1+e)}$$

$$= \frac{G\gamma_w}{1+e} \quad (5.22)$$

For saturated soils, S = 1 and e = wG (from Eqn. 5.19).

Putting the values in Eqn. (5.21), we get

$$\gamma_{sat} = G\gamma_w \frac{(1+w)}{(1+e)}$$

$$= \gamma_w \frac{(G+wG)}{(1+e)}$$

$$= \gamma_w \frac{(G+e)}{(1+e)} \quad (5.23)$$

Submerged unit weight (γ_{sub}) can be written as:

$$\gamma_{sub} = \gamma_{sat} - \gamma_w$$

$$= \gamma_w \frac{(G+e)}{(1+e)} - \gamma_w$$

$$= \gamma_w \frac{(G+e-1-e)}{(1+e)}$$

$$= \gamma_w \frac{(G-1)}{(1+e)} \quad (5.24)$$

Relation between γ_{sat}, γ_{bulk}, γ_{dry} and s we can write from equation

$$\gamma_{bulk} = G\gamma_w \frac{(1+w)}{(1+e)}$$

$$= \gamma_w \frac{(G+wG)}{(1+e)}$$

$$= \gamma_w \frac{(G+Se)}{(1+e)}$$

$$= \frac{G\gamma_w}{1+e} + S\frac{e\gamma_w}{1+e}$$

$$= \gamma_{dry} + S\left[\frac{(G+e)\gamma_w}{1+e} - \frac{G\gamma_w}{1+e}\right]$$

$$= \gamma_{dry} + S\left[\gamma_{sat} - \gamma_{dry}\right] \quad (5.25)$$

5.1.9. Soil Classification

Soils with comparable behavior are grouped together in a soil classification system, which is a universal language understood by all geotechnical engineers. Systematic and logical methods of classifying and characterizing these soils are then suggested. Any ambiguity in the description of the properties of the soil is eliminated by the use of such clear and standard terminology. Currently, a number of soil classification schemes are in use. The most widely used system is the Unified Soil Classification System (USCS), which is currently in use in many countries. In road construction, the American Association of State Highway and Transportation Officials (AASHTO) classification system is widely used. It groups soils based on whether they are suitable for use as subgrade or embankment materials. Most geotechnical works in Australia are conducted in accordance with AS 1726–1993.

Table 5.3. Soil Groups, Descriptors and Symbols

Major Soil Group	Descriptor
	Well graded (W)
Gravel (G)	Poorly graded (P)
Sand (S)	Silty (M)
	Clayey (C)
Silt (M)	Low plasticity (L)
Clay (C)	Medium plasticity (I)
Organic (O)	High plasticity (H)

Source: https://www.geoengineer.org/storage/education/10/general_file_collection/7906/siva-classification.pdf.

- **AS 1726 – 1993:** The Unified Soil Classification System and Australian Standards AS 1726 categorize soils into four main groups and provide standard descriptors for each. Table

5.3 displays the main soil groups, descriptors, and suggested symbols. According to AS 1726–1993, the soil should be described as "XY," where "X" stands for the major soil group and "Y" for the descriptor.

Clay or silt is assigned to fine-grained soil based on the Atterberg limits rather than the percentages. The PI-LL chart presented by Casagrande (1948) is depicted in Figure 5.4, where the Aline divides the silts and clays. A fine-grained soil is clay if its LL and PI values plot above the A-line. Plotting below the A-line indicates silty soil. A fine-grained soil is denoted by the letters "L," "I," and "O" when the LL value is less than 35, 35–50, or more than 50, respectively. For silts, AS1726–1993 does not advise using "I." Therefore, as Figure 5.4 illustrates, fine-grained soils can be CL, CI, CH, ML, or MH. An organic clay is designated with the symbols "OH" or "OL," plots below the A-line, and has low plasticity. Any one of the four characteristics listed in Table 5.3 can apply to coarse-grained soil, whether it contains sand or gravel.

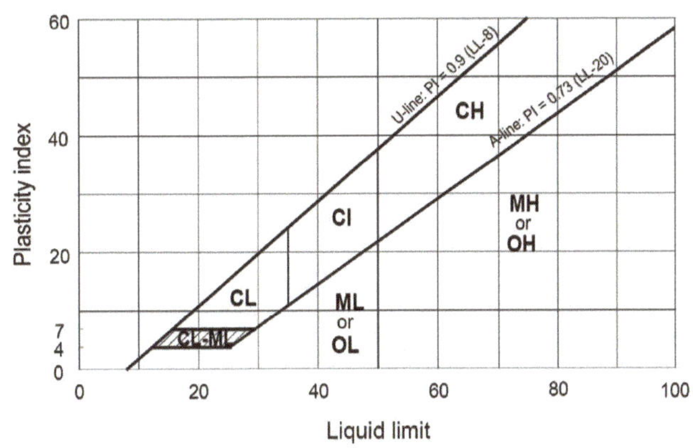

Figure 5.4. *Casagrande's PI-LL chart.*

Source: https://www.geoengineer.org/storage/education/10/general_file_collection/7906/siva-classification.pdf.

Take note that Casagrande's PI-LL chart has a U-line. This is roughly the PI-LL relation's upper bound. The Atterberg limit values plotting above this line should raise some concerns.

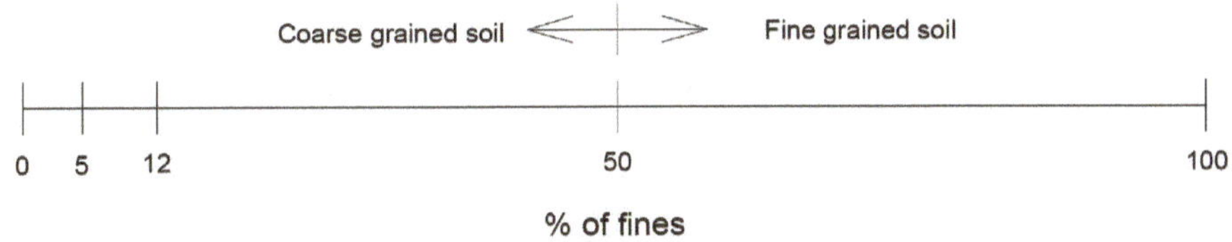

Figure 5.5. *AS1726–1993 soil classification cases.*

Source: https://www.geoengineer.org/storage/education/10/general_file_collection/7906/siva-classification.pdf.

The percentage of fine particles in soil has a big impact on its geotechnical properties. As a result, this needs to be taken into account when dividing up the soil. A soil is considered fine-grained when its fines percentage is higher than that of coarse-grained soils, and vice versa. If the amount of gravel in coarse-grained soil exceeds that of sand, it is gravel, and vice versa. Based on the fine content, AS 1726–1993 categorizes the soils into four cases (Figure 5.5).

A fine-grained soil with a fine content greater than 50% is denoted by a symbol like XY, where X stands for fines and Y for the descriptor (refer to Table 5.3). In this scenario, the soil could be ML, MH, CL, CI, CH, OL, or OH. The soil is assigned the dual symbol CL-ML if the plots of the PI and LL values occur in the hatched area.

When fines are less than 5%, they have little effect. In this case, the soil is designated with the symbol XY, where X represents the majority of the coarse-grained soil and Y denotes the gradation of the coarse-grained soil. In this instance, the soil could be GW, GP, SW, or SP.

If the percentage of fine particles in the soil falls between 12 and 50%, the soil has large amounts of fine particles. The major coarse-grained soil component, represented by the symbol X, and the fine-grained component, represented by the descriptor Y, are assigned to the soil. In this instance, the soil could be GM, GC, SM, or SC.

Both the fine content and the gradation of the coarse-grained soil are significant when the fine content is between 5 and 12%. The major coarse-grained soil present (G or S), the soil's gradation (W or P), and the fine-grained soil present (M or C) are represented by the symbols XY-XZ for the soil. Therefore, GW-GM, GW-GC, GP-GM, GP-GC, SW-SM, SW-SC, SP-SM, or SP-SC can all be considered soils in this situation.

Unified Soil Classification System

The boundary between sands and gravels in the Unified Soil Classification System (USCS) is 4.75 mm (keep in mind that it is 2.36 in AS 1726). The USCS has a special case where the fine content is between 45 and 55% when classifying the soils based on their fine content. The proportions of fine-grained and coarse-grained soils are roughly equal when the fine content is between 45 and 55%. In this case, the soil is given the symbol XY-YZ, where X and Y stand for the corresponding fine- and coarse-grained soils. Z is the fines' descriptor (L or H). The USCS advises against using the pronoun "I" to describe fines.

5.2. SITE INVESTIGATION TECHNIQUES

An essential component of geotechnical engineering is site investigation, which yields valuable data regarding the geological and geotechnical characteristics of a location. It entails gathering information to describe a site's subsurface features, including its geology, the characteristics of the rock and soil, the presence of groundwater, and other elements that could have an impact on a project's design and construction. Site investigation is a crucial step in the building process because it aids engineers and designers in comprehending the site's characteristics and creating foundation designs and construction techniques that are sustainable, affordable, and safe. It is also used to find possible risks or hazards that could compromise a project's performance or safety.

A site investigation is usually conducted in phases, such as a reconnaissance of the site, a preliminary investigation, and a detailed investigation. The size, complexity, and degree of risk posed by the site conditions will all influence the kind and extent of the investigation (Figure 5.6).

Figure 5.6. *Drilling and sampling.*

Source: https://geologyscience.com/geology-branches/engineering-geology/site-investigation-techniques/?amp.

5.2.1. Techniques for Site Investigation

Depending on the particulars of the project and the site, a variety of techniques can be applied to the site investigation process. Some common techniques include:

1. **Desk Study**: This involves gathering and reviewing all available data and information related to the site and the surrounding area, including geological maps, aerial photographs, historical records, and previous site investigation reports.
2. **Site Reconnaissance**: This involves visiting the site to make observations and take measurements, such as mapping surface features, assessing the condition of existing structures, and identifying potential hazards.
3. **Geophysical Surveys**: These involve using various techniques to measure and map physical properties of the subsurface, such as electrical resistivity, magnetic susceptibility, and seismic waves. Examples of geophysical techniques include ground-penetrating radar, seismic reflection and refraction, and electrical resistivity imaging.
4. **Drilling and Sampling**: This involves drilling boreholes or excavating test pits to obtain samples of soil and rock, which can be analyzed in the laboratory to determine their physical and mechanical properties.
5. **In-Situ Testing**: These tests are carried out on-site to determine the properties of soil and rock in their natural state. Examples of in-situ tests include Standard Penetration Test (SPT), Cone Penetration Test (CPT), and Pressure meter Test.
6. **Laboratory Testing**: This involves analyzing soil and rock samples in a laboratory to determine their physical, mechanical, and chemical properties. Examples of laboratory tests include grain size analysis, triaxial compression test, and direct shear test.
7. **Environmental Testing**: These tests are carried out to evaluate the environmental conditions of the site, including the presence of contaminants in soil, groundwater, and surface water. Examples of environmental tests include soil and groundwater sampling, and analysis for heavy metals, hydrocarbons, and other pollutants.
8. **Remote Sensing**: This involves using satellite or aerial imagery to map and analyze the site and surrounding areas. Remote sensing can be used to identify land use patterns, vegetation cover, topography, and other features that can affect site suitability and development.

These techniques can be used in combination to provide a comprehensive understanding of the site and its characteristics.

5.2.2. Geophysical Techniques

Without the need for excavation or drilling, a collection of techniques known as geophysical techniques can be used to investigate a site and provide information about the subsurface geological conditions. These methods entail measuring a range of subsurface physical characteristics, including seismic velocity, electrical conductivity, density, and magnetic susceptibility. Images and models of the subsurface are produced using the data gathered from geophysical surveys, and these can be used to identify geological structures like faults, fractures, and lithological changes. Some of the commonly used geophysical techniques in site investigation include:

1. **Seismic Reflection**: This technique involves generating seismic waves using a source and measuring the

reflected waves using sensors. The data collected can be used to create a 2D or 3D image of the subsurface.

2. **Electrical Resistivity Tomography (ERT)**: This technique measures the electrical resistivity of the subsurface materials by passing an electrical current through the ground and measuring the voltage difference. The data collected can be used to create a model of the subsurface.
3. **Ground-Penetrating Radar (GPR)**: This technique involves transmitting and receiving electromagnetic waves to and from the subsurface. The data collected can be used to create a cross-sectional image of the subsurface.
4. **Magnetic Survey**: This technique measures the magnetic properties of subsurface materials using a magnetometer. The data collected can be used to identify magnetic anomalies associated with certain geological structures.
5. **Gravity Survey**: This technique measures the gravitational field of the subsurface using a gravimeter. The data collected can be used to identify changes in the density of subsurface materials, which can indicate the presence of geological structures.
6. **Electromagnetic Survey**: This technique involves transmitting and receiving electromagnetic waves to and from the subsurface, which can be used to identify changes in subsurface conductivity. This can help in the identification of certain geological structures.

When combined with other techniques, like drilling and sampling, these geophysical methods can yield important information for site investigation and help to provide a thorough understanding of the subsurface geology.

5.2.3. Drilling and Sampling

In order to gather data about a site's subsurface conditions, geotechnical engineers and geologists rely heavily on drilling and sampling as essential methods of site investigation. The methods entail boring holes in the earth and gathering soil or rock samples at different depths.

The physical and engineering characteristics of the soil and rock, such as their composition, strength, permeability, and deformation characteristics, are ascertained using the data gathered from drilling and sampling. The design of foundations, tunnels, excavations, and other structures is then done using this information.

There are several types of drilling techniques, including:

1. **Auger Drilling**: This involves using a helical screw to penetrate the soil and collect samples. It is commonly used for shallow investigations and for soil types that are not too hard.
2. **Rotary Drilling**: This involves using a rotary drill bit to penetrate the soil or rock. It can be used for both shallow and deep investigations.
3. **Percussion Drilling**: This involves using a hammer to drive a drill bit into the soil or rock. It is commonly used for rock formations.

Once the borehole is drilled, samples of soil or rock can be collected using various sampling techniques, such as:

1. **Standard Penetration Testing (SPT)**: This involves driving a split

spoon sampler into the soil using a hammer and counting the number of blows required to drive the sampler a specified distance. This information is used to determine the soil density and strength.

2. **Shelby Tube Sampling**: This involves using a thin-walled tube to collect undisturbed soil samples from the borehole. This technique is commonly used for sampling cohesive soils.
3. **Rock Coring**: This involves using a diamond-tipped drill bit to collect rock samples. The samples can be either continuous or discontinuous, depending on the drilling technique used.

After being taken from the borehole, the samples are sent to a lab for examination and analysis. The findings are utilized to create a geotechnical report that offers details on the site's subsurface conditions as well as suggestions for the layout and building of the foundation.

5.2.4. In-situ Testing

When soil and rock samples are measured in-situ—that is, without taking them out of their natural environment—we are referring to the techniques used for this type of analysis. For engineering and construction projects, in-situ testing can yield valuable information because it gives a more precise understanding of the characteristics of the soil and rock materials that will be encountered during excavation, construction, or other activities.

There are several different types of in-situ testing techniques, including:

1. **Standard Penetration Test (SPT)**: This is a widely used method for determining the strength and density of soils. A cylindrical sampler is driven into the soil using a hammer, and the number of blows required to penetrate the soil a certain distance is recorded.
2. **Cone Penetration Test (CPT)**: This test involves pushing a cone-shaped penetrometer into the soil at a constant rate while measuring the resistance to penetration. The data can be used to determine soil strength, density, and other properties.
3. **Pressuremeter Test**: This test involves inflating a cylindrical probe inside a borehole and measuring the pressure required to expand the probe. The data can be used to determine the in-situ stress-strain properties of the soil or rock.
4. **Shear Wave Velocity Test**: This test involves measuring the speed of shear waves traveling through the soil or rock using seismic methods. The data can be used to determine the stiffness and density of the material.
5. **Cross-Hole Seismic Test**: This test involves creating seismic waves at one location and measuring the waves at another location using sensors installed in boreholes. The data can be used to determine the shear wave velocity and other properties of the soil or rock.
6. **Thermal Conductivity Test**: This test involves measuring the rate of heat flow through the soil or rock using a heat source and temperature sensors. The data can be used to determine the thermal properties of the material.
7. **Electrical Resistivity Test**: This test involves measuring the electrical

resistance of the soil or rock using probes inserted into the ground. The data can be used to determine the moisture content and other properties of the material.

These are but a handful of the numerous kinds of in-situ testing methods that are available for use in site investigation and geotechnical engineering. The particular characteristics of the soil or rock materials, the site's characteristics, and the investigation's goals all influence the technique choice.

5.2.5. Laboratory Testing

The physical and mechanical characteristics of soil and rock samples taken from the site are ascertained through laboratory testing, which is an essential component of site investigation. The engineering qualities of rock and soil, such as strength, permeability, compressibility, and deformation characteristics, are determined by these tests and are crucial for the planning and building of engineering structures.

There are several laboratory tests that can be conducted on soil and rock samples, including:

1. **Grain Size Analysis**: This test is used to determine the particle size distribution of soil. The test involves sieving the soil through a series of standard sieves and measuring the weight of soil retained on each sieve.
2. **Atterberg Limits**: This test is used to determine the plastic limit, liquid limit, and shrinkage limit of soil. The test involves measuring the moisture content of the soil at different stages of the test to determine the boundaries of the different states of consistency.
3. **Compaction Test**: This test is used to determine the maximum dry density and optimum moisture content of soil. The test involves compacting soil in a standard mold using standard compaction energy and measuring the resulting density.
4. **Shear Strength Test**: This test is used to determine the shear strength of soil or rock. The test involves applying a shear force to a soil or rock sample and measuring the resulting deformation.
5. **Permeability Test**: This test is used to determine the permeability of soil or rock. The test involves measuring the rate of flow of water through a soil or rock sample under a known hydraulic gradient.
6. **Consolidation Test**: This test is used to determine the rate and magnitude of settlement of soil. The test involves applying a load to a soil sample and measuring the resulting deformation over time.
7. **Rock Mechanics Testing**: This includes tests such as uniaxial compression, triaxial compression, and direct shear tests, which are used to determine the strength and deformation characteristics of rock samples.

Usually, these laboratory tests are carried out in compliance with industry-standard testing protocols set forth by associations like ASTM International and the International Society for Rock Mechanics.

5.2.6. Environmental Testing

Environmental testing is the process of examining environmental samples to ascertain whether contaminants or pollutants are present and at what concentration. Testing of this kind is necessary to evaluate how human activity affects the environment and to make sure environmental laws are being followed.

Some common types of environmental testing include:

1. **Water Testing**: This includes the analysis of surface water, groundwater, and wastewater to determine the presence and concentration of pollutants such as heavy metals, organic compounds, and pathogens.
2. **Air Testing**: This involves the collection and analysis of air samples to measure the levels of pollutants such as particulate matter, volatile organic compounds (VOCs), and air toxics.
3. **Soil Testing**: This includes the analysis of soil samples to determine the presence and concentration of pollutants such as heavy metals, pesticides, and petroleum hydrocarbons.
4. **Sediment Testing**: This involves the analysis of sediment samples from lakes, rivers, and other bodies of water to determine the presence and concentration of pollutants such as heavy metals and organic compounds.
5. **Biological Testing**: This includes the analysis of biological samples such as fish, shellfish, and other aquatic organisms to determine the presence and concentration of pollutants that may be harmful to human health.

Environmental consulting firms, government agencies, and other entities that focus on environmental monitoring and remediation are usually the ones that conduct environmental testing. Environmental testing results can be used to determine possible health risks, create plans for remediation, and make sure environmental regulations are followed.

5.2.7. Remote Sensing

The process of gathering data about a target or area without physically touching it is known as remote sensing. This data is usually collected from aircraft or satellites. The physical, chemical, and biological characteristics of the Earth's surface and atmosphere, as well as human activity and environmental changes over time, can all be learned from remote sensing data. Passive and active remote sensing are the two primary categories. Using sensors, passive remote sensing records the energy that a target or area, like sunlight, naturally emits or reflects. Infrared sensors that detect heat, radiometers that gauge an object's energy output, and cameras that record visible light are a few examples of passive remote sensing devices. On the other hand, active remote sensing uses sensors to measure the energy emitted or reflected after an energy source is directed toward a target or area. Devices used for active remote sensing include lidar and radar. Numerous disciplines, including geology, agriculture, forestry, and urban planning, can benefit from the application of remote sensing. In geology, mineral deposits, alterations in land use, and geological structures like folds and faults can all be found and mapped using remote sensing. It can also be used to keep an eye on calamities caused by nature, like earthquakes and landslides.

5.3. BEARING CAPACITY AND SETTLEMENT OF FOUNDATIONS

The portion of a structure that transfers the weight of the structure to the earth is called the foundation. Every building on land has foundations to support it. A foundation serves as a link between the actual structure and the supporting earth. The foundation soil's bearing strength properties are a key consideration in the design of civil engineering structures. Bearing capacity, as used in nontechnical engineering, refers to the soil's ability to withstand loads placed on it. The maximum average contact pressure between the foundation and the soil that prevents the soil from shearing is known as the bearing capacity of the soil. Allowable bearing capacity is calculated by dividing the ultimate bearing capacity by a safety factor. Ultimate bearing capacity is the highest theoretical pressure that can be supported without failure. Large settlements under loaded foundations without shear failure can occasionally happen on soft soil sites; in these situations, the maximum allowable settlement determines the allowable bearing capacity.

5.3.1. Bearing Capacity

The ability of soil to withstand loads applied to it is known as bearing capacity. The highest pressure that can be applied to the soil without causing failure is known as the allowable bearing pressure. On the other hand, the maximum pressure that can be applied to the soil before it fails is known as the ultimate bearing capacity. Bearing capacity is limited by three different types of failure: punching shear failure, local shear failure, and general shear failure. It is dependent on the soil's shear strength as well as the depth, form, and kind of foundation.

5.3.2. General Bearing Failure

A general bearing failure happens when the soil moves significantly on a shear failure surface that stretches upward and away from the footing due to the load on the footing. The size of the footing and the characteristics of the soil are taken into consideration when calculating the footing's general bearing capacity. Terzaghi developed the basic method; Meyerhof and Vesi⊠ added modifications and other factors. The case that is typically examined is the general shear failure scenario. Settlement calculations implicitly account for prevention against other failure modes. Ludwig Föppl (1941) and Gerhard Schubert (1942) discovered the stress distribution in elastic soils beneath foundations in a closed form. There are numerous approaches to determining when this failure will transpire.

5.3.3. Ultimate Bearing Capacity

Shear failure would occur under the least amount of pressure in the supporting soil that is next to and beneath a foundation. The maximum gross pressure intensity at the foundation base at which the soil does not fail in shear is known as the ultimate bearing capacity. The UBC may be interpreted as bearing capacity when the term is used.

- **Footing Size Calculation Using Ultimate Bearing Equation:** The maximum load that a foundation or soil can sustain before failing or experiencing excessive settlement is referred to as the ultimate bearing capacity. It is a crucial factor in geotechnical engineering and is ascertained using a variety of soil testing and analysis techniques.

  Soil ultimate bearing capacity is contingent upon a number of factors, such as soil type, moisture content, density, and foundation size and shape.

- **Theories and Methods to Calculate Ultimate Bearing Capacity:** There are different theories and equations used to estimate the ultimate bearing capacity, with some of the commonly used ones being:

  1. **Terzaghi's Bearing Capacity Theory:** This theory, developed by Karl Terzaghi, provides a method for calculating the ultimate bearing capacity of shallow foundations. It takes into account the cohesion of the soil (C), the effective overburden pressure (⊠'), and the angle of internal friction of the soil (⊠).

  2. **Meyerhof's Bearing Capacity Theory:** Proposed by Helmut Meyerhof, this theory extends Terzaghi's theory to consider the shape of the foundation and the inclination of the failure plane. It introduces shape factors and inclination factors to modify the bearing capacity equation.

  3. **Vesic's Bearing Capacity Theory:** Developed by Zden⊠k P. Bažant, this theory considers the effect of the foundation depth on the ultimate bearing capacity. It incorporates a depth factor to modify the bearing capacity equation based on the depth of the foundation.

Although these theories offer approximations for determining a foundation's ultimate bearing capacity, it's crucial to remember that they are predicated on idealized soil behavior and simplifying assumptions. In actuality, the ultimate bearing capacity for a project is ascertained through site-specific soil testing and analysis. Standard penetration tests (SPT), cone penetration tests (CPT), and plate load tests are examples of field tests that are commonly used in conjunction with laboratory testing of soil samples.

To precisely determine the ultimate bearing capacity and design foundations that can safely support the intended loads, it is imperative to consult a qualified geotechnical engineer or a professional with foundation design experience.

5.3.4. Allowable Bearing Capacity

It is the highest pressure that can be given to the soil while still meeting the two essential conditions: limiting settlement to a manageable level. It is the net loading intensity at which excessive settlement or soil failure in shear poses no risk to the structure and prevents soil shear failure.

The highest pressure would guarantee a factor of safety i.e., 2.5–3 of soil and preventing excessive settlement. The bearing capacity of a shallow foundation can be found by

1. Analytical method
2. In correlation with field test
3. In situ determination
4. Presemptive value

5.4. SLOPE STABILITY AND LANDSLIDES

The ability of a slope or hillside to withstand the collapse or downward movement of rock and soil materials is known as slope stability. One frequent type of slope failure is landslides, which can have a serious negative impact on the environment, cause fatalities, and seriously damage infrastructure and property. In engineering geology and geotechnical engineering, slope stability and landslides are crucial factors to take into account, especially when planning, designing, and constructing infrastructure projects like buildings, bridges, and roads.

Slope instability and landslides can result from a variety of factors, such as the kind of geological materials present, the aspect and gradient of the slope, the presence of groundwater, and the effects of both natural and man-made erosion. Slope instability can be caused by a variety of factors, such as excavation or construction activities removing support from the base of a slope, heavy rainfall or snowmelt, and changes in soil moisture content (Figure 5.7).

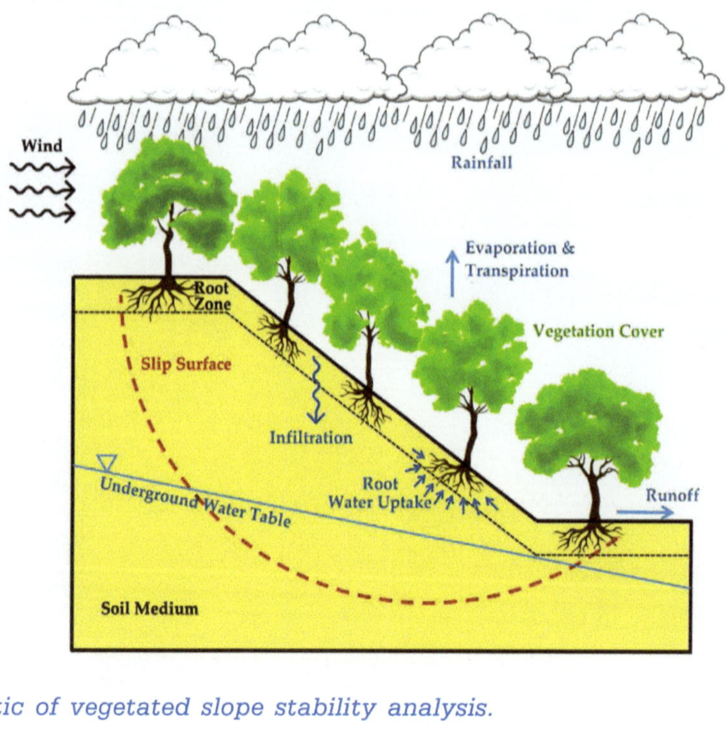

Figure 5.7. *Schematic of vegetated slope stability analysis.*

Source: https://geologyscience.com/geology-branches/engineering-geology/slope-stability-and-landslides/?amp.

Geologists and engineers use a range of techniques, such as field mapping and observation, geophysical surveys, drilling and sampling, and in-situ testing like the Standard Penetration Test (SPT) and Cone Penetration Test (CPT), to evaluate the potential for slope instability and landslides. Predicting how slopes will behave and possible failure mechanisms under various circumstances is another application for computer modeling and simulation.

Building retaining walls or stabilization structures, enhancing drainage and vegetation cover, and changing the slope geometry through grading or excavation are some typical techniques for reducing the risk of slope instability and landslides. Relocating residential areas or infrastructure away from high-risk areas may be necessary in certain situations. All things considered, the study of landslides and slope stability is a crucial component of geotechnical engineering and can support the sustainability and safety of human settlements and infrastructure projects in high-risk areas.

5.4.1. Causes of Slope Failure

Slope failure can occur due to various natural and human-induced factors. Some of the common causes of slope failure are:

1. **Geology and Soil Properties:** The type and properties of soil and rock underlying the slope can contribute to instability. For example, slopes with weak or weathered rock, clay soils, or soils with a high water content are more prone to failure.
2. **Hydrological Conditions:** Water is a significant factor in slope instability, and its presence can contribute to slope failure. Excessive rainfall, flooding, or changes in groundwater level can cause landslides and slope failure.
3. **Slope Geometry:** The angle of the slope and its height can contribute to instability. The steeper the slope, the greater the potential for failure.
4. **Seismic Activity:** Earthquakes and other seismic activities can trigger landslides by altering the stability of slopes.
5. **Human Activities:** Human activities like excavation, construction, mining, or logging can alter the stability of slopes and lead to instability and failure.
6. **Vegetation:** Removal of vegetation can cause instability and contribute to slope failure by reducing soil cohesion and increasing water flow.
7. **Climate Change:** Climate change-induced phenomena like heavy rainfall, droughts, and changes in temperature can contribute to slope failure.
8. **Other Factors**: Other factors that can contribute to slope failure include erosion, freeze-thaw cycles, and natural slope movement over time.

5.4.2. Types of Landslides

Landslides are categorized into various types according to the kind of material they involve and their movement. Some of the common types of landslides are:

1. **Rockfall:** This occurs when rocks or boulders detach from a steep slope and fall to the ground.
2. Rockslide: This occurs when a large block of rock slides downhill along a plane of weakness, such as a fault or joint.
3. **Debris Flow:** This occurs when a large volume of soil, rock, and

water flows downhill, usually in a channel.

4. **Mudflow:** This is similar to debris flow, but the material is mostly fine-grained soil and water.
5. **Earthflow:** This occurs when saturated soil moves downhill in a slow, viscous flow.
6. **Creep:** This is a slow, continuous movement of soil or rock downhill, usually caused by expansion and contraction of the material due to seasonal changes in temperature and moisture.
7. **Slump:** This occurs when a mass of soil or rock moves downhill along a curved surface, leaving a crescent-shaped scar on the slope.
8. **Complex Landslide:** This is a combination of two or more types of landslides, such as a rockslide that triggers a debris flow.

5.4.3. Slope Stability Analysis Techniques

There are several techniques used for slope stability analysis, including:

1. **Limit Equilibrium Analysis:** This method assumes that the slope fails along a failure plane, and the factor of safety is the ratio of the resisting forces to the driving forces along that plane. Different methods can be used for this type of analysis, such as Bishop's method, Janbu's method, and Spencer's method.
2. **Finite Element Analysis:** This method involves dividing the slope into a large number of small elements and analyzing the behavior of each element. This allows for the consideration of more complex geometries, soil behaviors, and load conditions.
3. **Shear Strength Reduction Analysis:** This method is used to assess the stability of a slope under different loading conditions. The shear strength of the soil is reduced incrementally until the slope fails, and the factor of safety is calculated.
4. **Probabilistic Analysis**: This method involves the use of statistical models to assess the probability of slope failure based on the variability of input parameters, such as soil properties and loading conditions.
5. **Empirical Methods:** These methods are based on experience and observation and are often used for preliminary analysis. Examples include the stability number method and the Swedish circle method.

Each of these methods has benefits and drawbacks, and they are suitable for varying soil types and slopes. The type of slope, the available data, and the required level of accuracy all play a role in choosing the best technique.

5.4.3.1. Limit Equilibrium Analysis

Limit equilibrium analysis is a widely used method for assessing a slope's stability. Its foundation is the equilibrium principle, which says that when the forces acting on a slope are balanced, the slope is stable. The slope is divided into several sections for analysis, and the stability of each section is taken into account independently (Figure 5.8).

The factor of safety (FS), in limit equilibrium analysis, is a metric for a slope's stability. The ratio of the driving and resisting forces operating on the slope is known as the safety factor. The slope is classified as stable if the factor of safety is more than one; and unstable if it is less than one.

There are various methods of limit equilibrium analysis, including:

1. **Bishop's Method:** This is a widely used method for analyzing slopes. It assumes that the shear strength of the soil increases linearly with depth and that the forces acting on the slope can be resolved in two perpendicular directions.
2. **Janbu's Method:** This method is similar to Bishop's method, but it considers the possibility of circular failure surfaces.
3. **Spencer's Method:** This method is used for analyzing complex slopes with irregular geometries. It considers the distribution of forces along the slope and uses a graphical approach to determine the factor of safety.
4. **Morgenstern-Price Method:** This method is based on the assumption that the shear strength of the soil varies along the failure surface, and uses numerical techniques to calculate the factor of safety.

Although limit equilibrium analysis is a commonly used method for assessing slope stability, it is not without limitations. It makes the assumption that the characteristics of the soil are uniform and isotropic, which may not always be the case. Additionally, it ignores the effects of pore water pressure, which has a big impact on slope stability. Therefore, the outcomes of limit equilibrium analysis may be supplemented by other analytic methods such as finite element analysis (FEA) or finite difference method (FDM).

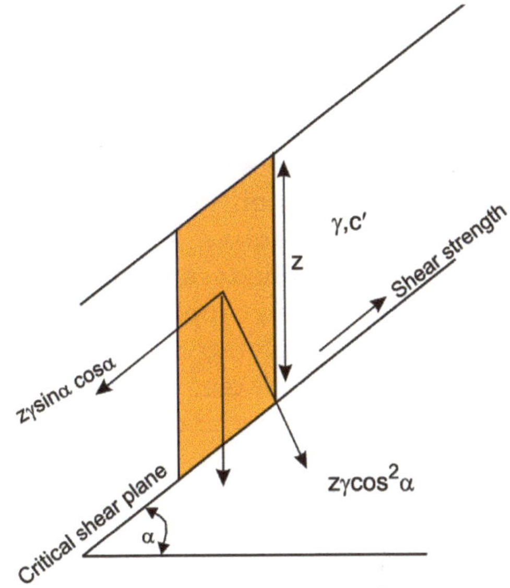

Figure 5.8. *Forces used in limit equilibrium analysis of a simple one-layered vertical soil column.*

Source: https://geologyscience.com/geology-branches/engineering-geology/slope-stability-and-landslides/?amp.

5.4.3.1.1. Bishop's Method

A slope stability analysis method called Bishop's method is used to calculate a slope's factor of safety (FoS) under various loading scenarios. The technique was created by W. W. Bishop in the 1950s and is frequently applied in the practice of geotechnical engineering.

The failure surface in a slope is assumed to be circular or partially circular by Bishop's method. The slope is divided into several slices for the analysis, each of which is taken to be a rigid block. After the forces acting on each slice are separated into their vertical and horizontal components, a force equilibrium equation is used to assess each slice's stability. The ratio of the total driving force to the total available resisting force is known as the slope's factor of safety.

The soil's weight, pore water pressure, and shear strength are all taken into consideration by Bishop's method. Depending on the slope's characteristics and the properties of the soil, the analysis can be carried out using either the total stress method or the effective stress method. Because of its simplicity and ease of use, the method is widely applied in practice; however, when applying it to real-world slope stability problems, one should be aware of its limitations and presumptions.

5.4.3.1.2. Janbu's Method

In geotechnical engineering, Janbu's method is a popular slope stability analysis technique. It is a limit equilibrium technique that examines the stability of slopes using circular failure surfaces. The Mohr-Coulomb failure criterion is assumed to control the soil's shear strength in this method.

The slope is divided into several vertical slices by Janbu's method, and the forces acting on each slice are examined using the statics principles. The technique considers how the properties of the soil change with depth as well as how pore water pressure affects the slope's stability.

The ratio of the driving forces to the resisting forces, or the factor of safety, is calculated as part of the analysis. A slope is considered stable if its factor of safety is greater than 1, and unstable if it is less than 1.

Because of its versatility and ability to work with a variety of slope geometries and soil types, Janbu's method is widely used. It does, however, have certain drawbacks, including the presumption of circular failure surfaces and the disregard for the effects of strain-hardening and strain-softening on soil shear strength.

5.4.3.1.3. Spencer's Method

One kind of limit equilibrium analysis used to assess the stability of slopes is Spencer's method. It bears Edmund H.'s name, who created it. Spencer. To assess the forces operating on a slope and establish its stability, the method makes use of the idea of "wedges."

The slope is divided into several possible failure wedges using Spencer's method, and each is assessed for stability. The weight of the soil above the wedge, the pore pressure in the soil, and any external forces operating on the slope are all taken into account by the method in addition to the weight of the wedge itself. A set of formulas that consider the forces acting on the wedge and the soil's shear strength are used to calculate each wedge's stability.

For the analysis of complex slopes, where there may be several failure surfaces, Spencer's method is especially helpful. Additionally, slopes with irregular geometry or soil with variable properties can have their stability assessed using this method. It does, however, have certain drawbacks, such as the assumption of a two-dimensional failure surface and the assumption that soil properties remain constant along the failure surface, just like other limit equilibrium techniques.

5.4.3.1.4. Morgenstern-Price Method

The pore water pressure created in the slope by water infiltration is taken into consideration by the Morgenstern-Price method, a slope stability analysis technique. Zden⊠k Morgenstern and William Allen Price, two Canadian geotechnical engineers, created this technique in the 1960s.

The approach is predicated on the idea that a slope can be segmented into multiple slices, each of which has a unique factor of safety against failure. The process entails figuring out the effective stresses in each slice, or the stresses on the soil particles left over after deducting the total stress from the pore water pressure. The shear strength of the soil and the shear stress acting on each slice are then compared to determine the factor of safety against failure for each slice.

Any type of slope can be analyzed using the Morgenstern-Price method, including ones with intricate soil profiles and geometries. It has been integrated into numerous software packages for slope stability analysis and is widely utilized in practice. The technique does have certain drawbacks, though, such as the assumption that pore water pressure and soil characteristics would remain constant throughout the slope, which may not always be the case in real-world situations.

5.4.3.2. Finite Element Analysis

A computational technique called finite element analysis (FEA) is used to examine and forecast the behavior of intricate engineering systems. It entails disassembling a system into smaller, more manageable components, known as finite elements, and modeling the behavior of each element using numerical techniques and mathematical equations. To find the solution for the entire system, the equations for each element are solved concurrently.

FEA is frequently used in geotechnical engineering to simulate how rock masses and soil behave, particularly under complicated geological circumstances. Among other things, FEA can be used to analyze issues with excavation, tunneling, foundation behavior, and slope stability. The geometry, boundary conditions, material characteristics, and loading conditions of the system under analysis must all be thoroughly understood for FEA to be performed. The complexity of the model and the accuracy of the input parameters determine how accurate the results are. Although FEA is an effective tool, it also necessitates a large amount of computational power, specialized software, and knowledge of computer programming and numerical techniques.

5.4.3.3. Shear Strength Reduction Analysis

A numerical technique called shear strength reduction analysis (SSRA) is used to assess the stability of embankments and slopes. It is also referred to as the c-method, shear strength reduction method, or stability reduction method.

In SSRA, a slope's factor of safety (FoS) is determined by gradually lowering the soil's shear strength until failure happens. The method is predicated on the idea that a slope fails when the soil's shear strength is reached at any given point in the slope's maximum shear stress.

When it is difficult to apply conventional techniques like limit equilibrium analysis due to complex soil properties or slope geometry, the SSRA method comes in handy. Nevertheless, SSRA is a computationally demanding method that necessitates the use of sophisticated software and potent computers in order to perform the required simulations.

In geotechnical engineering, slope stability has been extensively analyzed using SSRA for a variety of purposes, such as open-pit mining, dam construction, and roadways. It has also been used to look into how environmental elements like earthquakes, rainfall, and climate change affect slope stability.

5.4.3.4. Probabilistic Analysis

Slope stability analysis uses probabilistic analysis as a technique to determine the likelihood that a slope failure will occur. It entails allocating probabilities to various elements that may affect the slope's stability, including the soil's strength, the slope's geometry, and the amount and duration of loading.

Instead of assigning a single deterministic value to each factor, probabilistic analysis uses a range of values. Because it accounts for the inherent variability and uncertainty found in real-world conditions, this enables a more realistic assessment of the slope's stability.

In probabilistic analysis, Monte Carlo simulation is a frequently employed method. Numerous simulations must be conducted, each with a unique set of input values chosen at random from the designated probability distributions. Following that, the simulation results can be used to determine the most important parameters affecting the slope's stability as well as the likelihood that a slope failure will occur.

5.4.3.5. Empirical Methods

Slope stability analysis techniques based on historical slope behavior observations are known as empirical methods. They rely on empirical relationships derived from case histories of slope failures instead of mathematical models. These techniques come in handy when data is scarce or the geotechnical conditions are complicated and challenging to model.

The "Stability Number" approach, which is used to analyze slopes with planar failure surfaces, is an illustration of an empirical method. The soil's cohesion, friction angle, slope angle, and unit weight are used to calculate the stability number. The methodology is predicated on the finding that slopes are generally regarded as unstable when their Stability Number is less than 1.0, and as stable when it is greater than 1.0.

Another illustration is the semi-empirical "Swedish method," which is popular in Scandinavia. Using this method, the shear strength of the soil is compared with an analysis of the pore pressure distribution within the slope. The slope is deemed unstable if the pore pressure is greater than the shear strength.

In order to get more information about the stability of a slope, empirical methods are frequently used in conjunction with other analysis techniques. They are most frequently employed in scenarios where the geotechnical conditions are intricate and challenging to model, or when the amount of available data is small.

ROLE MODEL

Source: https://upload.wikimedia.org/wikipedia/commons/thumb/7/7c/KT_1926.jpg/800px-KT_1926.jpg.

KARL VON TERZAGHI

Karl von Terzaghi (October 2, 1883 – October 25, 1963) was an Austrian mechanical engineer, geotechnical engineer, and geologist known as the "father of soil mechanics and geotechnical engineering."

Early Life:

In 1883, he was born the first child of Army Lieutenant-Colonel Anton von Terzaghi, of Italian origin, and Amalia Eberle in Prague, in what is now the Czech Republic. Upon his father's retirement from the army, the family moved to Graz, Austria.

At 10, Terzaghi was sent to a military boarding school, where he developed an interest in astronomy and geography. At age fourteen, Terzaghi entered a different military school, in Hranice, the Crown of Bohemia. He was an excellent student, especially in geometry and mathematics, and graduated with honors at 17.

In 1900, Terzaghi entered the Technical University in Graz to study mechanical engineering, where he also developed an interest in theoretical mechanics. He was nearly expelled at one point but ended up graduating with honors in 1904. Terzaghi then fulfilled his compulsory one-year military service. While fulfilling his military obligations, Terzaghi translated a popular English geology field manual into German and greatly expanded it. He returned to the university for one year and combined the study of geology with courses on subjects such as highway and railway engineering. Shortly afterward, he published his first academic paper on the geology of terraces in southern Styria.

Early Professional Years

His first job was as a junior design engineer for the firm Adolph von Pittel, Vienna. The firm was becoming more involved in the relatively new field of hydroelectric power generation, and Karl became involved in the geological problems the firm faced. His responsibilities quickly increased, and by 1908, he was managing a construction site, workers, and the design and construction of steel-reinforced structures. He embarked on a project to construct a hydroelectric dam in Croatia. For six months in Russia, he developed some novel graphical methods for the design of industrial tanks, which he submitted as a thesis for his PhD at the university.

In 1912 he traveled to the United States and undertook an engineering tour of major dam construction sites in the West. He used the opportunity to gather reports and firsthand knowledge of the problems of many different projects, and he returned to Austria in December 1913. When World War I broke out, he found himself drafted into the army as an officer directing a 250-man engineering battalion. He eventually led 1,000 men, faced combat in Serbia, and witnessed the fall of Belgrade. After a short stint managing an airfield, he became a professor at the Royal Ottoman College of Engineering in Istanbul (now Istanbul Technical University).

There, he began a rigorous study of the properties of soils in an engineering context. Both his measurements and his analysis of the force on retaining walls were first published in English in 1919, and they were quickly recognized as an important new contribution to the scientific understanding of the fundamental behavior of soils.

After the war, he was forced to resign his post at the University but managed to find a new post at Robert College in Istanbul, where he switched his teaching language from French to English. He began studying experimental and quantitative aspects of the permeability of soils to water and produced theories to explain the observations. He invented novel equipment as part of the work. In 1924 he published *Erdbaumechanik auf Bodenphysikalischer Grundlage* (*The Mechanics of Earth Construction Based on Soil Physics*) which would have a profound impact on the field. That resulted in a job offer from the Massachusetts Institute of Technology (MIT) which he accepted.

Later Years

One of his first tasks in the United States was to bring his work to the attention of engineers. This he proceeded to do by writing a series of articles for the *Engineering News Record*, which were published in winter 1925, then as a small book in 1926. He found the facilities at MIT abominable and had to deal with obstruction from the administration. Brushing the obstacles aside, he once more set up a new laboratory geared towards making measurements on soils with instruments of his own devising. He entered a new phase of prolific publication and a rapidly growing and lucrative involvement as an engineering consultant on many large-scale projects.

In 1927, Aurelia Schober Plath, who would become the mother of the poet Sylvia Plath, worked as a secretary for Terzaghi. She was of Austrian descent and worked for him by translating a handwritten manuscript in German, dealing with new principles of soil mechanics. After work, they would have dinner together when Terzaghi's conversation led her to Greek drama, Russian literature, the works of Hermann Hesse, the poems of Rainer Maria Rilke as well as the writings of great world philosophers. She claims the experience affected her for the rest of her life and that she "realized how narrow my world had been and that self-education could be and should be an exciting lifelong adventure. It was the beginning of my dream for the ideal education of the children I hoped some day to have."

From 1926 to 1932, Arthur Casagrande, another pioneer of soil mechanics and geotechnical engineering, worked as Terzaghi's private assistant at MIT.

Terzaghi was much in demand as a dinner companion and was a fascinating conversationalist. His striking good looks and evident power was very attractive to women. In 1928, he met the young Harvard doctoral student in geology, Ruth Dogget, and fell deeply in love. That year, Terzaghi was finally fed up with MIT and its president and determined to return to Europe. He accepted a chair at the Vienna *Technische Hochshule* in the winter of 1929. He married Ruth, who became his editor and collaborator as well. A short consulting trip to the Soviet Union before taking up his post horrified him, and he came to oppose the communist system there, as a regime exemplified by its brutality and chaos. Using Austria as his base, he traveled ceaselessly throughout Europe, consulting, lecturing and making new professional contacts and collaborations. His teaching workload was now relatively light so he continued his experimental investigations and became especially interested in the problems of the settling of foundations, and of grouting. He began writing the manuscript for a much updated and expanded version of *Erdbaumechanik*, now set for two volumes. However, the political turmoil in Austria began to interfere with his work, and in 1935, he decided to take a leave from Vienna from 1935 to 1936.

He began his sabbatical with a short trip to consult with Todt and the architects of the proposed grandiose plans for immense buildings at the Nazi's Party Day Rally site in Nuremberg. That led to a conflict over the best way to lay a sound foundation, which led to a discussion with Hitler himself, who took an intense interest in all the details of the architecture. Terzaghi then returned to America, where he gave a plenary lecture at the International Conference on Soil Mechanics at Harvard University (the event led to the establishment of the International Society for Soil Mechanics and Geotechnical Engineering; Terzaghi was its first president). He made a lecture tour of many other universities but discovered that prospects for employment were dim. He returned to Vienna in September 1936, shortly after the birth of his first son Eric.

In Vienna, he returned to a nasty professional and political controversy (including an acrimonious dispute with Paul Fillunger), which he overcame only with some difficulty. He memorably stated, "The Fatherland denoted me as a Nazi, the Nazis as a Bolshevik, and the Bolsheviks as a conservative idealist. Certainly only one of the three could be right, and that one is the Bolsheviks." He escaped from Vienna frequently by extended consulting trips to major construction projects in England, Italy, France, Algeria, and Latvia, adding greatly to his engineering experience.

In 1938, he emigrated to the United States and took up a post at Harvard University. By the end of the war, he had consulted on the Chicago subway system, Newport News Shipways construction, and raising the Normandie, among others. He became an American citizen in March 1943. He was awarded the Frank P. Brown Medal in 1946.

He remained a part-timer at Harvard University until his retirement in 1953, at the mandatory age of 70.

In 1950, he published work on the geological aspects of soft ground tunneling including a classification system for the assessment of different soil types.

In July 1954, he became the chairman of the Consulting Board for the construction of the Aswan High Dam. He resigned from that post in 1959 after coming into conflict with the Soviet engineers in charge of the project but continued to consult on various hydroelectric projects, especially in British Columbia. He died in 1963, and his ashes were interred in South Waterford, Maine, near "Bear's Corner," the family retreat.

Legacy

The American Society of Civil Engineers established in 1960 the Karl Terzaghi Award to an "author of outstanding contributions to knowledge in the fields of soil mechanics, subsurface and earthwork engineering, and subsurface and earthwork construction." The Terzaghi and Peck Library, which is managed by the Norwegian Geotechnical Institute, in Oslo, Norway, holds an extensive collection of his papers.

In 1963, The American Society of Civil Engineers' Soil Mechanics and Foundations Division, along with and the friends of Terzaghi, established the Karl Terzaghi Lecture. The first lecture was delivered by Ralph B. Peck, a student of Terzaghi's, in 1963. In 2016, the Geo-Institute of the American Society of Civil Engineers started observing Terzaghi Day, held every year on Terzaghi's birthday.

In 1983, the Austrian Postal Service issued a commemorative stamp to mark the centenary of Terzaghi's birth. The Mission Dam in British Columbia, Canada, was renamed in his honor as the Terzaghi Dam in 1965.

As Professor Goodman describes him, Karl Terzaghi was a remarkable man and an impassioned engineer. As he put it himself, "All the modest achievements which I have to my credit can be described by a simple formula... Guided by common sense and casual observations, I recognized weak points in traditional procedures and tried to make them less weak. Sometimes I failed, but usually I succeeded."

CASE STUDY

GEOTECHNICAL

A geotechnical engineer investigated a site for a housing tract that contained a known active fault. Earth materials consisted of 3 m of alluvium overlying bedrock. The fault was shown on published maps, described in the open literature, and in reports of adjacent sites. The engineer's final report submitted to the permitting agency stated that "The fault was not exposed at ground surface and, therefore, did not exist at the site." He did not conduct trench studies. The reviewing agency's reply was that available information indicated the fault existed at the site and provided the consultant with published maps, reports of adjacent sites, and references to the consultant. The agency recommended a trench be placed across the strike of the fault. The engineer refused, stating that his investigation was conclusive. The agency persisted and the engineer agreed only because the agency stated it would not approve the report without documentation that the fault either did or did not exist at the site.

A 4-m deep trench was dug. The trench walls were not supported with shoring. The consultant's staff engineer entered the trench, identified the fault in bedrock, and traced it through overlying alluvium to the ground surface. He began to photograph and map the fault when the trench collapsed and killed him. Work stopped; the engineer did not complete the project.

The initial investigation did not identify the fault because it was "not obvious at the ground surface," being covered by grass and bushes. The engineer's decision to not conduct subsurface investigations was not consistent with the standards of practice for active fault investigations. The active fault was an existing adverse condition and posed a hazard. The agency correctly requested additional work be done and refused to grant a permit based on the information contained in the initial report.

The death resulted in criminal action against the engineer. At trial, the engineer blamed the collapse of the trench on "Evaporation of soil moisture; the surface tension of the pore water was reduced and no longer held the grains together." That argument was dismissed by the court as being not relevant; the trench collapsed because it did not have shoring. In addition, the engineer violated Federal and State laws by not shoring the trench. The court remanded the engineer's technical report to the Licensing Board. The Board's action has not been made public.

The permitting agency acted properly. The engineer did not follow standard practice for (a) not investigating subsurface conditions of a known active fault and (b) not shoring the trench. His argument that evaporation of soil moisture caused the trench to collapse was specious. Had he followed Federal and State laws to shore the trench, the death would have been avoided.

CLASS ACTIVITY

A partially saturated sample of soil from an earth fill has a natural moisture content of 22% and a bulk unit weight of 20.0 kN/m^3. Assuming the specific gravity of soil solids as 2.65, the unit weight of water as 10 kN/m^3, and the weight of solids in the soil as unity. Students draw the three-phase diagram and determine its degree of saturation and voids ratio. If subsequently the soil gets saturated, the student determine its unit weight.

SUMMARY

- Geotechnical engineering is the branch of civil engineering that studies soils, structures, or structural components residing on or inside these soils. The mechanics of soils provide the necessary physical attributes to be used to understand the mechanisms that play a key role in what happens to soils under external effects, such as those induced by structural loads or changes in the water content.
- Soil may be defined as a natural aggregate of mineral grains, loose or moderately cohesive, inorganic or organic in nature, that have the capacity to be separated by means of simple mechanical processes. It is useful to have an idea of different types of soils.
- Soil mechanics is a discipline of Civil Engineering involving the study of soil, its behavior, and its application as an engineering material. Soil mechanics is the application of laws of mechanics and hydraulics to engineering problems dealing with sediments and other unconsolidated accumulations of solid particles, which are produced by the mechanical and chemical disintegration of rocks, regardless of whether or not they contain an admixture of organic constituents.
- Residual soils are those that result from the disintegration of rock and stay at the place of their formation. Sands, silts, and clays belong to this category.
- Transported soils are those that result from the disintegration of rock and are carried away by forces of gravity, water, wind, and ice and deposited at another location.
- The soil mass consists of solids and voids. The solids consist of mineral matter. The voids may consist of water and air. The voids may be partially or fully filled with water or air. It has been observed that the solids and voids in a sample of soil do not occupy separate spaces, they can be represented as occupied separate volumes.

- Soil mass is composed of matters existing generally in three phases-solid, liquid, and gaseous. Soil particles form the solid phase, water forms the liquid phase, and air the gaseous phase.
- Site investigation is a critical aspect of geotechnical engineering, providing important information about the geological and geotechnical properties of a site. It involves the collection of data to characterize the subsurface conditions of a site, such as the geology, soil, and rock properties, groundwater conditions, and other factors that may affect the design and construction of a project.
- Geophysical techniques are a set of methods used in site investigation to provide information about subsurface geological conditions, without the need for excavation or drilling. These techniques involve measuring various physical properties of the subsurface, such as density, magnetic susceptibility, electrical conductivity, seismic velocity, and others.
- A foundation is the part of a structure that transmits the weight of the structure to the ground. All structures constructed on land are supported on foundations. A foundation is a connecting link between the structure proper and the ground which supports it. The bearing strength characteristics of foundation soil are major design criteria for civil engineering structures.
- Bearing capacity is the capacity of soil to support the loads applied to the ground. The bearing capacity of soil is the maximum average contact pressure between the foundation and the soil which should not produce shear failure in the soil. *Ultimate bearing capacity* is the theoretical maximum pressure that can be supported without failure; *allowable bearing capacity* is the ultimate bearing capacity divided by a factor of safety.
- Slope stability refers to the ability of a slope or hillside to resist the downward movement or collapse of soil and rock materials. Landslides are a common form of slope failure, which can result in significant damage to property and infrastructure, loss of life, and environmental impacts. Slope stability and landslides are important considerations in engineering geology and geotechnical engineering, particularly in the planning, design, and construction of infrastructure projects such as roads, bridges, and buildings.

REVIEW QUESTIONS

1. Define soil in a broad sense.
2. List the types of soils.
3. What is bearing capacity in geotechnical engineering?
4. What is allowable bearing capacity?
5. What do you mean by slope stability?

REFERENCES

1. Casagrande, A. (1948). Classification and identification of soils. *Transactions of the ASCE, 113*, 901–930.
2. Coduto, D. P. (2001). *Foundation Design: Principles and Practices* (2nd ed.). Prentice Hall. ISBN 0135897068.
3. Das, B. M. (1994). *Principles of Geotechnical Engineering* (3rd ed.). PWS Kent Publishers.
4. Das, B. M. (1999). *Shallow Foundations: Bearing Capacity and Settlement*. CRC Press. ISBN 0849311357.
5. Das, B. M. (2007). *Principles of Foundation Engineering* (6th ed.). Thomson. ISBN 978-0495082460.
6. Gardner, C. M. K., Laryea, K. B., & Unger, P. W. (1999). *Soil Physical Constraints to Plant Growth and Crop Production* (1st ed.). Food and Agriculture Organization of the United Nations. CiteSeerX 10.1.1.466.9332.
7. Herranz, J. M., Ferrandis, P., & Martínez-Sánchez, J. J. (1998). Influence of heat on seed germination of seven Mediterranean Leguminosae species. *Plant Ecology, 136*(1), 95–103. https://doi.org/10.1023/A:1009702318641.
8. McMichael, B. L., & Burke, J. J. (1998). Soil temperature and root growth. *HortScience, 33*(6), 947–951. https://doi.org/10.21273/HORTSCI.33.6.947.

CHAPTER 6

Construction Technology

LEARNING OBJECTIVES

After studying this chapter, you will be able to:

- Understand the modern building construction techniques;
- Discuss the architectural technology;
- Describe the tunneling technology;
- Explain the bridge construction technology; and
- Define the construction safety management.

KEY TERMS FROM THIS CHAPTER

3D printing technology	3D volumetric construction
Arch bridge	Automated construction equipment
Bridge construction technology	Building construction
Building information modeling (BIM)	Construction management software
Construction technology	Engineered wood products (EWPs)
Flat slabs	HVAC systems
Hybrid concrete construction	Insulating concrete formwork (ICF)
Internet of things (IoT)	Seismic technology
Structural insulated panels (SIPs)	Suspension bridge
Thin-joint masonry	Truss bridge

INTRODUCTION

Construction technology encompasses a wide array of tools, machinery, modifications, software, and systems used during the construction phase of a project that enables advancement in field construction methods, including semi-automated and automated construction equipment. This technology aims to improve the efficiency, productivity, quality, and safety of construction processes. Key aspects include Building Information Modeling (BIM), which provides digital representations of the physical and functional characteristics of places. BIM enhances collaboration among architects, engineers, and contractors, enabling more precise planning and execution. Another significant technology is prefabrication, where building components are manufactured off-site in a controlled environment before being transported and assembled on-site. This method reduces waste, speeds up construction timelines, and often results in higher-quality outcomes.

Drones, or unmanned aerial vehicles (UAVs), have revolutionized site surveying and monitoring by providing real-time, high-resolution images and data, allowing for better site management and progress tracking. Similarly, the advent of robotics and automation in construction, such as robotic bricklayers and 3D printers that create building components layer by layer, has led to increased precision and reduced labor costs. Wearable technology and sensors play a crucial role in enhancing safety and productivity on construction sites. Wearables, such as smart helmets, vests, and glasses, can monitor workers' health and movements, providing data to prevent accidents and improve overall site safety. Internet of Things (IoT) devices further contribute by connecting construction machinery, tools, and equipment to a central system, facilitating real-time monitoring and management of resources.

The use of green construction technologies is also rising, focusing on sustainability and reducing the environmental impact of buildings. Innovations such as energy-efficient HVAC systems, solar panels, and green roofs contribute to creating more sustainable structures. Advanced materials, including self-healing concrete and translucent wood, offer new possibilities for durable and energy-efficient buildings. Construction management software has become indispensable for project management, offering tools for scheduling, budgeting, and resource allocation, enhancing project coordination, and reducing the likelihood of delays and cost overruns. Overall, construction technology is driving significant improvements in how buildings and infrastructure are designed, constructed, and maintained. It supports the development of smarter, more sustainable, and safer construction practices, ensuring that projects are completed on time, within budget, and to the highest quality standards. As technology continues to evolve, its integration into the construction industry promises even greater advancements, further transforming how construction projects are approached and executed.

6.1. MODERN BUILDING CONSTRUCTION TECHNIQUES

There are many different construction methods that can be used when building a home. Some of the most popular ones are stick-built wood homes, while others are more traditional and are only used in specialized projects. Light-gauge steel framing and earthen block construction are examples of techniques that are gradually gaining popularity. The techniques used in the construction of buildings have greatly improved recently, thanks to the best possible utilization of cutting-edge contemporary technologies. Working closely with architects and builders is essential when using the newest materials to construct any kind of house. There are numerous ways to construct a completely new home, particularly when using prefabricated materials like light gauge steel, modular pieces, and innovative concrete designs (Figure 6.1).

Figure 6.1. *Building construction techniques.*

Source: https://www.constructionplacements.com/wp-content/uploads/2021/09/What-are-Construction-Techniques-.jpg.

A custom plan will make it easy to frame several items on-site and complete them in accordance with the highest industry standards when building a home. There are numerous standard techniques used in contemporary home building to produce improvements for your new residence. When the time comes to build a new house, you must be aware of these cutting-edge techniques. When attempting to determine which method of home construction would be best for your house in the future, it can frequently be useful.

- **How to Determine the Construction Type?:** When it comes to designing a structure, the construction type is chosen based on several things, such:
  - **Purpose of the Structure:** Assume for the moment that you wish to construct a detached garage to house a boat, lawnmower, and additional vehicle. It wouldn't require a structure to support several stories when it was designed, though you could add a room upstairs to avoid the need for labor-intensive building techniques.
  - **Size of the Structure:** A heavier construction type would be used when constructing something larger than a garage or a house, such as a grocery store or auto dealership showroom. We're talking about expansive roofs supported by metal trusses and open floor plans devoid of inner walls or columns.
  - **Load Bearing Constraints:** Building height increases require an increase in structural loads. In light of this, the type of construction needs to be robust enough to support these loads. In this instance, wood is probably not the best material for a building with multiple floors.
  - **Environmental Factors:** If you have ever driven near a coastline, you have undoubtedly noticed that the buildings are raised in order to protect them from tidal waves. Buildings in seismically active regions must also be constructed with the right kind of materials. A system that is overly rigid prevents it from flexing during tremors.
  - **Necessary Speed of Construction:** Construction types can expedite the building process if you need to have your building up quickly. By using pre-engineered techniques, these construction methods enable large components to be delivered to the construction site and swiftly erected.
  - **Cost of Materials:** f course, when deciding on a construction method, material costs are a factor. Investing money in an overly complex structure is not necessary.

6.1.1. Benefits of Modern Construction Techniques

Modern building techniques accelerate project timelines and help to reduce construction costs. However, there are other tangible benefits of using modern building methods.

- **Eco-Friendly**: In general, MMCs have a lower environmental impact than conventional building techniques. When compared to the manufacturing, assembly, and installation of conventional materials, the MMC production process produces fewer emissions. Not to mention, many MMCs' energy-efficient insulation can contribute to a decrease in carbon footprints.
- **Energy-Efficient**: Modern building methods make use of incredibly

energy-efficient insulation, which lowers the energy needed to heat and cool interior spaces. MMCs offer more sustainable energy solutions integrated into every construction project, a response to the global increase in energy costs.

- **Less Risk**: There is significantly less risk in the building process when MMCs are manufactured in a controlled manufacturing environment and are pre-planned using sophisticated software. Builders typically don't have to operate at an elevated height, which significantly lowers their risk of injury from falls. Additionally, risks such as moving construction vehicles are removed.
- **Higher Quality**: Conventional construction is prone to unanticipated difficulties and delays. Construction projects can be completed more quickly and with greater quality thanks to the standardization and consistency of MMC products produced in off-site facilities.
- **Comfortable**: Most importantly, MMC-built buildings are both structurally sound and comfortable. Warmer in the winter and colder in the summer, interior temperatures are maintained with airtight insulation.

6.1.2. Types of Construction Techniques/Methods

The kind of construction that is taking place will usually determine the kind of construction method that is used. The following building construction types are commonly used:

- **Pre-Engineered/Precast Flat Panel System/Precast Foundations:** Repetitive cellular projects benefit greatly from the sturdy structures created when floor and wall units are manufactured off-site in a factory and assembled on-site. Panels may comprise finishes, windows, doors, and services. As load-bearing components, building envelope panels with ornamental cladding and insulation installed at the factory are also options. This offers both the speed of on-site erection and factory quality and accuracy. Systems for precast concrete can be utilized to build foundations quickly. The elements are typically custom-made and cast in a factory setting, providing assurance for the final product. The foundations are frequently connected and supported by concrete piles. These systems reduce the need for excavation, which is especially helpful when working with the contaminated ground and increase productivity, especially in bad weather.
- **3D Volumetric Construction:** Three-dimensional units are produced in controlled factory conditions prior to being transported to the construction site, a process known as 3D volumetric construction (also called modular construction). From a simple structure to one with all internal and external finishes and services installed and ready for assembly, modules can be delivered to the site in a variety of configurations. When producing units that require a lot of repetition and quick on-site assembly, casting modules take advantage of factory conditions to produce highly desirable units for

service. This contemporary building technique provides the on-site speed of erection along with the inherent advantages of concrete, including thermal mass, sound, fire resistance, factory quality, and accuracy.

- **Tunnel Formwork Systems:** A tunnel form system is a type of formwork that enables a contractor to construct slabs and walls in a single operation on a daily basis. The flexibility and economy of cast-in-situ construction are combined with the speed, quality, and accuracy of ready-mixed concrete and formwork produced offsite or in a factory. For repeating cellular projects like hotels, apartment buildings, and dorms, this expedited building technique is appropriate. In addition to utilizing the natural advantages of concrete, like its resistance to fire and sound, it provides economy, speed, quality, and accuracy. The tunnel form's formwork sections are large and must be swung out from the building's side by a crane when concrete is struck. This indicates that it is inappropriate for small spaces.
- **Flat Slabs:** Flat slabs can be constructed more quickly because modern formwork is reduced and simplified. Panelized formwork and early striking are combined to provide a quick turnaround. Because there are continuous service zones beneath the floor slab, prefabricated services can be maximized. As a result, flat slab construction allows for faster overall construction and easier service installation. Flat slab construction not only reduces construction time but also allows for unrestricted placement of horizontal services and partitions. The occupier can easily modify internal layouts to accommodate changes in the structure's use, giving them a great deal of flexibility. Flat slab post-tensioning allows for thinner, longer slabs with less reinforcing, which has a major positive impact on labor and the program.
- **Hybrid Concrete Construction:** The advantages of both cast-in-situ construction and pre-casting are combined in hybrid concrete construction. When the two are combined to create a hybrid frame, construction efficiency, quality, and speed are all increased. By offering straightforward, competitively priced, and easily constructed structures with reliable performance and quality, hybrid concrete construction can meet client demands for reduced costs and improved quality (Figure 6.2).
- **Thin-Joint Masonry:** Thin-Joint Masonry makes it possible to reduce the mortar's depth from 10 mm to just 3 mm or less, which speeds up wall laying and increases productivity—especially for lengthy wall runs. Large-format concrete blocks, which have a face size equal to two regular concrete blocks, can accelerate construction even more—by 13.5%. More courses can be laid each day because the mortar cures quickly, reaching full bond strength in one to two hours and doing away with the "floating" issue.

Figure 6.2. *Hybrid concrete construction technique.*

Source: https://i0.wp.com/theconstructor.org/wp-content/uploads/2014/03/hybrid-concrete-construction.jpg?fit=644%2C394&ssl=1.

- **Insulating Concrete Formwork:** Expanded polystyrene panels or blocks with twin walls are used in Insulating Concrete Formwork (ICF) systems, which can be rapidly assembled to construct formwork for a building's walls. In order to create a sturdy structure, this formwork is then filled with factory-produced, quality-assured ready-mixed concrete. High levels of thermal insulation are still provided by the expanded polystyrene blocks, and robustness and good levels of sound insulation are provided by the concrete core.
- **Wood Frame (Engineered Wood Products (EWPs)):** Framed home construction frequently uses Engineered Wood Products (EWPs), which are materials made by binding fibers, particles, or wood veneers together with adhesives. Plywood and oriented strand board (OSB) are the most widely used and recognized EWPs; these wood panels are frequently fastened to the wall studs of a framed wall to offer extra structural support and wall surfaces. For unique projects, engineered wood products (EWPs) can also be utilized in place of timber beams, providing stronger beams. Structural insulated panels (SIPs) are a type of EWP that consists of two sheets of OSBs or another EWP and an inner layer of insulating foam core. SIP panels can be used as a covering for a home's framing or as the entire structural support in place of a frame. For some types of home design, their size and strength can be advantageous over standard platform framing.
- **Light Gauge Steel Construction:** The use of studs, rafters, plates, and joists in building structures is comparable to that of wood construction. Light steel is employed in lieu of wood. S-shaped cross-sections are also available, but C-shaped cross-sections are the most common for steel members. Similar to how wood construction techniques are used, this kind of building makes material carrying simple. On-site cutting and erecting of the materials is another option. By using screw guns and metal cutting instruments instead of saws and nails as in wood construction, this technique of construction is frequently applied in commercial buildings.
- **Joisted or Load Bearing Masonry Construction:** In this construction method, load-bearing walls are constructed using concrete bricks or blocks. This method produces heavy structures, which is undesirable in regions that are prone to earthquakes. The word "joistered Masonry" refers to the use of

wooden joists in the construction of floors and ceilings. Because every brick or block needs to be a handset, this method requires a lot of work.

- **Hybrid Concrete Building Technique:** This method integrates the benefits of in-situ building with concrete pre-casting, thereby reducing construction turnaround time. Construction costs decline while quality rises. Concrete hybrid structures are competitive, simple to construct, and have reliable performance.
- **Insulating Concrete Formwork (ICF) Technique:** Polystyrene blocks with twin walls that are quick to assemble are used in the ICF technique to create building wall formwork. Superior ready-mixed concrete is subsequently pumped into the formwork. Building construction becomes error-free, and the finished structure has excellent thermal and acoustic insulation. The technologies that underpin building construction methods have advanced to a point where they have matured significantly. Astute builders are using these techniques to maximize your return on investment. These techniques provide buildings with outstanding quality assurance, cost competitiveness, and superior results.

6.1.3. Innovative Building Construction Techniques

Many of the building methods still in use today date back hundreds of years. Some are a little more contemporary, and although we value the tried-and-true methods that helped the sector reach its current state, it's fascinating to observe how fresh innovations and methods are changing the way we do things. You can be sure that all facets of building construction methods—from modular to insulation panels—are impacted by these advancements.

- **Augmented Reality-Assisted Building:** Using augmented reality (AR) in the construction process is one of the most eagerly awaited innovations in modern construction methods. Through the integration of Building Information Modeling (BIM) with contemporary augmented reality wearables, developers can view fully-rendered visualizations of the final product, a construction project. Apart from the apparent advantages of pre-construction planning for specific projects, this technology will facilitate innovative construction approaches in all facets of urban planning, including traffic routing, utility management, housing development zoning, and more. AR can also help with other construction-related decisions like what kind of materials are used on the job site and if insulating concrete is needed for the building.
- **Raised Access Flooring:** These days, there are many options for raised flooring to select from, based on the requirements of your facility, available space, HVAC system, and cable routing requirements. Raised access flooring systems give you more options for service distribution throughout your building and give you access to more contemporary methods of building facilities, like underfloor air distribution, which uses less energy and provides better airflow

than traditional HVAC systems. By keeping reconfiguration costs down over the course of the building's life cycle, these modifications are crucial for future-proofing a facility.

- **Self-Healing Concrete:** It's a little surprising that it took us this long to get to this point, considering that concrete is the material that is used the most everywhere in the world. However, self-healing concrete has finally developed to the point where it will soon be a competitive substitute for conventional materials. These materials offer a great long-term solution for concrete degradation by automatically repairing any cracks or fissures that may appear over time by utilizing bacteria that are present within the concrete building itself.
- **Kinetic Footfall Energy Harvesting:** It's no secret that utilizing alternative energy in various forms is one of the main areas of new construction techniques in buildings. By installing ground sensors in busy locations, kinetic footfall energy harvesting captures and stores thermal mass energy from pedestrian footfall. This is an inventive approach to incorporating more sustainable off-grid energy solutions into long-term planning. Similar to solar roadways, these kinetic solutions are also being tested on roads right now, but thus far, the results have been mixed. It might take some time before the technology becomes widely used.
- **3D Printed Buildings:** The use of 3D printing technology on construction sites has streamlined the process recently. These days, businesses use 3D printing to create materials on location in almost exact replicas of manufacturing environments. This lowers the need for extensive supply chains and lowers the overall cost of the project. However, there are now more widespread use cases in which 3D printing software receives architectural designs directly. This paves the way for a quicker, more intelligent development process that requires less time and effort from all stakeholders by enabling builders to automate as much of the construction process as possible. For a new building system, not too bad.

6.2. ARCHITECTURAL TECHNOLOGY

The majority of primitive buildings are pit houses, which are essentially made-up of earthen pits covered in shingles. These structures have been discovered all over the world. Using natural materials like grass, wood, stone, mud, and animal pelts, people built rain-proof homes thousands of years ago using simple tools they had learned to make and use. As civilization advanced, structures with symbolic purposes beyond simple habitation—like religious, hierarchical, or memorial—began to be constructed. Among those structures are theaters, palaces, and temples, with the pyramids serving as one of their representatives. Throughout history, a wide range of structures with diverse architectural forms have been constructed, depending on the natural circumstances that gave rise to the civilization as well as social factors like hierarchy, religion, and ways of thinking. As new building materials and techniques were developed, these buildings grew larger both laterally and vertically. Medieval churches and temples that still stand in Europe are good examples. In Europe during the 19th century, the advent of modern architecture coincided with the industrial revolution. The Industrial Revolution pushed the economy to unprecedented heights and expedited industrialization. The need for factories and other industrial facilities to be more economical, efficient, and functional resulted in a significant shift in the demand for buildings. Steel, cement, and glass are examples of modern building materials used in reinforced concrete and steel-reinforced concrete structures, which have become commonplace. In line with this, there was a greater need for dimensional expansion in both height and space.

6.2.1. Super High-rise Building Technology

Advanced structural analysis techniques and materials technology support super high-rise buildings. Super high rises that reach heights of more than 200 meters were originally mostly built using steel frames. Reinforced concrete, which outperforms steel-frame structures in constructability and cost-effectiveness, is now used in many of the extremely tall buildings we see today. The development of high-strength concrete technology, whose strength is several times higher than that of regular concrete, is responsible for the structural excellence of these buildings. In those new constructions, concrete with extremely high strength—80 to 130 N/mm^2—is used under stringent quality control. Furthermore, the CFT technique was created to achieve even more cost-effective super high rises. This method involves filling concrete into steel columns to increase compressive strength. Earthquake-resistant engineering is one of the amazing technological components that supports extremely tall buildings; this topic will be covered in the section after this one. By using vibration damping structures, such as braces and reaction-generating pendulums to set off an external force due to seismic action or wind pressure, it creates safe and comfortable living spaces even in very tall buildings.

6.2.2. Seismic Technology

Buildings that are constructed to be either rigid enough to withstand seismic force or flexible enough to withstand it are considered to be earthquake resistant according to conventional standards. In the event of a significant earthquake, buildings built using either concept will withstand the damage better, but the interior installations, furnishings, and equipment will be severely damaged. The goal of the most advanced building engineering concept is to control and mitigate the seismic force acting on the building in order to lessen vibration and safeguard both the structure and its contents. These methods can be divided into two categories: seismic isolation and vibration control. A vibration-controlled structure is classified as either "active" or "passive" control based on how seismic force is generated.

It is intended to trigger seismic force by providing a counterforce in the opposite direction. By providing a reaction force produced, for example, by computer-controlled hydraulic power, active damping manages vibration energy by detecting seismic vibration using sensors. Vibration energy with inertial force produced, for example, by pendulums, is absorbed by passive damping. In today's ultra-high rises, vibration damping is an essential technological advancement. Rubber or sliding bearings are inserted in the isolation layer of a seismic isolation structure to absorb and lessen seismic force transmission to floors above the isolation layer, thereby mitigating vibrations. Usually located on the foundation, the isolation layer can also be found, depending on the design of the building, in the middle-height floors of some structures. This method works particularly well for low- to mid-rise structures.

6.2.2.1. Origin and Measurement of Earthquakes

1. **Plate Tectonics, the Cause of Earthquakes:** The earth's surface rolls, trembles, or shocks suddenly during an earthquake. In essence, the crust of the Earth is made up of a number of plates that are constantly moving (at a rate of 2 to 130 millimeters per year), spreading outward from the center, sinking at the edges, and renewing themselves. As a result of stresses created by friction from plates colliding, extending, or subducting (one plate sliding under the other), an earthquake can cause complex wave motions that radiate through the crust and result in ground failure, including landslides, liquefaction, subsidence, and tsunamis. Consequently, the built environment in the vicinity of the earthquake may suffer minor damages or complete destruction.
2. **Measuring Seismic Forces:** In order to characterize or measure the effect of an earthquake on the ground (a.k.a. ground motion), the following definitions are commonly used:

- *Acceleration* is the rate of change of speed, measured in "g"s at 980 cm/sec^2 or 1.00 g.
    - o For example,
        - 0.001 g or 1 cm/sec^2 is perceptible by people
        - 0.02 g or 20 cm/sec^2 causes people to lose their balance

  - 0.50 g is very high but buildings can survive it if the duration is short and if the mass and configuration have enough damping

- *Velocity* (or speed) is the rate of change of position, measured in centimeters per second.
- *Displacement* is the distance from the point of rest, measured in centimeters.
- *Duration* is the length of time the shock cycles persist.
- *Magnitude* is the "size" of the earthquake, measured by the Richter scale, which ranges from 1–10. The Richter scale is based on the maximum amplitude of certain seismic waves, and seismologists estimate that each unit of the Richter scale is a 31 times increase in energy. Moment Magnitude Scale is a recent measure that is becoming more frequently used.

  The power of destruction can be defined by combining the duration and the acceleration level. Generally speaking, the building can withstand less acceleration a longer the duration. A building's ability to endure extremely high acceleration for brief periods of time is directly correlated with the amount of damping that the structure contains.

The magnitude of the earthquake's local damage is known as its intensity, and it is measured using the 12-level Modified Mercalli Scale (MM), where each level corresponds to a specific level of destruction caused by an acceleration of the ground. Damage from earthquakes varies according to factors like construction type, local soil conditions, and distance from the epicenter.

6.2.2.2. Effects of Earthquakes on Buildings

The forces that earthquakes apply to buildings are calculated using the seismic measures mentioned above. The majority of seismic damage is caused by internal forces within buildings known as the Inertial Force (F_{Inertial}), which are created when the ground shakes (pushes back and forth, sideways, up and down).

$$F_{Inertial} = \text{Mass (M) X Acceleration (A)}.$$

The internal inertial forces generated increase with the building's mass (weight). In seismic design, lighter, less massy construction is usually advantageous. Greater mass results in greater lateral forces, which raises the risk of columns buckling, shifting, or going out of plumb when subjected to a vertical load (P delta Effect).

Earthquakes generate waves that may be slow and long, or short and abrupt. The length of a full cycle in seconds is the *Period* of the wave and is the inverse of the Frequency. All objects, including buildings, have a natural or fundamental period at which they vibrate if jolted by a shock.

The natural period is a primary consideration for seismic design, although other aspects of the building design may also contribute to a lesser degree to the mitigation measures. If the period of the shock wave and the natural period of the building coincide, then the building will resonate and its vibration will increase or "amplify" several times.

The soil also has a period varying between 0.4 and 1.5 sec., very soft soil being 2.0 sec. Soft soils generally have a tendency to increase shaking as much as 2 to 6 times as compared to rock.

Also, the period of the soil coinciding

with the natural period of the building can greatly amplify the acceleration of the building and is therefore a design consideration.

6.2.2.3. Seismic Design Factors

The building's design is influenced by and affects the following elements. The design team must be aware of these factors and take appropriate action during the design process.

- **Torsion:** Every building and object has a center of mass, which is the point at which rotation cannot occur and the object can be balanced. The geometric center of the floor and the center of mass may coincide if the mass is uniformly distributed. Stress concentrations are produced by torsion, which is caused by an uneven mass distribution that places the center of mass outside of the geometric center. Every building design will inevitably contain some torsion. However, a symmetrical mass arrangement will maintain torsion within a reasonable range and produce balanced stiffness against either direction.
- **Damping:** Buildings in general are poor resonators of dynamic shock and dissipate vibration by absorbing it. Damping is the rate at which natural vibration is absorbed.
- **Ductility:** A material's ability to bend, flex, or move is known as its ductility; it is a property of steel that stops working only after significant deformation has taken place. Non-ductile materials crumble quickly and break easily, like badly reinforced concrete. Carefully designed joints can have good ductility.
- **Strength and Stiffness:** A material's ability to withstand and withstand applied forces within a safe limit is known as its strength. The degree of a material's resistance to deflection or drift—drift being a horizontal relative displacement from story to story—is known as its stiffness.
- **Building Configuration:** This term describes the dimensions, shape, and structural and nonstructural components of a building. The distribution of seismic forces within a structure, as well as their relative magnitude and potential design concerns, are determined by the configuration of the building.
    - *Regular Configuration* buildings have Shear Walls or Moment-Resistant Frames or Braced Frames and generally have:
        - Low Height to Base Ratios
        - Equal Floor Heights
        - Symmetrical Plans
        - Uniform Sections and Elevations
        - Maximum Torsional Resistance
        - Short Spans and Redundancy
        - Direct Load Paths
    - *Irregular Configuration* buildings are those that differ from the "Regular" definition and have problematic stress concentrations and torsion.

Soft First Story is a discontinuity of strength and stiffness for lateral load at the ground level.

Discontinuous Shear Walls do not line up consistently one upon the other causing "soft" levels.

Variation in *Perimeter Strength* and *Stiffness* such as an open front on the ground level usually causes eccentricity or torsion.

Reentrant Corners with the forms H, L, T, U, +, or [] cause torsion and stress concentration at the reentrant corner. Reentrant corners should be sufficiently isolated or reinforced by seismic designs. Understanding the period, torsion, damping, ductility, strength, stiffness, and configuration of the building can be helpful in choosing the best seismic design elements and mitigation techniques to use.

6.2.2.4. Seismic Design Strategies and Devices

- **Moment Frames:** The frame, a basic engineering structure, is a two-dimensional arrangement of connected pieces that are joined together. The members can be freely joined at any point along their length and are not always straight. Moment frames in two orthogonal directions are frequently joined to create a three-dimensional frame of columns and beams in real constructions. Although moment frames are intended to support both vertical and horizontal loads in the same plane, they can also be used to resist horizontal loads that are outside of the frame's plane. They are utilized in a wide range of building materials and applications. A moment frame is a unique kind of frame in which all of its constituent parts are connected rigidly. Because of the inherent bending moment and shear strength in its members and connecting joints, this configuration can withstand lateral and overturning forces. Consequently, in seismic design, the stiffness and strength of the moment frame rely on the strength and stiffness of its members. Moment frames permit larger movements during an earthquake because they can be more flexible than other solutions, like shear walls. To prevent damage, non-flexible parts that are fastened to the frame, like the cladding, must be made to fit the extra movement.
- **Braced Frames:** Bracing is another essential engineering concept that entails adding extra components to a frame to improve its resistance to lateral loads. Centric and eccentric braced frames are the two primary types.
- **Concentric Bracing:** Diagonal braces positioned in the frame's plane make up concentrated bracing. To create a truss and a stiff frame, the brace's two ends join at the ends of other framing members. Concentric bracing can be set up in a variety of ways, including X, K, and one-directional diagonal bracing. The bracing members can be made to act in compression, tension, or both directions. For medium-rise buildings, balanced diagonal bracing is most frequently used because it offers equal strength in both directions. It is challenging to dissipate energy efficiently in concentrically braced frames.
- **Eccentric Bracing:** Diagonal braces that are positioned in the frame's plane and where one or both of their ends do not join at the ends

of other framing members are known as eccentric bracing. In essence, the system minimizes the drawbacks of both moment frames and concentrically braced frames by combining their best features. An eccentric brace transfers lateral forces via shear to a vertical column or to another brace because of its eccentric connection to the frame. Eccentric braced frames have the potential to be more ductile and capable of dissipating more energy than concentric braced frames made of the same material when they are proportionately balanced.

- **Shear Panels:** A shear panel is a solid sheet element that is inserted into a frame to strengthen it against moments of overturning and lateral loads. Shear panels distribute vertical loads down to the building's foundation or lateral forces from one area of the frame to another. A shear wall can be created by joining multiple shear panels in the same plane. These could also be monolithic systems that reach the entire height of the building or multiple stories. Large-scale concrete structures made of a single piece are known as shear walls in most commercial buildings (such as tall shear-wall buildings). They can also be made from a collection of components that resemble panels and are fastened together rigidly (like in a warehouse). When windows, doors, and other building features are incorporated into the design, shear panels may lose some of their effectiveness.
- **Horizontal Diaphragms:** A horizontal diaphragm is a horizontal truss (in a roof plane) or solid sheet element (in a floor), similar to a shear panel. It is positioned in between vertical components, such as moment frames, vertical trusses, or shear panels, to transfer lateral loads to the vertical elements. For instance, in timber floors, the in-plane bending forces are carried by the perimeter joists and plates, and the shear forces are carried by the floor sheeting. Vertical forces, as opposed to lateral ones, will frequently determine the thickness of sheeting and the size of joists. In most modern timber buildings, plywood or particleboard are used as floor diaphragms. A diaphragm of some kind is used in the majority of building designs on the roof and on every floor. One example of a horizontal diaphragm is a floor plate. In many designs, a diaphragm also functions as the building's roof or floor element.
- **Energy-Dissipating Devices:** Increasing the resistance of the building structure will cause more shaking, which could harm the building's contents or functionality. In order to reduce shaking, energy-dissipating devices are employed. If ductile materials undergo controlled deformation, energy will be released. Eccentric Bracing serves as an illustration of how energy can be dispersed through the controlled deformation of framing members. This won't, however, stop or lessen damage to the building's contents. Using energy-dissipating devices, which work similarly to shock absorbers in a moving car, is a more straightforward solution. The building will withstand shaking for

a longer amount of time and within a bearable range.

- **Base Isolation:** This seismic design approach serves as a shock absorber and involves separating the building from the foundation. The building moves more slowly when the ground moves because most of the shock is absorbed by the isolators. In order to prevent overturning, the building needs to be conceived as a single, inflexible unit, akin to a rigid box, with flexible utility connections that allow for movement at the base of the structure. Designing new construction with Base Isolation in mind is the simplest task. The Base Isolators may need to be inserted into existing buildings, which may need to have their foundations split from the superstructure to make them more rigid. The entire building will move back and forth a full foot or more, so more room—a moat—must be made available for horizontal displacement. A costly operation, base isolation retrofit is typically appropriate in high-asset-value facilities and may necessitate the full or partial removal of building occupants during installation. Elastomeric isolators are made of natural rubber, high-damping rubber, or another elastomer combined with metal components. Frictive Isolators are also employed, and they are mainly composed of metal components.

Tall structures cannot be base-isolated without causing them to topple. They require control over their horizontal displacement because they are more flexible than low-rise buildings. Dampers, which absorb a significant portion of the energy and make the displacement bearable, can be used to achieve this. Dampers are frequently a simpler retrofit option for existing buildings than base isolators, particularly when the application is external or doesn't interfere with occupants' daily lives.

There are many types of dampers used to mitigate seismic effects, including:

- o Hysteric dampers utilize the deformation of metal parts
- o Visco-elastic dampers stretch an elastomer in combination with metal parts
- o Frictive dampers use metal or other surfaces in friction
- o Viscous dampers compress a fluid in a piston-like device
- o Hybrid dampers utilize the combination of elastomeric and metal or other parts

6.2.2.5. Nonstructural Damage Control

All items, which are not part of the structural system, are considered as "nonstructural," and include such building elements as:

- Exterior cladding and curtain walls
- Parapet walls
- Canopies and marquees
- Chimneys and stacks
- Partitions, doors, windows
- Suspended ceilings
- Routes of exit and entrance
- Mechanical, Plumbing, Electrical and Communications equipment
- Elevators
- Furniture and equipment

To stop damage or complete destruction, bracing must be used to stabilize these items. Seismic isolating devices, which are adapted variants of standard vibration isolators, can be installed on building machinery and equipment. A loss from nonstructural damage may exceed a loss from structural damage by a factor of multiple. Due to both structural and nonstructural seismic damages, the likelihood of a business closing or failing completely increased in the wake of the Loma Prieta, Northridge, and Kobe earthquakes.

6.2.2.6. Design Seismic Forces Combined with Other Loads

In line with the load combinations specified by the code, design seismic forces must be combined with other loads, primarily dead and live loads as well as loads resulting from lateral earth pressures and fluids.

It provides three sets of load combinations:

- Strength (LRFD) load combinations;
- Basic ASD load combinations; and
- Alternative ASD load combinations.

The strength load combinations are meant to be applied in conjunction with LRFD techniques. With ASD, one can use either the standard or custom load combinations. The main distinction between the alternative and basic ASD load combinations is that the latter allow a one-third stress increase for load combinations including transient loads, like seismic or wind. The former do not. Steel structure designs that are similar to the strength load combinations for LRFD will be produced by using the fundamental ASD load combinations. The other ASD load combinations might not, though. Because the alternative ASD load combinations do not offer the same intrinsic protection against failure as the strength load combinations for LRFD, some engineers have been pushing for their elimination for a while.

6.2.2.7. Application

The principles and strategies of seismic design and construction are applied in a systematic approach that matches an appropriate response to specific conditions through the following major steps:

1. **Analyze Site Conditions:** The location and physical properties of the site are the primary influences of the entire design process. The following questions can serve as a checklist to identify seismic design objectives.
   - i. Where is the location of the nearest fault?
   - ii. Are there unconsolidated natural or man-made fills present?
   - iii. Is there a potential for landslide or liquefaction on or near the site?
   - iv. Are there vulnerable transportation, communication, and utilities connections?
   - v. Are there any hazardous materials on the site to be protected?
   - vi. Is there potential for battering by adjacent buildings?
   - vii. Is there exposure to potential flood from tsunami, seiche, or dam failure?

   Take into account the potential for seismic activity to disrupt business operations or mission critical sites nearby. This could result in the project site becoming inaccessible, causing utility outages, posing a fire risk, or releasing toxic materials onto the site. Investigate below the surface to find any uncontrolled fill or loose soils that might exacerbate ground motion. Solid dense rock makes for the most reliable and seismically secure building foundation, while hard, dense soils continue to be more stable.

2. **Establish Seismic Design Objectives:** It is advised to create seismic design objectives using a performance-based methodology. This establishes a degree of predictable building behavior by reacting to the highest seismic activity that is taken into account. The level of performance required for the building project can be determined using a risk analysis and threat/vulnerability assessment. Some suggested seismic design performance goals are:
   - Conform to local building codes providing "Life Safety," meaning that the building may collapse eventually but not during the earthquake.
   - Design for repairable structural damage, required evacuation of the building, and acceptable loss of business for a stipulated number of days.
   - Design for repairable nonstructural damage, partial or full evacuation, and acceptable loss of business for a stipulated number of days due to repair.
   - Design for repairable structural damage, no evacuation required, and acceptable loss of business for a stipulated number of days due to repair.
   - No structural damage, repairable nonstructural damage, no evacuation, and acceptable loss of business for a stipulated number of days due to repair.
   - No structural or nonstructural damage, and no loss of business caused by either (excluding damage to tenants' own equipment such as file cabinets, bookshelves, furniture, office equipment, etc., if not properly anchored).

   Regarding the magnitude of the earthquake, it may also be stipulated as "Low," "Moderate," or "Large" as another matrix of grading threat and establishing corresponding building performance goals.

3. **Select/Design Appropriate Structural Systems:** The choice of the best structural system and related building systems for the project can be significantly influenced by the seismic design objectives. Some construction-type options, and corresponding seismic properties, are:
   - Wood or timber frame (good energy absorption, lightweight, framing connections are critical).
   - Reinforced masonry walls (good energy absorption if walls and floors are well integrated; proportion of spandrels and piers are critical to avoid cracking)
   - Reinforced concrete walls (good energy absorption if walls and floors are well integrated; proportion of spandrels and piers are critical to avoid cracking)
   - Steel frame with masonry fill-in walls (good energy absorption if bay sizes are small and the building plan is uniform)

- Steel frame, braced (extensive bracing, detailing, and proportions are important)
- Steel frame, moment-resisting (good energy absorption, connections are critical)
- Steel frame, eccentrically braced (excellent energy absorption, connections are critical)
- Pre-cast concrete frame (poor performer without special energy-absorbing connections)

To guarantee ductility and natural damping and to limit damages to a manageable range, structural and architectural detailing as well as construction quality control are crucial. It is unlikely that the possibility of both structural and nonstructural damage will disappear if energy-dissipating devices are not used carefully. Energy-dissipating device additions run between 1% and 2% of the total cost of the structure. This is a small figure, especially considering the building's life-cycle cost. The cost is minimal over the course of a 30- to 50-year life cycle.

6.2.3. Large Space Technology

The Pantheon, which was built in Rome at the start of the second century to house Roman gods, is the oldest large-scale structure still standing in the world. Large spaces are in high demand in a variety of buildings these days, including commercial complexes, stadiums, train stations, and airports. Large space technology can be broadly divided into three categories: simulation, erection, and structure techniques. Large-space roof support structures are the focus of structural technique. A semi-sphere-shaped geodesic dome structure created by Richard Buckminster Fuller, a beam string structure using pretensioned cables, a membrane structure in which a roof is supported by the increasing internal pressure of a building over external pressure and a 3D truss structure made of triangles are among the regular plans that have been proposed thus far. An increasing number of recently constructed stadiums have movable roofs, indicating that larger venues are utilizing increasingly complex mechanisms. Additionally, there are now sliding, folding, and rotating varieties of openable mechanisms. The erection technique takes into account how to build long-span roofs for large spaces so that, once the entire roof structure is finished, the roofs have adequate strength. The lift-up method and the temporary-support method are the fundamental ideas. The typical process for providing temporary support involves forming temporary supports and scaffolds to hold each piece of the roof in place while it is being assembled. Once the roof is fully assembled and the necessary strength of the roof structure is verified, the temporary installations are taken down. In contrast, the entire roof structure is assembled on the ground during the lifting-up process, and the necessary strength of the roof is verified. Next, computer control is used to raise the entire roof after hydraulic jacks are attached to the columns. The lift-up method offers faster and safer construction and can save more money, though this depends on the size and structure. The determines how to precisely guarantee large spaces' environmental comfort, including lighting and air conditioning. High levels of comfort and energy savings in real structures are realized through computational simulations for temperature, humidity, air current, illumination intensity, and acoustics under expected conditions, assuming different shapes of roofs with open or closed states.

6.3. TUNNELING TECHNOLOGY

Tunneling technology involves a variety of methods, equipment, and engineering practices essential for excavating tunnels through diverse soil and rock types, playing a critical role in transportation, utilities, and infrastructure projects. Traditional methods like cut-and-cover tunneling are straightforward and cost-effective for shallow tunnels but can be disruptive to surface activities. Bored tunneling, using tunnel boring machines (TBMs), minimizes surface disruption and is suitable for long, deep tunnels, although it requires significant initial investment and planning. The drill and blast method, effective for hard rock conditions, involves drilling holes, filling them with explosives, and detonating to break the rock, but it generates vibration, noise, and dust. The New Austrian Tunneling Method (NATM) utilizes the natural strength of surrounding rock for stability, with flexible support adjustments as needed, requiring detailed monitoring and skilled labor. Immersed tube tunneling, used for underwater crossings, involves sinking pre-fabricated tunnel sections, a process demanding precise engineering.

Key equipment includes TBMs, which bore through soil and rock while installing support segments, drilling rigs for precision in the drill and blast method, shotcrete machines for immediate wall support, and conveyors and muck cars for efficient removal of excavated material. Essential engineering practices begin with geotechnical investigations to understand soil and rock conditions, crucial for choosing suitable methods and equipment. Ground support systems, such as rock bolts, steel ribs, and shotcrete, stabilize tunnels, while tunnel ventilation systems ensure air quality and dust control. Waterproofing and drainage prevent water ingress, using membranes, drainage pipes, and grouting. Automation and real-time monitoring enhance safety and efficiency, employing sensors for ground movement and automated systems.

Tunneling technology's applications are vast, from creating subways, railway tunnels, and highway tunnels to improve transportation, to constructing utility tunnels for water supply, sewage, and conduits. In mining, tunnels facilitate mineral extraction and transportation, while in water management, they serve flood control, irrigation, and hydroelectric projects. Advances in tunneling technology have made it possible to build complex, durable tunnels in varied geological conditions, significantly contributing to efficient, safe, and sustainable infrastructure development globally.

6.3.1. Crushing Techniques of Rock

Throughout history, tools such as pickaxes, hammers, shovels, and chisels have been used to dig holes in the ground. Deer horns were discovered in some ancient human ruins; these horns were most likely used as pickaxes. Pickaxes and shovels were used by the people of modern-day Persia to begin building irrigation canals, known as qanat, around the eighth century BC. Because their route followed ground that was sufficiently firm for nonsupport excavation, the qanats ran unsupported for five to ten kilometers, and on occasion even up to seventy kilometers. During the Roman era, man-made tunnels were also constructed for military routes and water supplies. They attempted to soften hard rock by heating it with fire and cooling it with water when they came across it. Although the modern blasting technique was developed in the latter half of the 19th century with the invention of dynamite and the development of rock drills, gunpowder was first used to explode rocks into pieces.

Improvement of machines to dig tunnels was accelerated when the Industrial Revolution promoted the expansion of the railroad network and the construction of long tunnels for railroads over the Alps was started. Alfred Nobel invented dynamite in 1866 and the blasting cap in 1867. Improvement of rock drills started in the early part of the 19th century with a rotary drill using steam developed by Richard Trevithick in 1813, and a hammer drill by the Singer Brothers in 1838. Browton made use of compressed air in 1844, and a compressed air hammer drill that rotates a chisel to hit and break rock was then developed by Fowl in 1851. In 1861, Germain Sommeiller started to use improved drilling machines and he built the Mont Cenis tunnel (Figure 6.3) in the Alps between Italy and France in 1870. In 1897, J. G. Rheiner invented the "water liner" method, in which compressed air is sent through hollow steel chisels to the drilled hole bottom and blow off crushed muck to remove. Combined use of the improved rock drills and dynamite led to the construction of the Simplon Tunnel to its completion, a 20 km long transalpine tunnel connecting Italy and Switzerland, in 1905.

Figure 6.3. *The Mont Cenis tunnel.*

Source: https://todayinsci.com/Events/Tunnels/SommeillerBoringMachines(Fig178)750px.jpg.

A jumbo was introduced at the beginning of the 20th century and it was outfitted with numerous large compressed air rock drills. The advent of hydraulic rock drills in 1970 led to an improvement in drilling performance compared to compressed air drills. These drills have higher pressure and greater hammering rotation. The newest rock drill models are great for a variety of uses and have better drilling performance. A few have hydraulic and angle sensors installed to guarantee precise positioning, angle setting, and parallel drilling. Because automation techniques have improved, computer-aided jumbos (Figure 6.4) with automatic drilling management systems have also been developed and are in widespread use.

Figure 6.4. *Computer-aided jumbo.*

Source: https://www.tunnel-online.info/imgs/100834309_75f714b325.jpg.

6.3.2. Muck Transport Technique and Space Supporting

In the past, animals like horses and cows were used to pull crushed stone, also known as muck or debris, out of tunnels through manual labor. Following the development of the locomotive, tunnels were equipped with rails to allow trolleys pulled by locomotives to be used to remove dirt. Following World War II, steel supports took center stage. Because steel supports are more robust than wood ones, they can preserve more space, which enables the use of larger drilling equipment. Large dump trucks or containers are used to remove muck from tunnels with a large cross-section when they are excavated. Belt conveyors and crushing equipment are being used more often to remove muck from long tunnels where blasting is used. In the 1960s, a novel approach to tunnel support was presented, known as the New Australian Tunneling Method (NATM), which relies on rock bolts and sprayed concrete to support the tunnel space above the natural ground. The foundation of this new approach is the notion that the natural ground itself is capable of supporting the tunnel space and that, through measuring and tracking the behavior of the natural ground, appropriate actions (changes) should be made to the tunnel support in a timely manner.

The idea of excavating the entire cross-section of a rock tunnel with a tunneling machine instead of blasting was born in Italy in the middle of the 19th century. It was in 1952 when a machine close to today's TBM in terms of mechanism was produced and applied to a drainage canal tunneling project for the Oahe Dam, South Dakota, U.S.A. It was in those days that the rock tunneling technique using TBM started to develop in the USA as well as in Europe. In the USA and Europe, where the ground was suitable for TBM tunneling, application of TBM rapidly increased, and technical standards were upgraded in the fields of excavation speed, applicability to various types of geology, and upsizing. TBM was thus more recognized and appreciated eventually. In the 1990s, an improved version of TBM tuned to softer ground was developed. Incorporating new functions, TBM has gradually increased the applicable range of geology types. Today, TBM is frequently used to drill pilot tunnels for road tunnels, large water channel tunnels, and railroad tunnels of over 8 m in diameter. The world's largest TBM of 12.8 m in diameter, has also been in use for drilling two-lane road tunnels.

6.4. BRIDGE CONSTRUCTION TECHNOLOGY

Bridge construction technology involves advanced methods, materials, and engineering techniques aimed at creating durable, efficient, and aesthetically pleasing structures. Modern bridge construction benefits from significant advancements in materials science, such as the use of high-strength concrete, advanced steel alloys, and composite materials, which enhance the strength and longevity of bridges while allowing for innovative designs. Precast concrete segments, which are manufactured off-site in controlled environments, ensure consistent quality and enable faster assembly on-site, reducing construction time and costs.

One of the most significant advancements in bridge construction technology is the use of cable-stayed and suspension bridge designs. Cable-stayed bridges, with their cables directly connected to towers, and suspension bridges, where cables are hung between towers and anchored at both ends, allow for the construction of long spans that can cross vast distances over water or rugged terrain. These designs not only improve structural efficiency but also reduce the environmental impact by minimizing the number of supports needed in sensitive areas.

Incremental launching is another innovative construction method where the bridge is built in sections behind an abutment and then slid into place. This technique is particularly advantageous in environments where traditional construction would be challenging, such as over water or in urban areas where minimizing disruption is crucial. Similarly, the balanced cantilever method, where segments are built symmetrically from piers, is used for constructing bridges over obstacles like deep valleys and wide rivers, ensuring stability and reducing the need for false work.

Geotechnical engineering plays a critical role in bridge construction, ensuring that foundations are designed to handle the specific soil and rock conditions of the site. This includes the use of deep foundations such as piles and drilled shafts to reach stable soil layers, especially in areas prone to seismic activity. Advanced software tools, including Building Information Modeling (BIM), enable detailed planning and visualization, improving the accuracy and coordination among architects, engineers, and construction teams, and leading to more efficient project execution.

Maintenance and monitoring technologies have also evolved, with the integration of structural health monitoring systems that use sensors to continuously assess the condition of a bridge. These sensors can detect stress, strain, temperature changes, and vibrations, providing real-time data that can predict potential issues before they become critical, thereby extending the lifespan of the bridge and ensuring safety. Smart materials, like self-healing concrete, further contribute to reducing maintenance needs by automatically addressing micro-cracks and other minor damages. Environmental considerations are increasingly integral to bridge construction technology. Designs now often incorporate features that reduce ecological disruption, such as wildlife corridors and measures to prevent water pollution during construction. Sustainable practices, including the use of recycled materials and energy-efficient construction processes, are also prioritized to minimize the environmental footprint of bridge projects.

Bridge construction technology encompasses a range of sophisticated methods and materials that address modern infrastructure challenges. Through the use of high-performance materials, innovative construction techniques, advanced planning and monitoring tools, and sustainable practices, the field aims to build bridges that are not only functional and aesthetically pleasing but also resilient, safe, and environmentally responsible.

6.4.1. Bridges in Ancient Times

Bridges in ancient times were marvels of engineering and ingenuity, reflecting the technological and architectural advancements of their respective civilizations. These structures were crucial for facilitating trade, transportation, and communication, and often held significant strategic and cultural importance. The materials and construction methods used in ancient bridges varied greatly depending on the local resources and technological capabilities of the builders.

One of the earliest known bridges was constructed by the ancient Sumerians in Mesopotamia around 4000 BCE. These early bridges were typically made of simple materials like timber and mud bricks. As civilizations advanced, so did their engineering techniques. The ancient Egyptians, for example, built bridges primarily of stone, utilizing their extensive knowledge of stone masonry developed through the construction of monumental structures like pyramids and temples.

The ancient Greeks made significant contributions to bridge construction, particularly through their development of the arch. The use of arches allowed for the construction of longer and more durable bridges. One notable example is the Arkadiko Bridge, dating back to the Mycenaean period (circa 1300 BCE). This bridge, constructed from stone and using a corbel arch design, is one of the oldest surviving arch bridges in the world and demonstrates the advanced engineering skills of the Greeks.

The Romans further revolutionized bridge construction with their extensive use of concrete and the true arch, which provided even greater strength and stability. Roman bridges were remarkable for their durability and the scale at which they were built. The Pont du Gard in France, an ancient Roman aqueduct bridge constructed in the first century CE, is a prime example. This three-tiered bridge, made of precisely cut stones without mortar, spans the Gardon River and has withstood the test of time due to its robust construction and sophisticated design. In ancient China, bridge construction reached new heights with the development of sophisticated techniques and materials. The Chinese were among the first to use iron chains in bridge construction, as seen in the Anlan Bridge, also known as the Bridge of the God of Death, built during the Han Dynasty (202 BCE – 220 CE). Additionally, the Zhaozhou Bridge, constructed during the Sui Dynasty (581–618 CE), is the world's oldest open-spandrel stone segmental arch bridge. Its innovative design, using a stone arch with hollow spandrels, reduced the weight and material needed, demonstrating advanced engineering knowledge.

In the Indian subcontinent, ancient bridges like the Mahadeo Temple Bridge in the Chamba region, constructed during the 11th century, showcase the use of cantilevering techniques and stone construction. These bridges were often adorned with intricate carvings and inscriptions, reflecting the rich cultural heritage of the region.

Overall, bridges in ancient times were a testament to the ingenuity and engineering prowess of their builders. They utilized available materials and developed innovative construction techniques that laid the groundwork for modern bridge engineering. These structures not only facilitated the movement of people and goods but also stood as enduring symbols of the civilizations that built them, many of which continue to inspire awe and admiration today.

6.4.2. Elements of Bridge Design

There are six fundamental types of bridges: cantilever, suspension, beam, truss, arch, and cable-stay.

1. **Beam Bridge:** A beam bridge, where the green lines indicate compression forces and the red lines indicate tension forces. The most prevalent type of bridge is the beam bridge. A beam bends to support vertical loads. The top of the beam bridge experiences horizontal compression as it bends. Horizontal tension is applied to the bottom of the beam at the same time. By compressing the beam vertically, the supports transfer the loads to the foundations. A bridge is referred to as a simply supported beam bridge when its beams are supported by just two elements. The bridge becomes continuous if two or more beams are fixedly joined over supports.
2. **Truss Bridge:** Because a single-span truss bridge bends to support vertical loads, it functions similarly to a simply supported beam. Bending causes tension in the bottom chords, compression in the top chords (or horizontal members), and, depending on how they are oriented, either compression or tension in the vertical and diagonal members. Because they can support relatively large loads with a relatively small amount of material, trusses are widely used.
3. **Arch Bridge:** The primary method of load carrying for an arch bridge is compression, which applies both vertical and horizontal forces to the foundation. Therefore, arch foundations need to stop both horizontal sliding and vertical settling. The structure itself typically requires less material than a beam bridge of the same span, despite the more intricate foundation design.
4. **Suspension Bridge:** Vertical loads are supported by curved cables that are under tension in a suspension bridge. These loads are distributed to the anchorages, which have

to withstand the inward and occasionally vertical pull of the cables, as well as the towers, which carry them to the ground by vertical compression. With the towers in compression, the suspension bridge resembles an upside-down arch in tension. The deck is suspended in the air, so it needs to be careful not to move too much when something is loaded. For this reason, the deck needs to be stiff, heavy, or both.

5. **Cantilever:** A beam is said to be cantilevered when it projects outward, supported only at one end. A cantilever bridge is generally made with three spans, of which the outer spans are both anchored down at the shore and cantilever out over the channel to be crossed. The central span rests on the cantilevered arms extending from the outer spans; it carries vertical loads like a simply supported beam or a truss—that is, by tension forces in the lower chords and compression in the upper chords. The cantilevers carry their loads by tension in the upper chords and compression in the lower ones. Inner towers carry those forces by compression to the foundation, and outer towers carry the forces by tension to the far foundations.
6. **Cable-Stay:** The vertical main-span loads of cable-stayed bridges are supported by tensioned, nearly straight diagonal cables. Through vertical compression, the towers transmit the cable forces to the foundations. The deck is likewise compressed horizontally by the tensile forces in the cables.

6.4.3. Materials Used for Bridges

Wood, stone, iron, and concrete have historically been the four main building materials used for bridges. Iron has influenced modern bridges the most out of all of these. Prestressed and reinforced concrete are made with steel, which is derived from iron. Steel, reinforced concrete, and prestressed concrete are nearly always used in the construction of modern bridges.

1. **Wood and Stone:** Wood has always been readily available and reasonably priced, despite its relative weakness in both compression and tension. Wood has proven to be an effective material for small, lightweight bridges like footbridges. Modern bridges sometimes feature laminated wooden arches and beams designed by engineers. Stone is weak in tension but strong in compression. It has mostly been used in abutments, piers, and arches.
2. **Iron and Steel:** Cast iron, which is strong in compression but weak in tension, was the first type of iron used during the Industrial Revolution. While cast iron is stronger in compression, wrought iron has a significantly higher tensile strength. Steel is an even more advanced form of iron, stronger than any other iron in both tension and compression. Steel can be produced in a variety of alloys, with some being five times stronger than others. These are what the engineer calls high-strength steel.
3. **Concrete:** An artificial stone called concrete is created by combining water, sand, gravel, and a binder like cement. It is weak in tension

and strong in compression, just like stone. Reinforced concrete is made of concrete that has steel bars embedded in it. Because the steel bears all of the tension and the concrete shields it from fire and corrosion, reinforcement permits the use of less concrete. One significant kind of reinforced concrete is prestressed concrete. Post-tensioned prestressing is a common process that entails casting concrete beams with longitudinal holes for steel tendons, which are similar to reinforced concrete but with upward curves from end to end. Once the tendons are fitted inside the beams, they are stretched and then secured at the ends. The beam is compressed as a result of the tendons' extreme tension pulling the two anchored ends together. Furthermore, the upward force exerted by the curved tendons can be adjusted by the designer to offset a significant portion of the downward load anticipated to be supported by the beam. Prestressed concrete allows for lighter, frequently less expensive designs than reinforced concrete designs by reducing the amount of steel and concrete required in a structure.

6.4.4. Performance in Service

Bridges are built to support traffic, or live loads, in addition to their own permanent weight, or dead load, and to withstand natural forces like earthquakes and winds.

6.4.4.1. Live Load and Dead Load

A bridge's main purpose is to support trains, automobiles, and large trucks. Traffic loading needs to be estimated by engineers. The maximum possible load may be carried across short spans; that is, four heavy trucks may cross simultaneously, two in each direction, on spans shorter than thirty meters (100 feet). The cost of designing for the maximum possible load on longer spans of a thousand meters or more is excessive because it is such a remote possibility (imagine the Golden Gate Bridge with only heavy trucks crossing at the same time in both directions). As a result, engineers base their designs on likely loads. The structure needs weight to support traffic, and on short spans, this weight is typically less than the live loads. However, dead load is greater than live loads on longer spans; therefore, it becomes more crucial to design forms that minimize dead load as spans get longer. Longer spans are typically constructed using cantilevers, cable-stays, and suspension forms, whereas shorter spans are typically constructed using beams, hollow boxes, trusses, arches, and continuous versions of the same. Questions of form, materials, and shape become more significant as spans get longer. In order to provide longer spans with greater strength from less material, new forms have developed.

6.4.4.2. Forces of Nature

While natural forces can be either vertical or horizontal, dead and live weight are basically vertical loads. Two significant loads are caused by wind: static and dynamic loads. The horizontal force attempting to push a bridge sideways is known as the static wind load. Oscillations can occur in any direction due to vertical motion caused by dynamic wind load. Oscillations are vibrations that have the potential to fail a bridge, much like the breaking of an overused violin string. A thin deck that is improperly shaped and supported could twist or move vertically, which could be dangerous.

The use of expansion joints in the deck and bearings at the abutments and tops of piers has reduced the expansion and contraction of bridge materials due to heat and cold. The bridge can adjust to changes in temperature thanks to its bearings without putting undue strain on the material. Engineers occasionally create hinges in arches to lessen stresses brought on by temperature changes. In addition, natural calamities like earthquakes and tropical cyclones must not destroy contemporary bridges. Structures with as little dead weight as possible are generally better able to withstand earthquakes because the horizontal forces resulting from ground accelerations are proportional to the structure's weight. The basic Newtonian law, which states that force equals mass times acceleration, explains this phenomenon. In general, it is preferable for cyclones to pass through or around a bridge that is aerodynamically designed with minimal solid material facing the winds, allowing them to do so without creating hazardous oscillations.

6.5. CONSTRUCTION SAFETY MANAGEMENT

Planning, organizing, coordinating, directing, controlling, and supervising any industrial or business project or activity with responsibility for others are executive functions that are specifically included in the definition of management. A structured management approach to safety, construction safety management considers the unique organizational structures and operational safety-related processes, such as accountabilities, policies, and procedures. Establishing the organizational safety policy is the first step in this. The safety organization moves forward with planning, organizing, staffing, coordinating, communicating, and budgeting for safety in accordance with the management's safety policy and strategy (Figure 6.5).

Figure 6.5. *Construction safety management cycle.*

Source: https://labor.delhi.gov.in/sites/default/files/Labor/important-news/safety_manual_for_construction_workers.pdf.

- **Planning for Safety**
  - o Planning is the most fundamental and the first function of the management process and other functions rest on it.
  - o The need for planning is increasing because of changes in technology, materials, methods, processes, demands, law, government policy, procedure, and competition.
  - o It includes setting safety objectives, formulating safety policy, safety

programming, budgeting, and determining safe or standard procedures.

- o Good planning at the design stage always helps.

6.5.1. Purpose/Areas of Safety Planning

- To draft health and safety policy and environmental policy for the management.
- To decide safety committee, its objectives, and effective functioning.
- To decide safety targets like zero/ minimum accidents, maximum working hours without accidents, safety education, training, and awareness programs for the workers and the public.
- Areas of hazards and their detection, inspection, audit, risk assessment, and measures for their removal or minimization.
- Accident investigation, analysis, cost calculation, and introducing safety measures to prevent their recurrence.
- Organization structure, staff, key persons, and their roles for safety.
- Standards for safety equipment, tools, permissible exposures, etc.
- Preparation, rehearsal, and updating of on-site emergency plans.

OSH objectives Consistent with the OSH policy, measurable OSH objectives should be established, which are:

- Specific to the organization, and appropriate to and according to its size and nature of activity;
- Consistent with the relevant and applicable national laws and regulations, and the technical and business obligations of the organization with regard to OSH;
- Focused on continually improving workers' OSH protection to achieve the best OSH performance;
- Realistic and achievable; and
- documented, and communicated to all relevant functions and levels of the Organization; and periodically evaluated and if necessary updated.

6.5.1.1. Hazard Prevention and Control Measures

Hazards and risks to workers' safety and health should be identified and assessed on an ongoing basis. Preventive and protective measures should be implemented in the following order of priority:

- Eliminate the hazard/risk;
- Control the hazard/risk at source, through the use of engineering controls or organizational measures;
- Minimize the hazard/risk by the design of safe work systems, which include administrative control measures; and
- Where residual hazards/risks cannot be controlled by collective measures, the employer should provide appropriate personal protective equipment, including clothing, at no cost, and should implement measures to ensure its use and maintenance.

6.5.1.2. Controlling Risk

Risk control is the process of continually assessing the condition of the project and developing options to permit alternative solutions. Project managers should take care to identify consequences that are likely to occur and any indicators of the start of the problem. The following are some suggestions for risk control:

- Continually update the risk management plan.
- Implement risk avoidance actions.
- Implement risk contingency actions.
- Report on each risk issue.

6.5.2. Organizing for Safety

The term 'organization' is used in management in different ways:

- It refers an activity, process or function of management i.e., organizing.
- It is used in a dynamic way referring to a process by which the structure is created, maintained and used.
- It is used in a static way referring a static structure (skeleton) of responsibilities and authorities i.e., relationships among individuals and positions in an organization.

6.5.2.1. Safety Organization

Safety organization can be defined as the structure and process by which groups of people are divided into sections or departments, each section or department is assigned specific safety function or duty.

- Authority and responsibility of everybody are clearly defined and interrelationship between them is specified for the accomplishment of organizational safety goals.
- A large unit may have a safety department which may have groups of people for the division of such safety functions and responsibilities.
- But in a small unit (majority) if such division is not possible and only a few persons are available for safety work, they will be assigned specific duties and other departmental heads (production, purchase, personnel, etc.) will be explained their role and responsibility towards safety goals.
- All supervisors shall be integrated with safety as part of their duty.

1. Directing for Safety

- It is a part of the management process that guides, inspires, instructs, and harnesses people to work effectively and efficiently to achieve the goals.
- Once the objectives are decided and planning is made on how to achieve them, the organization structure is designed by staff, appointing qualified and experienced persons and the organization can be said ready for action. But this action cannot begin until orders and instructions are issued i.e., the employees are 'directed' to carry the orders. This process of inspiring and guiding people on what to do and how to do the best of their ability is known as 'directing.'

2. Controlling for Safety: To complete the management cycle it is important to know about the function of controlling after knowing about planning, organizing, staffing and directing.

Thus controlling is aimed at:

- Monitoring the outcome of activities
- Reviewing feedback about this outcome and
- If necessary, take corrective action to achieve the outcome according to the plan.

The control function is closely connected to planning. It can be said an effective counterpart to planning. Using these definitions, controlling for safety can be defined as "a process that verifies and guides activities towards predetermined (planned) safety goals and takes necessary action, if required, to achieve the goals.

i. Its main purpose is:
   - To measure progress
   - To uncover deviations or change, delegation, mistakes, complexity, etc.
   - To indicate corrective action.

ii. Benefits: It offers
   - Guide to operations
   - Policy verification
   - Managerial accountability
   - Employee morale
   - Psychological pressure
   - Co-ordination in action.

6.5.3. Job Site Safety

Various measures are available to improve job site safety in construction industries. Several such measures are supposed to occur before construction is undertaken. These include

- Design;
- Choice of technology; and
- Education.

Certain structures can be made safer or more dangerous to build by changing their designs. For instance, rather than having to be as high as required by building codes, parapets can be designed to the proper heights for the safety of construction workers. The choice of technology can also have a significant impact on a job site's safety. Inbuilt safety features in machinery can alert operators to issues or stop injuries. For instance, basic switches can stop equipment from working in the absence of protective shields. e. The moment the machine guards are removed, the machine shuts down instantly. Jobsite safety may be directly impacted by providing workers with the necessary knowledge and training about risks and appropriate procedures. By providing safety training:

- You can avoid workplace accidents and health issues.
- You can understand the financial costs of accidents in the workplace involved in construction injuries
- You will avoid damaged products, loss in site, and injured staff.

The majority of job sites now routinely conduct safety meetings and inspections. Another significant chance to increase safety is through pre-qualification of contractors and subcontractors in terms of safety. Enough safety is guaranteed on the part of contractors if they are only asked to submit a bid if they have a respectable safety record. The most crucial safety precautions to take during construction are to make sure managers, inspectors, and workers are alert and cooperative. Being vigilant entails weighing the dangers associated with various work practices. Maintaining temporary physical barriers like guy lines, barricades, braces, railings, toe boards, and the like is also part of it.

Improving worksite safety is very important in construction sites due to the following reasons:

- Job site is continually changing as construction progresses.
- Workers do not have fixed worksites and must move around a structure under construction.
- The worker's familiarity in site is less compared to manufacturing site the employer employee relationship are less settled compared to manufacturing site as the tenure of a worker on a site is short.

CASE STUDY

ALICE TECHNOLOGIES: OPTIMIZATION WITH CONSTRUCTION OPTIONEERING

ALICE Technologies is a tech startup that's transforming the way the AEC industry approaches project management. They've used the power of artificial intelligence to optimize construction scheduling.

Their technology allows construction managers to input their project parameters, and then it generates thousands of potential schedules. Through analysis and simulation, ALICE will predict the impact of each option on time, cost, and quality. It will then evaluate and rank the best options based on simulation, identifying the most optimal choices.

This is a revolutionary approach that makes the most efficient use of time and resources, compared to a traditional way of construction planning.

According to their metrics, ALICE significantly reduces project duration and also greatly cuts labor and equipment costs.

17%	14%	12%
Project duration reduced	Labor cost savings	Equipment cost savings

SCS JV: Saving £2M on the London Tunnels of the UK's High Speed 2 Railway.

Three major European general contractors – Skanska, Costain, and STRABAG – formed a joint venture, SCS JV, to optimize the construction of the London tunnels of HS2, a national high-speed railway linking London, Birmingham and the north of the country.

The question at hand for SCS JV was whether it could use ALICE to find efficient sequencing options that would enable an earlier construction start for the Euston cavern shaft, which provides critical access to the main tunnels leading to Euston station.

Through the "what if" analyzes with ALICE, the SCS JV team was able to identify a number of ways to re-sequence and improve the build time for the pre-shaft sinking activities.

Results: With the ALICE platform, the SCS JV team was able to trim the build time for the shaft by 86 working days, a savings worth £2M in associated overhead expenses.

Align JV: Saving Four Months on the HS2's Colne Valley Viaduct

Three major European general contractors – Bouygues Travaux Publics, Sir Robert McAlpine, and VolkerFitzpatrick – formed a joint enture, Align JV, to collaborate on a project on Phase One of HS2. The construction schedule of this project had already been built over the previous three years.

Align wanted to use the ALICE construction optioneering platform to test its schedule assumptions and look for improvements. The Align team focused on using ALICE on the schedule for the viaduct, a complicated and critical part of the project.

- **Results:** Through the "what if" analyzes, the Align team optimized the sequences, resources and calendars for the viaduct construction with ALICE. Through this work, Align identified more than four months of savings on the finishes work that were implemented in the schedule to mitigate the risk of delays.

Ananda: Slashing Costs by 32% on the Elio Del Nest Project

Ananda Development wanted to improve efficiency and profitability of Elio Del Nest, a large residential high rise project it was planning to develop in Bangkok. Ananda partnered with ALICE Technologies to explore different construction scenarios for the project and identify the best possible paths.

Using ALICE and the power of its computational AI, Ananda experimented with a number of variables and challenged its own core assumptions on formwork to come out with the most efficient way in which to build Elio Del Nest.

- **Results:** Ananda was able to decrease the duration of its project by 208 days. And in doing so, it would decrease costs by 32% over a plan that didn't include overtime pay.

Advanced Manufacturing Centre (AMC): Reducing Concrete Waste

The Advanced Manufacturing Centre (AMC) project in Tseung Kwan O Industrial Area was facing the risk of project delays, concrete waste, and the need to innovate their construction tech stack.

They turned to Converge's ConcreteDNA, which integrates concrete sensor data with building information modeling (BIM) and provides in-depth material performance analytics. This formed the first real-time digital twin of in-situ concreting.

- **Results:** The implementation of Converge's technology led to significant improvements in the AMC project. Gammon Construction was able to save over

110 tons of concrete waste, reducing both environmental impact and project costs. The need for cube testing was reduced by approximately 14,000 cubes on the project, saving time and resources.

White Hart Lane: Project Early Finish with Converge Concrete Strength Data

The White Hart Lane station upgrade project was a complex engineering challenge that required a precise construction sequence. The project was part of the improvements needed to accommodate the increased matchday peak throughput at the Tottenham Hotspur Stadium, London's largest Premier League stadium.

Before constructing the underpasses, the archway piers needed underpinning to prevent any movement that might compromise the upper structure. The process involved excavating a bay, filling it with rapid-curing concrete, and waiting for the concrete to reach a strength of 2MPa before injecting grout to fill any small remaining spaces between the surfaces of the concrete and original foundations.

To prevent ground movement, the previously concreted bay needed to have achieved at least 15MPa before excavating the next. With 40 bays to complete and two concrete strength hold points for each, concrete curing was without question on the critical path.

Converge provided the site team with real-time concrete strength data, which unlocked a 'day shift – evening shift' work sequence.

- **Results:** The use of Converge's real-time data allowed the team to work continuously without ever having to wait for the right strength result. This led to all the underpins being completed 2 weeks early, significantly accelerating the project timeline and enhancing productivity.

London City Airport: Accuracy Under Variable Weather Conditions

The London City Airport extension project was a major civils project designed to deliver a new taxiway which increased the airport's capacity. The project was unique in nature as the taxiway was constructed over open water where weather conditions are variable.

This meant that typical methods to estimating curing time, such as cubes, were unrepresentative compared to actual in-situ concrete curing which is greatly affected by ambient temperature. The piling and decking team faced several challenges given the unique nature of the project. With a high traffic area and a small project footprint, the majority of works could only be performed at night and outside of usual working hours for most suppliers.

The project made use of Converge's Concrete Strength Predictions engine; an intelligent algorithm that combines Converge's proprietary concrete performance dataset with local weather feeds, allowing the team to see with >95% accuracy when their pour will reach strength.

- **Results:** The use of Converge's Concrete Strength Predictions Engine not only allowed the site team to strike at the earliest, safest opportunity but also to proactively plan the resources needed to complete critical path activities in advance. This led to a significant reduction in build time, keeping the project on schedule and reducing costs.

CLASS ACTIVITY

CONSTRUCTION MATERIALS AND TECHNOLOGY

Consider a project having activities and their associated time estimates as given in Table:

Activity	Predecessor activity	Time (Days)		
		Optimistic	Most likely	Pessimistic
A	-	2	4	6
B	A	8	12	16
C	A	14	16	30
D	B	4	10	16
E	B,C	6	12	18
F	E	6	8	22
G	D	18	18	30
H	F,G	8	14	32

a. Draw the network diagram.

b. Identify the critical path and compute the expected project completion time.

SUMMARY

- Construction is one of the branches of civil engineering that is concerned directly with common people, as everyone wants to have beautiful dwellings. Buildings were built long ago in history but the difference is in technology as early buildings were simple and just for the purpose of shelter.
- The construction industry includes a wide range of constructions suitable for all classes of society. Commercial construction, domestic construction, industrial construction, and heavy or civil construction are a few examples that are now displaying master pieces of construction technologies.
- Building construction is the process of adding structure to real property. The vast majority of building construction projects are small renovations, such as the addition of a room, or renovation of a bathroom. Often, the owner of the property acts as a laborer, paymaster, and design team for the entire project.
- Tunnels are dug in types of materials varying from soft clay to hard rock. The method of tunnel construction depends on such factors as the ground conditions, the groundwater conditions, the length and diameter of the tunnel drive, the depth of the tunnel, the logistics of supporting the tunnel excavation, the final use and shape of the tunnel, and appropriate risk management.
- Timber frame construction uses a range of structural frames that have been made off-site in a factory, such as external and internal walls, floors, and roofs to help form a combined structure that can be clad in another material such as brick on-site.
- Flat slab construction involves placing flat slabs of concrete that are placed and reinforced with concrete columns, and in doing so removing the need for beams. A key benefit of this method is that it offers flexibility in layout, such as no restrictions on height between floors, and reduces the time needed for construction.
- Precast panels are where floor and wall units are made off-site and transported to be erected on-site to form a solid structure which can be ideal for things such as new builds where repetitive projects are made.
- Twin wall technology is a method that combines both pre-made concrete and in-situ concrete, which helps with both the speed and strength of the structure. Twin wall technology occurs when two walls that have been pre-made are joined and reinforced before the gaps are filled with concrete.

REVIEW QUESTIONS

1. How to determine the construction type?
2. What are the primary benefits of using modern construction techniques compared to traditional methods?
3. Describe the process and benefits of using 3D printing in construction.
4. Discuss the role of Building Information Modeling (BIM) in innovative construction.

5. Explain the role of wind engineering in super high-rise building design.
6. What are the fundamental principles of seismic technology in construction?
7. How do building codes address the combination of seismic forces with other loads?
8. Explain the benefits of using mobile crushers in rock crushing operations.
9. Define the importance of efficient muck transport in underground construction projects.
10. Discuss the importance of corrosion resistance in bridge materials.

REFERENCES

1. Allen, E., & Iano, J. (2019). *Fundamentals of Building Construction: Materials and Methods*. John Wiley & Sons.
2. American Concrete Institute (ACI). (2005). *Building Code Requirements for Reinforced Concrete* (ACI 318-05). American Concrete Institute.
3. American Society of Civil Engineers (ASCE). (2003). *Seismic Evaluation of Existing Buildings* (ASCE/SEI 31-03). American Society of Civil Engineers.
4. American Society of Civil Engineers (ASCE). (2005). *Minimum Design Loads for Buildings and Other Structures* (SEI/ASCE 7-05). American Society of Civil Engineers.
5. American Society of Civil Engineers (ASCE). (2007). *Seismic Rehabilitation of Existing Buildings* (ASCE/SEI 41-06). American Society of Civil Engineers.
6. Bonowitz, D. (2000). *State of Art Report on Past Performance of Steel Buildings in Earthquakes* (FEMA 355F). Federal Emergency Management Agency.
7. Bowles, J. E. (2011). *Foundation Analysis and Design*. McGraw-Hill Education.
8. Ching, F. D. K. (2014). *Building Construction Illustrated*. John Wiley & Sons.
9. Construction Industry Institute. (2009). *Construction Productivity: Measurement and Improvement*. Construction Industry Institute.
10. Cornell, A., & Krawinkler, H. (2000). Progress and challenges in seismic performance assessment. *PEER Center News, 3*(2), 4.
11. Federal Emergency Management Agency (FEMA). (2000). *Prestandard and Commentary for the Seismic Rehabilitation of Buildings* (FEMA 356 Report). Federal Emergency Management Agency.
12. Federal Emergency Management Agency (FEMA). (2006). *Next-Generation Performance-Based Seismic Design Guidelines–Program Plan for New and Existing Buildings* (FEMA 445). Federal Emergency Management Agency.
13. Federal Emergency Management Agency (FEMA). (2012). *Seismic Performance Assessment of Buildings, Methodology and Implementation* (FEMA P-58). Federal Emergency Management Agency.
14. Ghali, A., & Neville, A. M. (1978). *Structural Analysis: A Unified Classical and Matrix Approach*. Chapman and Hall.

15. Gordon, J. E. (1978). *Structures–Or Why Things Don't Fall Down*. Pelican Books.
16. Greeno, R., & Chudley, R. (2016). *Building Construction Handbook*. Routledge.
17. Heintz, J. A., Hamburger, R. O., & Mahoney, M. (2014). FEMA P-58 Phase 2–Development of performance-based seismic design criteria. In *10th US National Conference on Earthquake Engineering–Frontiers of Earthquake Engineering*, Anchorage, Alaska, July 21–25, 2014.
18. Horne, M., & Petrequin, J. R. (2013). *Construction Contracts: Law and Management*. Routledge.
19. Jennings, A. (1977). *Matrix Computation for Engineers and Scientists*. John Wiley and Sons.
20. Jha, N. K., & Jha, R. S. (2014). *Construction Project Management: Theory and Practice*. Prentice Hall.
21. Kanok-Nukulchai, W. (2002). *Analysis Interpretive Computer Package–1993 Version*. Asian Institute of Technology.
22. Liska, M., et al. (2010). *Construction Materials: Their Nature and Behavior*. CRC Press.
23. McKinney, C. (2010). *Modern Construction: Lean Project Delivery and Integrated Practices*. CRC Press.
24. Nawy, E. G. (2008). *Concrete Construction Engineering Handbook*. CRC Press.
25. New Zealand National Society for Earthquake Engineering (NZNSEE). (2000). *Earthquake Risk Buildings–Recommendations and Guidelines for Classifying, Interim Securing and Strengthening*. New Zealand National Society for Earthquake Engineering.
26. New Zealand Society for Earthquake Engineering (NZSEE). (2006). *Assessment and Improvement of the Structural Performance of Buildings in Earthquakes* (Incl. Corrigenda 1–4). New Zealand Society for Earthquake Engineering.
27. Nikolic, D. (2001). *Building Design and Construction Handbook*. McGraw-Hill Education.
28. O'Brien, J. J., & Bachman, R. W. (2013). *Building Construction: Principles, Materials, and Systems*. Pearson.
29. Ricketts, S. L., & Stein, B. (2019). *Building Information Modeling: Planning and Managing Construction Projects with 4D CAD and Simulations*. McGraw-Hill Education.
30. SAC Joint Venture. (2000). *Recommended Post-Earthquake Evaluation and Repair Criteria for Welded Steel Moment-Frame Buildings* (FEMA 352). Federal Emergency Management Agency.
31. SAC Joint Venture. (2000). *Recommended Seismic Design Criteria for New Steel Moment-Frame Buildings* (FEMA 350). Federal Emergency Management Agency.
32. Salvadori, M. G., & Heller, R. A. (1963). *Structure in Architecture*. Prentice-Hall.

CHAPTER 7

Transportation Engineering

LEARNING OBJECTIVES

After studying this chapter, you will be able to:

- Understand transportation systems;
- Discuss the geometric design of roads; and
- Describe traffic flow theory.

KEY TERMS FROM THIS CHAPTER

Crest vertical curves
Engineering design
Highway engineering
Hub-and-spoke networks
Road slopes
Traffic controllers
Transportation engineering
Vehicular control
Cross slope
Geometric design
Horizontal alignment
Road geometry
Sag vertical curves
Transport system
Transportation infrastructures

INTRODUCTION

Transportation engineering involves using technology and scientific principles to plan, design, operate, and manage transportation facilities for various modes of transportation. The goal is to ensure safe, efficient, fast, comfortable, convenient, cost-effective, and environmentally friendly transportation of people and goods.

Transportation engineering planning involves urban planning elements and includes technical forecasting decisions and political factors. Technical forecasting for passenger travel typically utilizes an urban transportation planning model to estimate trip generation, distribution, mode choice, and route assignment. Advanced forecasting may also consider traveler decisions like auto ownership, trip chaining, and land use forecasting. Passenger trips are important in transportation engineering as they often represent the highest demand on transportation systems.

Transportation engineering encompasses the planning, design, construction, maintenance, and operation of various transportation facilities that cater to air, highway, railroad, pipeline, water, and space transportation. Design considerations involve determining the size, materials, and capacity of transportation facilities, as well as designing pavement thickness and geometry for optimal alignment.

Prior to any planning, an engineer must conduct an inventory of the area or existing system. This database should include details on population, land use, economic activity, transportation infrastructure, travel patterns, laws and regulations, financial resources, and community values. This inventory helps the engineer develop accurate forecasts for future system conditions and create effective business models.

Traffic engineering is essential for ensuring the smooth movement of vehicles on roads and tracks. Traditional methods like signs, signals, markings, and tolling have been used in the past, while newer technologies such as intelligent transportation systems offer advanced solutions like variable message signs and ramp meters. Human factors play a crucial role in transportation engineering, focusing on the interaction between drivers and vehicles, as well as the usability of road signs, signals, and markings.

7.1. TRANSPORTATION SYSTEMS

Transportation systems are essential for society to function, enabling the movement of people and essential goods. Without them, emergency responders would be unable to reach those in need, and store shelves would remain empty.

7.1.1. Transport Systems

We describe the fundamental principles and characteristics of transportation systems and their components and introduce a number of basic analysis tools and models that can be used to address transportation systems problems.

7.1.1.1. Systems and Their Characteristics

A system consists of interconnected components that work together to accomplish a shared objective, ultimately functioning as a unified entity that remains stable and operates as a cohesive unit. The behavior of a system is determined by the relationships between its individual parts rather than the parts themselves. Systems exhibit various fundamental traits.

For a system to operate effectively, it requires all its components to be present and organized in a specific manner. Consequently, systems possess properties that extend beyond their individual components. Changing one element within a system can lead to unintended consequences. For instance, enhancing public transportation in a city could decrease the number of vehicles on nearby roads as more individuals opt for public transit over driving. While widening a road might temporarily alleviate traffic congestion, it could ultimately attract more drivers and trips, potentially worsening conditions in the long term.

Additionally, systems typically serve distinct purposes within the broader system they operate in, providing coherence and stability. In the case of transportation systems, the primary objective is to facilitate the safe and efficient movement of people and goods.

Another key aspect of systems is their ability to receive and transmit feedback. This exchange of information is essential for the functioning of systems and the process of systems thinking. In the case of transportation systems, there is a feedback loop between transportation and land use systems. The land use system influences travel demand by shaping the distribution of activities such as where people live, work, and shop. On the other hand, the transportation system also impacts land use patterns by stimulating development through the construction of new infrastructure like roads, transit lines, and airports (Nelson & Sopasakis, 1998).

7.1.1.2. Components of Transportation Systems

A transportation system consists of three components: (1) physical elements, (2) human resources, and (3) operating rules.

7.1.1.2.1. Physical Elements

Physical elements include (1) infrastructure; (2) vehicles; (3) equipment; and (4) control, communications, and location systems.

Infrastructure consists of the unchanging elements of a transportation system, such as roads, bridges, and tunnels. These fixed facilities make up the network of connections within the system and are designed by civil engineers. This includes designing the soil and foundation, structure, drainage systems, and overall layout of the elements that make up the infrastructure.

These include travel ways, terminals, and stations.

Travel ways vary depending upon the transportation vehicle or mode. For example;

- Highways and roads are travel ways for automobiles and trucks.
- Rail transportation requires railroad tracks, and
- Air transportation utilizes specified air corridors.

Terminals are essential for buses, trains, planes, trucks, and boats. They control the arrival and departure of vehicles, as well as provide storage for vehicles and cargo. Terminals are key locations where users can board or exit the transportation system, and they allow for intermodal transfers between different modes of transportation. In comparison, stations only provide some of the functions of terminals, mainly serving as entry or exit points for the transportation system.

- Examples include bus, subway, and railway stations.
- A parking garage or a regional airport also serves as a station.

In a transportation system, vehicles are the moving components that travel along the roadways. Flow entities refer to the different units that move through the established infrastructure. For roads, the infrastructure must be able to accommodate a diverse range of vehicles, from bicycles to large tractor-trailers.

- Examples of vehicles include automobiles, buses, locomotives, railroad cars, ships, and airplanes. Some vehicles are self-propelled, such as automobiles, locomotives, ships, and aircraft, while others do not have propulsion, like railroad cars, barges, and truck trailers.

Equipment is physical tools used primarily to aid in the process of transporting objects.

- Some examples are snowplows, maintenance vehicles for railroad tracks, and baggage-handling conveyor belts at airports.

Control encompasses the necessary components for determining right-of-way. It is comprised of both vehicular control and flow control systems. The flow control system specifically allows for the effective and seamless movement of vehicles while minimizing conflicts between them.

- Vehicular control is the technological process of guiding individual vehicles on fixed facilities, which can be either manually operated or automated.
- Determining right-of-way involves the use of air traffic control centers, traffic signals, travelway signs and markings, signal systems, and

established rules of operation.

- Communication systems connect traffic control centers with various travelway equipment including variable message signs, traffic signals, transit vehicles, air traffic controllers, and pilots.
- The use of location systems enables the real-time identification of individual vehicles, thanks to the utilization of GPS technology. This helps track various types of vehicles like transit vehicles, trucks, and emergency vehicles, ultimately improving routing efficiency.
- For highway facilities with manually controlled vehicles, factors like the driver's reaction time to different stimuli are important. Automated systems also have their own well-defined response times.

7.1.1.2.2. Human Resources

Human resources are vital for the functioning of transportation systems. This includes vehicle operators like truck drivers, bus drivers, and railroad engineers, as well as airline pilots, maintenance workers, and transportation managers. Professionals in this field use their knowledge and information to improve transportation operations. Transportation managers encompass roles such as strategic planners, marketing personnel, maintenance managers, operations analysts, and administrators.

7.1.1.2.3. Operating Rules

The operating rules consist of schedules, crew assignments, connection patterns, trade-offs between cost and level of service, and contingency plans.

- **Schedules:** Transportation schedules outline the times when vehicles will arrive and depart from various terminals and stations. Adhering to the schedule is crucial for ensuring the quality of service provided by a transportation mode.
- **Crew Assignment:** Connection patterns involve the allocation of operators to various vehicles, such as assigning bus drivers to buses or assigning pilots and flight attendants to specific flights in a transit agency or airline. This task is complex as it requires meeting several constraints for each assignment. These constraints include adhering to the maximum allowable continuous work hours for each individual, ensuring that operators are matched with vehicles they are certified to operate, and minimizing costs.
- **Connection Patterns:** Connection patterns are the way in which services are structured across a transportation system or network. One common example is the "hub-and-spoke" system, where passengers and goods are transported from multiple cities to a central hub location, where they are then grouped together based on their final destination. This system presents various operational challenges, such as ensuring efficient transfer times between vehicles, adhering to strict schedules, and dealing with the system's vulnerability to external disruptions like accidents or bad weather conditions.
- **Cost/Level of Service Tradeoff:** Balancing cost and level of service is crucial in managing transportation systems. This tradeoff involves

establishing operational guidelines that consider the expenses incurred versus the quality of service provided. For instance, increasing the number of buses on a route for a transit agency enhances passenger experience but also raises operational costs. Similarly, a state highway department must decide between constructing a wider highway for a better driver experience or opting for a more cost-effective option. Airlines also face this tradeoff when deciding to offer direct flights between cities, as this improves traveler convenience but may increase operating costs. This balance between cost and service quality is a key aspect of effectively operating transportation systems.

- **Contingency Plans:** When issues arise with the transportation system, contingency plans are put into action. This includes having a plan in place for redirecting traffic if a major highway is shut down due to an accident or construction, evacuating coastal regions during a hurricane, and managing increased traffic flow during events. Developing an effective contingency plan may require investing extra resources and entails a balance between costs and service levels.

7.1.2. Analysis Tools of a Transport System

The tools and models utilized in analyzing transportation systems consist of fundamental traffic analysis tools, regression techniques, fundamental probability theory, queuing theory, and optimization tools. These tools play a vital role in resolving issues within transportation systems.

7.1.3. The Geography of Transportation Networks

7.1.3.1. Transport Networks

Networks are frequently used as a metaphor to depict the structure and movement within transportation systems. Nodes, representing locations, are connected by routes, or links, forming a larger framework. Routes can be physical, like roads or rails, or abstract, such as air or sea pathways (Siam, Nasrin, Hadiuzzaman et al., 2017). The economic interactions within a region shape its territorial structure, with networks forming as a result of continuous improvements and changing conditions. These networks are influenced by strategies aimed at enhancing access and mobility, as well as technological advancements. Transport networks, whether comprised of permanent infrastructure like roads and railways or scheduled services like airlines and trains, play a crucial role in facilitating movement within a region. Different types of transport structures are identified in transport geography, with network structures ranging from centripetal to centrifugal in terms of location accessibility. Centripetal networks prioritize a select few locations, while centrifugal networks do not provide any specific locational advantages.

In the past few years, transport hubs have become a popular network structure for various types of transportation services, particularly in the aviation industry. While hub-and-spoke networks can increase network efficiency, they are also susceptible to disruptions and delays at hubs due to the lack of direct connections (Paveri-Fontana, 1975).

Hubs, as a network structure, provide increased flexibility in the transport system by centralizing flows.

The main advantages of hubs are:

- Economies of scale on connections by offering a high frequency of services. For instance, instead of one service per day between any two pairs in a point-to-point network, four services per day could be possible.
- Economies of scale at the hubs, enabling the potential development of an efficient distribution system since the hubs handle larger quantities of traffic.
- Economies of scope in the use of shared transshipment facilities. This can take several dimensions such as lower costs for the users as well as higher quality infrastructures.

Numerous transportation services have implemented a hub-and-spoke model. This structure is commonly seen in air passenger and freight services on global, national, and regional scales, with companies like UPS, FedEx, and DHL utilizing it. Despite its benefits, there may be drawbacks associated with this model as well.

- As additional transshipment as fewer point-to-point services are offered, which for some connections may involve delays and potential congestion as the hub becomes the major point of transhipment.

7.1.4. Transport Supply and Demand

What are the differences between a Boeing 747, an oil tanker, a car, and a bicycle? Many indeed, but they each share the common goal of fulfilling a derived transport demand, and they thus all fill the purpose of supporting mobility. Transportation is a service that must be used immediately and cannot be stored. Mobility relies on transport infrastructure to provide a means of transportation. Sometimes, the simplest form of transportation is walking, but in other cases, complex and expensive infrastructure is needed, such as for international air travel. An economic system that involves various activities in different locations creates the need for a transport system to support movement. Without movement, infrastructure would be pointless, and without infrastructure, movement would not be possible or efficient.

- **Transport Supply**: This refers to the ability of transportation infrastructures and modes to operate within a defined geographical area and timeframe. Supply is measured in terms of infrastructure capacity, service frequency, and network coverage. The quantity of passengers, volume (for liquids or containerized goods), or weight (for freight) that can be transported per unit of time and space is typically used to gage transport supply.
- **Transport Demand**: This represents the requirements for transportation, whether they are completely met, partially met, or not met at all. Like transportation supply, these needs are quantified in terms of the number of individuals, volume, or weight per specific unit of time and space.

A simple statistical method to measure transportation supply and demand for passengers or freight is through the use of passenger-km (or passenger-mile) and ton-km (or ton-mile). These measures compare the quantity of passengers or freight transported over a certain distance. While typically used to measure realized demand, they can also be applied to assess transport supply. For example, a Boeing 747–400 flight between New York and London may have a transport supply of 426 passengers over

5,500 kilometers, totaling 2,343,000 passenger-km. However, the actual demand for that flight could be 450 passengers or 2,465,000 passenger-km, suggesting a system where demand exceeds capacity at 105%.

Transport demand comes from the economy, which includes people, institutions, and industries that move people and goods. These movements create patterns of mobility and accessibility in space. The location of resources, factories, distribution centers, and markets is linked to the movement of goods. There are two main reasons why transport demand can change: an increase in the quantity of passengers or goods being transported, or an increase in the distance over which these passengers or goods are moved. Differences in geography and transport costs can lead to variations in freight transport demand between countries. The movement of passengers is influenced by where people live, work, and shop. The demand for transport, measured in passenger-km or ton-km, can increase due to a rise in the number of passengers or goods being transported, or an increase in the distance they are moved. Factors such as industrial movements, economic specialization, and suburbanization can contribute to these changes in transport demand. Often, these factors occur together, leading to an increase in both passengers and goods being transported over longer distances.

Simplifying transport supply can be achieved by identifying key functions that represent the primary variables affecting the capacity of transportation systems. These variables vary depending on the mode of transportation. For road, rail, and telecommunications, transport supply primarily relies on the capacity of routes and vehicles (modal supply), whereas, for air and maritime transport, supply is largely determined by terminal capacity (intermodal supply).

- **Modal Supply:** The availability of one mode of transportation can impact the availability of others, particularly in areas where different modes compete for the same infrastructure, like roads in heavily congested areas. For example, the supply of transportation for cars and trucks is interdependent because they both rely on the same road network.
- **Intermodal Supply:** Transport supply is also dependent on the transshipment capacity of intermodal infrastructures. For instance, the maximum number of flights per day between Mombasa and Nairobi cannot be superior to the daily capacity of the airports of Mombasa and Nairobi, even though the air corridor has potentially a very high capacity.

The need for transportation is often correlated with economic and social activities and is typically stable and predictable in many cases. This makes it easier to plan and provide services effectively. However, there are instances where transportation demand is unpredictable and fluctuates, making it challenging to offer appropriate services. For example, commuting is a regular and expected pattern of movement, but emergency vehicles like ambulances must deal with sudden and unpredictable demands. Transport demand functions vary according to the nature of what is to be transported:

- **Passengers:** For the road and air transport of passengers, demand is a function of demographic attributes of the population such as income, age, standard of living, race, and sex, as well as modal preferences.

- **Freight:** For freight transportation, the demand is a function of the nature and the importance of economic activities (GDP, commercial surface, number of tons of ore extracted, etc.) and of modal preferences. Freight transportation demand is more complex to evaluate than passengers.
- **Information:** For telecommunications, the demand can be a function of several criteria including the population (telephone calls) and the volume of financial activities (stock exchange). The standard of living and education levels are also factors to be considered.

7.2. GEOMETRIC DESIGN OF ROADS

Highway engineering's geometric design involves determining the optimal positioning of road elements based on standards and limitations. The main goals are to maximize efficiency and safety while minimizing expenses and environmental impact. The concept of geometric design also plays a role in promoting a new objective known as livability. This objective involves designing roads in a way that supports various community goals, such as enabling access to employment, schools, businesses, and homes. Additionally, it involves accommodating different modes of transportation like walking, biking, public transit, and cars, while also reducing fuel consumption, emissions, and environmental harm.

There are three main components of geometric roadway design: alignment, profile, and cross-section. When combined, they create a three-dimensional layout for a roadway.

- The alignment is the route of the road, defined as a series of horizontal tangents and curves.
- The profile is the vertical aspect of the road, including crest and sag curves, and the straight grade lines connecting them.
- The cross section displays the placement and quantity of vehicle and bicycle lanes, as well as sidewalks, including their cross slope or banking. Additionally, cross sections illustrate drainage elements, pavement structure, and other features beyond geometric design.

Roads are planned according to specific design guidelines and standards that are approved by national and sub-national authorities such as states, provinces, territories, and municipalities. These guidelines consider factors such as speed, vehicle type, road slope, view obstructions, and stopping distance. By following these guidelines and using sound engineering judgment, an engineer can create a roadway that is safe, comfortable, and aesthetically pleasing (Chen, Song, & Ma, 2019).

7.2.1. Profile of a Road

A road profile includes road slopes known as grades that are linked by parabolic vertical curves. These curves are designed to create a gradual transition between different road slopes, allowing vehicles to navigate grade changes smoothly while traveling.

Sag vertical curves are characterized by a steeper slope at the end of the curve compared to the beginning. When driving on a road, a sag curve resembles a valley, with the vehicle descending before reaching the lowest point of the curve and then ascending or remaining level.

Vertical crest curves are characterized by a decrease in slope at the end compared to the start of the curve. When traversing this type of curve, drivers will experience a gradual incline followed by reaching the peak of the curve and descending downhill.

The profile also affects road drainage.

7.2.1.1. Sag Curves

Vertical sag curves are curves that appear concave upwards when viewed from the side. This includes curves located at valley bottoms, as well as areas where an uphill slope increases in steepness or a downhill slope decrease.

The primary consideration in designing these curves is ensuring sufficient headlight sight distance. When a driver is navigating a sag curve at night, the ability to see ahead is limited by the steep grade in front of the vehicle. It is essential for this distance to be long enough so that the driver can easily identify and react to any obstacles on the road within the headlight sight distance. The headlight sight distance (S) is influenced by the headlight angle and the slope of the tangent at the end of the curve. By calculating the headlight sight distance (S) first and then determining the curve length (L) using the equations provided, the appropriate curve length can be established. If the S < L result shows that the curve length is greater than the headlight sight distance, then that value is acceptable. Conversely, if the S>L result indicates that the curve length is shorter than the headlight sight distance, then that value is acceptable.

Units	**Sight Distance < Curve Length ($S < L$)**	**Sight Distance > Curve Length ($S > L$)**
Metric (S in meters)	$L = \frac{AS^2}{120 + 3.5S}$	$L = 2S - \frac{120 + 3.5S}{A}$
US customary (S in feet)	$L = \frac{AS^2}{400 + 3.5S}$	$L = 2S - \frac{400 + 3.5S}{A}$

These equations are based on the assumption that the headlights are positioned 600 millimeters (2.0 ft) above the ground, and the headlight beam spreads 1 degree above the vehicle's longitudinal axis.

7.2.1.2. Crest Curves

Vertical crest curves are curves that are upwardly convex when seen from the side. They can be found at hill crests, as well as areas where an incline becomes less steep or a decline becomes steeper.

The primary consideration when designing these curves is ensuring adequate stopping sight distance for drivers. This refers to the distance a driver can see ahead over the top of the curve. If visibility is limited and a driver cannot spot potential hazards like a stalled vehicle or an animal on the road, there is a higher risk of accidents. The necessary stopping sight distance (S) is determined by the speed of traffic on the road. By calculating the stopping sight distance (S) and then using the equations provided to solve for the curve length (L), the appropriate length for the curve can be determined. The specific equation to use depends on whether the vertical curve is shorter or longer than the available sight distance. It is common practice to solve both equations and then compare the results to find the correct curve length.

sight distance > curve length (S > L)

$$L = 2S - \frac{200(\sqrt{h_1} + \sqrt{h_2})^2}{A}$$

sight distance < curve length (S < L)

$$L = \frac{AS^2}{200(\sqrt{h_1} + \sqrt{h_2})^2}$$

In the US, the driver's eye height is set at 1080 mm (3.5 ft) above the pavement, while the height of the object the driver must see is 600 mm (2.0 ft), similar to the taillight height of typical passenger cars.

In bike facilities, we assume the cyclist's eye height is at 1.4 meters (4.5 feet) and the object height is 0 inches. This is because a pavement defect can result in a cyclist falling or losing control.

7.2.2. Alignment

Road design involves aligning straight sections of the road with circular horizontal curves. These curves are defined by their radius and deflection angle, determining their tightness and extent. The design process includes establishing a minimum radius based on the speed limit, determining curve length, and considering any obstacles that may obstruct a driver's view.

An engineer utilizes AASHTO standards to create a road that prioritizes safety and comfort. Higher speed and tighter radius horizontal curves require increased superelevation for safety. When obstacles obstruct visibility around corners or curves, the engineer must ensure drivers have adequate sightlines to prevent accidents and effectively merge into traffic.

- **Terminology**
  - o BC = beginning of the curve
  - o EC = end of curve
  - o R = radius
  - o PC = point of curvature (point at which the curve begins)
  - o PT = point of tangent (point at which the curve ends)
  - o PI = point of intersection (point at which the two tangents intersect)
  - o *T* = tangent length
  - o *C* = long chord length (straight line between PC and PT)
  - o *L* = curve length
  - o *M* = middle ordinate, now known as HSO – horizontal sightline offset (distance from sight-obstructing object to the middle of the outside lane)
  - o *E* = external distance
  - o f_s = coefficient of side friction
  - o u = vehicle speed
  - o Δ = deflection angle
- **Geometry**

$$T = R\tan\left(\frac{\Delta}{2}\right)$$

$$C = 2R\sin\left(\frac{\Delta}{2}\right)$$

$$L = R\pi\frac{\Delta}{180}$$

$$M = R\left(1-\cos\left(\frac{\Delta}{2}\right)\right)$$

$$E = R\left(\frac{1}{\cos\left(\frac{\Delta}{2}\right)}-1\right)$$

The quantity 1 - COSq is the versine and 1/COSq - 1 is the exsecant.

- Curve Sight Distance

$$M = R\left(1-\cos\left(\frac{28.65?}{R}\right)\right)$$

7.2.3. Cross Section

The cross-section of a roadway is a visual representation of what would be visible if a trench were dug across the roadway by an excavator. It shows the number of lanes, their widths, and cross slopes, as well as the presence or absence of shoulders, curbs, sidewalks, drains, ditches, and other roadway features. The shape of the road surface in relation to managing runoff is known as the "crown."

7.2.3.1. Lane Width

The selection of lane width affects the safety, maximum capacity, and cost, of a highway.

For optimal safety in urban settings, lanes should be between 3.0 and 3.1 meters (9.8 to 10.2 ft) wide. Narrow lanes (less than 2.8 meters (9 ft 2 in)) and wide lanes (over 3.1 meters (10 ft)) have higher crash risks. Lanes wider than 3.3~3.4 meters (10.8–11.2 ft) are linked to 33% higher impact speeds and more severe collisions, as well as increased crash rates.

The ideal width for carrying capacity is 3.0 to 3.1 meters (9.8 to 10.2 ft) for both motor traffic and bicycles. The maximum throughput occurs at a speed of 18 miles per hour (29 km/h). As the lane width decreases to 3.0 to 3.1 meters (9.8 to 10.2 ft), traffic speed and vehicle intervals decrease.

As lanes narrow, pedestrian volume also increases, and intersections with narrower lanes are able to accommodate more bicycles.

Building and maintaining narrow lanes is usually more cost-effective. They can shorten the time it takes to walk across and minimize stormwater runoff.

In North America, lane widths are often wider than ideal, typically ranging from 9.8 feet to 12 feet. Roads with higher speeds and traffic volume, as well as a heavy presence of trucks and large vehicles, typically feature wider lanes and shoulders.

7.2.3.2. Cross Slope

Cross slope is the angle of a roadway that is perpendicular to the centerline. If a road was completely flat, water would not drain properly, leading to potential issues like hydroplaning and ice accumulation during colder temperatures.

In straight sections of roads, the cross slope of the road surface is typically 1–2% in order to facilitate drainage. This type of slope, known as a "normal crown," is usually not noticeable to drivers as it is applied evenly in both directions along the center of the roadway.

On curved sections, the outer edge of the road is elevated higher than the centerline to create a sloped surface. This slope allows gravity to pull the vehicle towards the inside of the curve, reducing the dependence on tire friction for steering. As a result, a larger portion of centripetal force is used to navigate the curve.

Superelevation slopes ranging from 4 to 10% are implemented to assist drivers in safely navigating curved sections of the road while maintaining a consistent speed. A maximum slope of 12% was chosen to accommodate construction and maintenance requirements, as well as to minimize the challenges of driving on steeply sloped curves at low speeds. In regions prone to snow and ice, most transportation agencies opt for cross slopes of 6 to 8%. Steeper cross slopes can make navigating icy surfaces at low speeds difficult, while lower cross slopes increase the risk of losing control at high speeds, particularly on icy roads.

Sharp curves are safer when designers choose up to 8% superelevation instead of 4%, as high-speed skidding has worse consequences than sliding inward at low speeds. Urban roadways typically use a 4% slope to prevent the outside road edge from being higher than the surrounding terrain, which is more suitable for lower speeds. The formula for the preferred curve radius, as depicted below, considers both the speed and superelevation (e). By rearranging this equation algebraically, one can determine the desired superelevation rate by providing the designated speed and curve radius of the roadway as inputs.

$$R = \frac{u^2}{15(e + f_s)}$$

Recent studies have revealed that the equation mentioned above underestimates the necessary cross-slope values for heavy vehicles such as semitrailers and buses due to their high center of gravity and increased rollover risk.

7.2.4. Safety Effects of Road Geometry

The safety performance of a road is significantly impacted by its geometry. Research on the causes of road accidents indicates that human factors are the leading cause, followed by roadway factors and then vehicle factors.

7.2.4.1. Design Consistency

Accidents are more likely to occur in areas where a sudden change in road conditions catches drivers off guard. A common scenario is a sharp curve appearing unexpectedly after a long straight section of road. Design consistency aims to prevent these incidents by evaluating neighboring road segments and identifying potential problem areas where drivers may be caught by surprise. Locations with significant changes in speed should receive extra attention in the design process. For instance, a sharp curve with a smaller radius than previous curves may require additional signage for better visibility. This approach goes beyond simply setting a minimum design speed, as it takes into account the overall road layout to ensure a smooth and safe driving experience. In the example mentioned earlier, a long straight followed by a sharp curve would be deemed acceptable if a design speed of 30 mph was selected, but a design consistency analysis would highlight the need to address the sudden change in speed at the curve.

7.2.4.2. Safety Effects of Alignment

The safety of a horizontal curve depends on several factors including the curve length, radius, use of spiral transition curves, and the roadway's superelevation. Curves with smaller radii are more prone to crashes while incorporating spiral transitions can reduce the risk. Insufficient superelevation can also increase the likelihood of accidents.

A safety performance function to model curve performance on two-lane roads is:

$$AMF = \frac{1.55L_c + \frac{80.2}{R} - 0.012S}{1.55L_c}$$

Where; AMF is the Accident modification factor, a multiplier that describes how many more crashes are likely to occur on the curve compared to a straight road; L_c is the Length of the horizontal curve in miles;

R is the Radius of the curve in feet; S = 1 if spiral transition curves are present.

7.2.4.3. Safety Effects of Cross-Section

The safety performance of a road is influenced by its cross slope and lane width. Roads with narrow lanes have a higher likelihood of experiencing certain types of crashes, known as "lane departure crashes," such as run-off-road collisions, sideswipes, and head-on collisions.

7.2.5. Sight Distance

The geometry of the road impacts the amount of sight available to the driver. Sight distance, when it comes to designing roads, is the distance of the road ahead that is visible to the driver. It is the distance a road user, typically a vehicle driver can see before their view is obstructed by a hill crest, obstruction inside a curve, or at an intersection. Inadequate sight distance can have negative effects on the safety and efficiency of a road or intersection.

In order to accurately assess the required sight distance for a specific situation, it is necessary to consider the total distance covered during two key phases of a driving maneuver: perception-reaction time (PRT) and maneuver time (MT). PRT refers to the duration it takes for a driver to recognize the need for a reaction to a road condition, determine the appropriate maneuver, and initiate the maneuver. MT, on the other hand, represents the time needed to successfully execute the maneuver. The combined distance traveled during both PRT and MT determines the overall sight distance required for the given scenario.

Highway engineers analyze the sight distance available during highway design and traffic safety investigations by comparing it to the required sight distance for the specific situation. Depending on the situation, one of three types of sight distances will be used:

7.2.5.1. Stopping Sight Distance

The stopping sight distance is the total distance covered while the driver perceives a need to stop, reacts by applying the brakes, decelerates, and comes to a complete stop. Factors such as road conditions, vehicle weight, road incline, and other variables impact the actual stopping distance. To design safe roadways, a conservative estimate of stopping distance is necessary to ensure that a vehicle traveling at the design speed can stop before reaching any obstacles in its path. Design sight distance is typically determined so that even a below-average driver can stop in time to prevent a collision.

7.2.5.2. Decision Sight Distance

Decision sight distance is necessary for drivers facing decisions beyond simple stop-or-go scenarios. It is greater than stopping sight distance to accommodate the additional distance needed to make a more intricate decision. This type of sight distance refers to the distance a driver needs to identify a challenging or unexpected hazard on the road, understand the level of threat it poses, choose an appropriate speed and route, and execute the necessary maneuver safely and effectively, especially in visually crowded environments. In an ideal scenario, roads are designed to accommodate the decision sight distance, allowing for 6 to 10 seconds for perception-reaction time and 4 to 5 seconds to complete the necessary maneuver.

7.2.5.3. Intersection Sight Distance

Intersection sight distance refers to the necessary distance for safely navigating through an intersection. The required distance varies depending on the type of traffic control present (such as uncontrolled, yield sign, stop sign, or signal) and the type of maneuver being made (left turn, right turn, or proceeding straight). All-way stop intersections typically require the least sight distance, while uncontrolled intersections typically require the most. The availability of intersection sight distance plays a crucial role in determining whether no control or yield control can be safely used, or if more restrictive control is necessary.

7.2.5.3.1. Corner Sight Distance

Corner sight distance (CSD) is the specified road alignment that ensures a clear line of sight for drivers, bicyclists, and pedestrians at intersections. This allows them to safely anticipate approaching vehicles and provides enough time for them to cross all lanes of traffic, make a left turn, or turn right without disrupting the flow of through traffic.

7.2.5.3.2. Uncontrolled and Yield Controlled Intersections

Uncontrolled and yield-controlled intersections need to have clear sight lines without obstructions for safe operation. At uncontrolled intersections, drivers must follow basic right-of-way rules (yielding to the vehicle on the right or following the boulevard rule). Drivers need to be able to see approaching traffic in order to adjust their speed or stop to yield before reaching the intersection. Clear visibility is not the only factor considered when determining the type of intersection control. Converting an intersection to stop control is often done in response to safety concerns.

7.2.5.3.3. Two-Way Stop Control

When calculating corner sight distance, it is important to consider the distance that the vehicle waiting at the crossroad should be set back. Some state MUTCDs and design manuals have established guidelines for the setback distance for the driver of the vehicle on the crossroad, which should be a minimum of 10 feet plus the shoulder width of the major road, but not less than 15 feet. The Federal MUTCD requires that a stop line, if utilized, should be at least 4 feet from the nearest travel lane. The line of sight for corner sight distance should be determined from a 3 and 1/2-foot eye height at the vehicle driver's location on the minor road to a 4 and 1/4-foot object height at the center of the approaching lane of the major road.

Corner sight distance, D_{CSD}, is equivalent to a specified time gap, t_g, at the design speed, V_{DS}, required for a stopped vehicle to turn right or left:

$$D_{CSD} = V_{DS} t_g$$

At two-lane intersections, passenger vehicles should maintain a distance of 7.5 seconds at the design speed. Trucks and buses, as well as multilane roads, require longer gaps. It is important for the public right-of-way to have and maintain a clear line of sight.

7.2.5.3.4. All-Way Stop Control and Signalized Intersections

At intersections where there is all-way stop control or traffic signals, drivers require the minimum amount of sight distance. At all-way stops, drivers must have visibility of vehicles halted at other approaches. At signal-controlled intersections, drivers approaching must have a clear view of the signal heads. In areas where right turns on red are permitted, drivers at right lane stop control require the same sight distance as two-way stop control. While not necessary under usual circumstances, it is advisable to have extra sight distance available in case of signal malfunctions or power outages.

7.2.5.4. Effects of Insufficient Sight Distance

Many older roads were built before current sight distance standards were in place, making it difficult for jurisdictions to meet the financial demands of acquiring additional right-of-way, redesigning roadbeds, or implementing new projects in challenging terrain or environmentally sensitive areas. In these situations, it is important to ensure that the corner sight distance is at least equal to stopping sight distance. While drivers should ideally have a corner sight distance that exceeds the braking distance at the design speed, they are still responsible for maintaining control and a safe speed to be able to stop within the assured clear distance ahead (ACDA), following the basic speed rule. Jurisdictions often offer some level of design immunity in government claims actions in these cases.

Warning signs are commonly utilized when there is not enough sight distance. According to the Manual on Uniform Traffic Control Devices, intersections must have Stop Ahead, Yield Ahead, or Signal Ahead signs when the traffic control device cannot be seen from a distance equal to the stopping sight distance for the speed of approaching traffic. Hill Blocks View signs may be used in areas where sight distance is restricted due to crest vertical curves. Nonetheless, numerous jurisdictions continue to require drivers to exercise caution when faced with obvious conditions, even in the absence of a sign. The vigilance and attention normally expected of drivers may need to be heightened on roads with lower functional classifications. The likelihood of sudden changes in traffic flow rises as the number of access points increases, a fact that drivers should be aware of even when a particular access point is not visible. As a result, full visibility around corners is typically not mandated for individual driveways in densely populated urban areas, where street parking within the right-of-way is often allowed.

7.3. TRAFFIC FLOW THEORY

Traffic flow theories aim to accurately illustrate the relationships between vehicles, drivers, and infrastructure through mathematical analysis. The infrastructure encompasses the highway system along with its various components such as control devices, signage, and markings. These theories are essential components of traffic models and analysis tools utilized in the planning and management of transportation networks.

Research on traffic flow began in the 1930s when probability theory was first used to analyze road traffic. Bruce D. Greenshields conducted early studies at the Yale Bureau of Highway Traffic focusing on models that linked traffic volume and speed, as well as the behavior of traffic at intersections. Following World War II, the study of traffic flow grew exponentially due to the rise in automobile usage and the expansion of highway systems. This period saw an increase in research on traffic characteristics and the development of various traffic-flow theories.

The first International Symposium on the Theory of Traffic Flow took place in December 1959 at the General Motors Research Laboratories in Warren, Michigan. This event marked the beginning of a series of triennial symposia focusing on the theory of traffic flow and transportation. Reviewing the proceedings of these symposia offers valuable insight into the significant advancements made in understanding and managing traffic flow processes over the past four decades.

The field of traffic-flow theory and transportation has expanded to the point where it cannot be adequately addressed by any single type of gathering. Many specialized conferences and symposia focusing on various traffic-related themes are now held regularly. Despite advancements in computational technology leading to a deeper comprehension of traffic-flow theory, the basic principles remain essential. They serve as the basis for the theories, methods, and practices used in designing, managing, and enhancing modern transportation systems.

Three specific systems stand out: a physical system, a transportation system, and a social system, as illustrated in Figure 7.1. The physical system typically includes non-living objects whose movements and interactions are governed by physical laws like Newton's laws of motion. On the other hand, the social system involves living beings such as humans whose behaviors can vary widely but generally follow some loosely defined guidelines (e.g., seeking benefits and avoiding losses). Consequently, physical science is considered "hard" for being more objective, rigorous, and precise, while social science is viewed as "soft" due to its subjectivity, flexibility, and lack of exactness. The transportation system combines aspects of both, involving both living beings (human drivers) and non-living elements (roads and vehicles). Therefore, transportation science could be seen as "firm" (in the absence of an appropriate term between "hard" and "soft") as it deals with both physical laws and social norms. It leans towards the "soft" end when it comes to strategic planning but moves towards the "hard" end when focusing on tactical decisions and operational control.

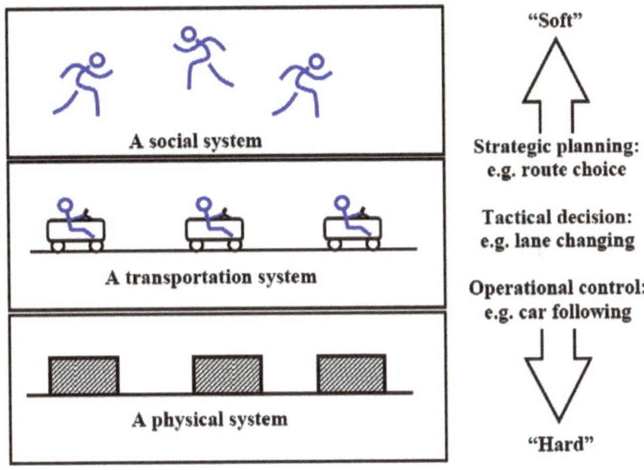

Figure 7.1. *Three systems.*

Source: Daiheng Ni, A Unified Perspective on Traffic Flow Theory Part I: The Field Theory, https://doi.org/10.1061/41186(421)420.

By blending physical laws and societal norms, a theory is proposed that defines the driving surroundings as a field perceived by a driver. Within this field, various objects like roadways and vehicles are considered individual component fields, collectively representing the overall danger that the driver aims to steer clear of. Therefore, the process of modeling vehicle movement involves choosing the safest path by moving through the field along its lowest point.

7.3.1. Physical Basis of Traffic Flow

Many traffic flow phenomena are analogous to those in the physical system, yet the transportation system has something special to distinguish itself.

7.3.1.1. Mechanics Phenomena

Forces in Physics are responsible for causing changes in motion. These forces can be measured and their effects can be replicated. For instance, Newton's second law states that the velocity of an object will change when it is acted upon by an external force. Newton's third law explains that every action has an equal and opposite reaction. Similarly, forces play a role in traffic flow, but these forces are perceived by individual drivers and can be subjective. Lewin and Gibson have noted that the difference between a perceived force and a physical force is that the former comes from within the driver, through motivation, rather than being externally imposed. For example, a fast driver feels a "force" (a stress in the driver's mind) when approaching a slow vehicle. Conversely, the slow driver may or may not be subject to the "reaction force" depending on whether the driver pays attention and responds to it, i.e., Newton's third law may or may not take effect in this case. More examples of mechanics phenomena are provided below:

- **M1: Directional Flow**: Traffic always flows in a pre-determined direction much like free objects always fall to the ground. The reason why free objects fall is that they are constantly subject to earth gravity. Similarly, it is reasonable to imagine that vehicles in the traffic are subject to a "gravity" along the roadway. Such a roadway gravity is, again, a subjective matter since it exists mentally in drivers and is not measurable, but it is recognized that the gravity is related to factors such as driver personalities (e.g., aggressiveness), vehicle properties (e.g., engine power), and road conditions (e.g., freeways vs. streets).
- **M2: Free Flow:** An object experiencing free fall will reach a stable speed caused by air resistance, just like a vehicle traveling freely on the road. However, the "resistance" in this scenario is the driver's decision to follow traffic regulations, such as speed limits, rather than factors like incline or air resistance. Unlike the predictable and consistent speed of free fall, the speed of a vehicle on the road is subjective and based on the driver's choices. When faced with the same conditions, drivers' choices may vary depending on the individual or the time. Different roads also have different free speeds. To prevent confusion, a driver's chosen free speed is referred to as their "desired speed," while the collective free speed of a group of vehicles is known as the "free-flow speed." Generally, desired speed is influenced by driver personalities and road conditions, while free-flow speed is impacted by road conditions and the population of drivers.
- **M3: Stopping at a Red Light:** Similar to an object coming to a halt when it hits a barrier, a vehicle slows down and stops at a red light. In this scenario, the force pushing back against the motion comes from the driver. Failing to obey the red light can lead to serious consequences such as accidents or fines. Unlike the predictable manner in which an object consistently stops, drivers have the freedom to slow down at their own pace before stopping. However, there may be instances where drivers may unintentionally forget to stop altogether.
- **M4: Road Barriers:** Vehicles on a roadway are separated by lane lines to prevent collisions with cars in adjacent lanes. Drivers should imagine barriers on both sides of their lane to stay in their designated lane. If a driver drifts out of their lane, they may feel a sense of pressure to correct their course, as if being pushed back in line by an invisible force. If a driver encounters a slower vehicle, they may feel the urge to change lanes, as if they are being lifted above a barrier to cross into the next lane. Barriers at the road edges are taller than lane barriers to prevent vehicles from running off the road. The center line barrier is even higher to deter vehicles

from crossing into oncoming traffic. These barriers are not physical but are instead perceived by drivers as a way to guide their movements on the road (Figure 7.2).

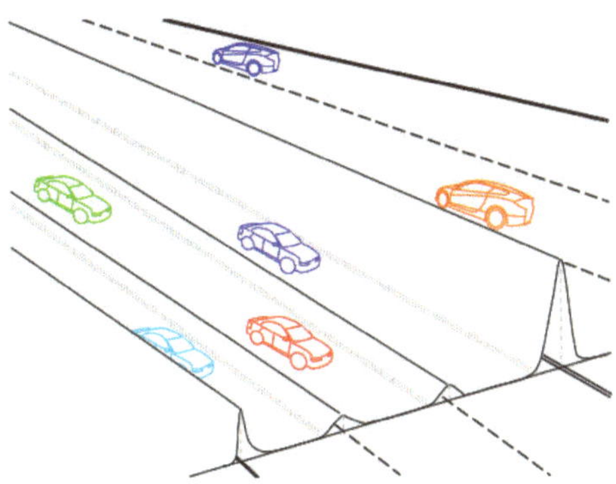

Figure 7.2. *Road barriers.*

Source: Daiheng Ni, A Unified Perspective on Traffic Flow Theory Part I: The Field Theory, https://doi.org/10.1061/41186(421)420.

7.3.1.2. Electromagnetic Phenomena

There are two ways in which an object can exert a force on another object: through collision or action at a distance. For instance, hitting a ball with a bat demonstrates the first method, while using a magnet to find a needle illustrates the second. Though collisions are not uncommon on highways, action at a distance is what vehicles normally interact with each other and examples of this kind include some of the above mechanics phenomena as well as the following:

- **E1: Car Following:** When a fast vehicle catches up with a slow vehicle, the fast driver realizes a potential collision if they maintain their current speed. The fear and potential cost of a collision prompt the fast driver to take evasive action. If changing lanes is not an option and the slow driver does not speed up, the fast driver must slow down and match the speed of the slow vehicle at a safe following distance. This situation can be compared to moving two like electric charges towards each other, as described by Coulomb's law. The force between the two charges is determined by the product of their charges and the inverse of the square of their distance. Similarly, in a car following, the stress on the fast driver increases as they approach the slow vehicle more quickly and shorten the distance between them. The slow driver may not experience the same force, as they may not be aware of the approaching vehicle.
- **E2: Tailgating:** Building on the scenario mentioned earlier, if the fast vehicle is tailgating (following at a dangerously close distance), the slow driver is likely to feel pressured to either speed up or yield to the faster vehicle. In this situation, it is similar to charge A driving charge B in the analogy, demonstrating Newton's third law. For a force to have an impact on a driver, it must first be perceived by them. The ability of a driver to perceive these forces depends on where they are scanning and how often they are checking their surroundings.
- **E3: Shying Away:** If two vehicles drive side by side, the drivers may feel intimidated and try to avoid a potential side collision by spreading out in space either lengthways or sideways. This apprehension becomes more pronounced when one of the vehicles is a large truck.

7.3.1.3. Wave Phenomena

W1: Harmonic wave – A line of vehicles on a road can be compared to a harmonic wave, with characteristics such as flow (vehicles per hour), traffic speed (km per hour), and density (vehicles per km). Similarly, a wave is defined by frequency (Hz or cycles per second), wave speed (meters per second), and wavelength (meters). It is clear that flow is equivalent to frequency, traffic speed is equivalent to wave speed, and spacing (the inverse of density) is equivalent to wavelength. In Figure 7.3, the top portion illustrates a line of vehicles as a harmonic wave.

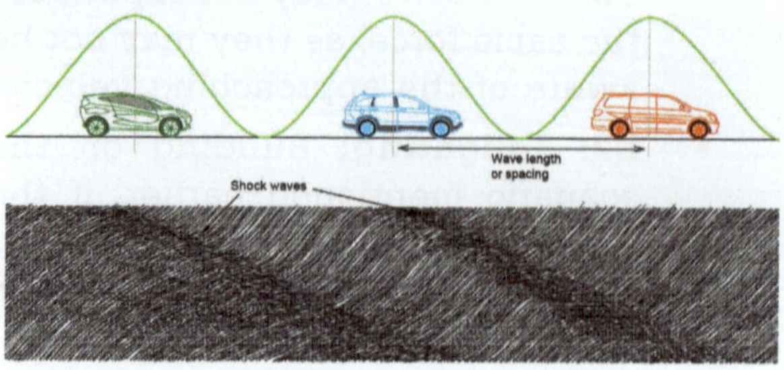

Figure 7.3. *Traffic and waves.*

Source: Daiheng Ni, A Unified Perspective on Traffic Flow Theory Part I: The Field Theory, https://doi.org/10.1061/41186(421)420.

- **W2: Signal Propagation:** The signal in this context does not pertain to a traffic signal, but rather to any variable that clearly indicates the location and speed of a disturbance in a medium. For example, when the front vehicle in a compact group suddenly slows down, a kinematic wave is created and moves through the group, with the signal being the momentary decrease in speed. Similarly, when a group of fast vehicles catches up to a group of slower vehicles, a shock wave is produced and moves through the traffic, with the signal being the boundary between the fast and slow vehicles. The lower portion of Figure 7.3 showcases various shock waves observed in vehicle paths.
- **W3: Wave-Particle Duality:** All objects, especially small objects, display characteristics of both waves and particles. The particle-like behavior is noticeable when focusing on individual particles (such as the photoelectric effect), while the wave-like behavior emerges when observing the collective behavior of multiple particles (like wave diffraction). In traffic flow, single vehicles behave like particles (such as following cars and changing lanes), whereas groups of vehicles act like waves (e.g., kinematic waves and shock waves).

7.3.1.4. Statistical Mechanics Phenomena

Many have modeled traffic flow as a one-dimensional compressible fluid, akin to gas. Gas molecules follow a Maxwell-Boltzmann distribution, with average speed and variance increasing with temperature. In traffic flow, however, a different trend emerges. Empirical observations show that traffic speed variance peaks at optimal density and decreases at lower and higher densities. This distribution, illustrated in Part 3 of Figure 7.4, contrasts the Maxwell-Boltzmann distribution and is further detailed in a three-dimensional model in Part 4, which serves as the basis for a stochastic fundamental diagram (Munjal & Pahl, 1969).

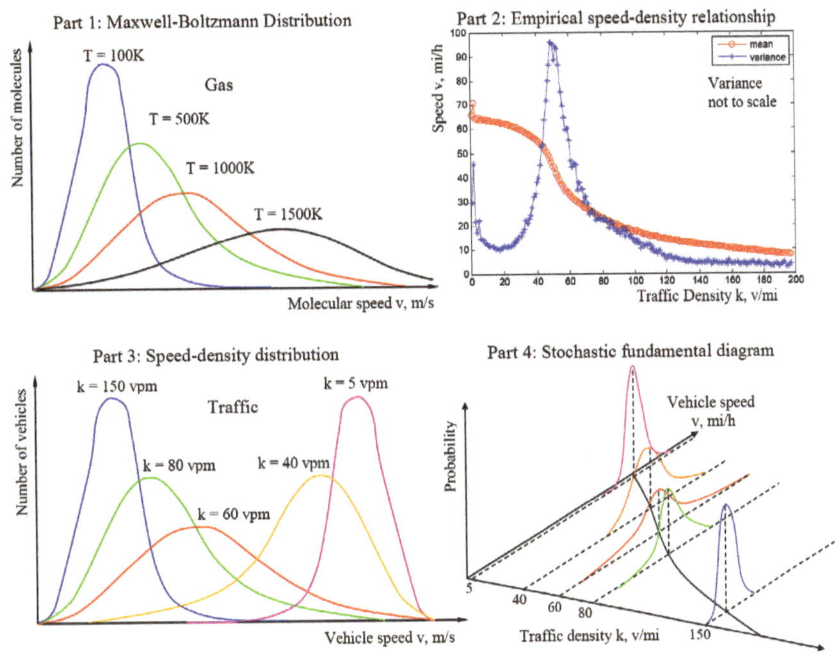

Figure 7.4. *Maxwell-Boltzmann distribution in traffic flow.*

Source: Daiheng Ni, A Unified Perspective on Traffic Flow Theory Part I: The Field Theory, https://doi.org/10.1061/41186(421)420.

7.3.2. The Field Theory

Lewin established the concept of the "field at a given time" within the realm of psychology, while Gibson applied this idea to the context of automobile driving. However, their observations lacked mathematical precision and were primarily qualitative. Therefore, the focus of this section is to develop a Field Theory for traffic flow. Given that the transportation system involves interactions between living beings (such as human drivers) and inanimate objects (like roadways and vehicles), it is subject to both physical principles and societal norms. Consequently, the Field Theory is being systematically formulated based on a series of postulates, with two pertaining to physical aspects (1 and 3) and the other two relating to social factors (2 and 4) (Kurt Lewin, 1943).

7.3.2.1. Postulate 1: A Road is a Physical Field

Postulate 1 is motivated by phenomena M1, M2, and M4. In the longitudinal (x) direction, a driver-vehicle unit i is subject to gravity along the road:

$$G_i = m_i g_i \qquad (1)$$

where G_i is the roadway gravity acting on the unit, mi is the mass of the unit, and

g_i is the acceleration of roadway gravity perceived by driver i. As discussed in M1, g_i is a function of driver personalities, vehicle properties, and road conditions. Meanwhile, the unit is also subject to a resistance R_i perceived by the driver due to his or her willingness to observe traffic rules (e.g., speed limits). As discussed in M2, R_i is related to the driver's perceived difference between his or her actual speed $\dot{x}_i$, and desired speed v_i which in term is related to the free-flow speed of the road v_f (subscript f here means free flow, not a vehicle ID), i.e., $R_i = R_i(\dot{x}_i, v_i(v_f))$. Therefore, the net force acting on unit i in the longitudinal direction can be expressed as:

$$m_i\ddot{x}_i = G_i - R_i$$

(2)

where $\ddot{x}_i$ is the acceleration of unit i.

The amount of net force on the right-hand side symbolizes the potential for acceleration of the unit, representing the driver's longing for movement. As the unit accelerates, the right-hand side diminishes due to an increase in R_i. Eventually, the right-hand side disappears when the unit reaches its desired speed v_i. If by chance some external disturbance causes the unit's speed to exceed v_i, the right-hand side turns negative. In such a scenario, the unit slows down and eventually returns to v_i.

In the lateral direction of the road, there are lane lines, road edges, and center lines to guide and separate traffic. As discussed in M4, these cross-section elements of the road can be mapped into a roadway potential field U_i^R perceived by the driver. When the unit deviates from its lane, the unit is subject to a correction force Ni which can be interpreted as the stress on the driver to keep his or her lane, see illustrations in Figures 7.2 and 7.6. The effect of such a force is to push the unit back to the center of the current lane. Based on physical principles, the force can be determined as the derivative of the roadway field, U_i^R, with respect to the unit's lateral displacement y_i:

$$N_i = -\frac{\partial U_i^R}{\partial y_i}$$

7.3.2.2. Postulate 2: A Driver Responds to His Surroundings Anisotropically

There are two main differences between the interaction of two driver-vehicle units and the collision of two objects. One involves Newton's third law of motion, which will be explained further, and the other relates to non-contact forces, which will be covered in the following postulate.

In classical physics, Newton's third law of motion applies when two objects collide, however, this law does not necessarily hold true in the interaction between two vehicles with drivers. For instance, when a fast-moving vehicle is approaching a slower vehicle, the fast driver experiences a sense of distance closing and may interpret it as a "repelling force" or stress. The faster the gap closes, the stronger this force may feel. The slow driver, on the other hand, may not perceive this force depending on their awareness of the situation and their willingness to react. As drivers are all facing forward while driving, the responsibility of maintaining a safe distance falls on the driver behind, who would be blamed for a rear-end collision if one were to occur. This can lead to situations where the lead driver may not be aware of what is happening behind them, like an approaching fast vehicle.

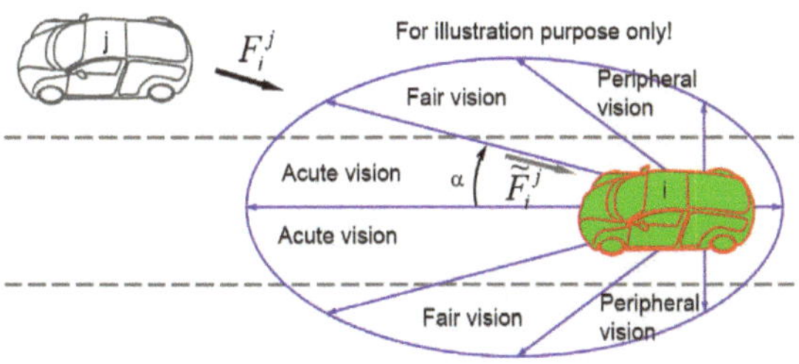

Figure 7.5. *Distribution of driver's attention.*

Source: Daiheng Ni, A Unified Perspective on Traffic Flow Theory Part I: The Field Theory, https://doi.org/10.1061/41186(421)420.

In general, a driver's awareness and reaction to their surroundings depend on their field of view and how often they scan their environment, as shown in Figure 7.5. For instance, the space directly in front of the driver, especially in their own lane, is within their immediate sight and requires constant monitoring and quick responses. The driver also pays attention to the front areas of the adjacent lanes within their reasonable sight as vehicles in those lanes may switch over, necessitating caution during lane changes. On the other hand, the driver scans less frequently towards the sides of their vehicle (considered their peripheral vision zones) unless a lane change or lane-adjacent maneuver is required. The last and least attended area is the rear of the vehicle not only because it is difficult to access (indirectly by means of side or rear mirrors) but also because the responsibility rests on the drivers behind. Therefore, it is reasonable to assume that the driver's directional response to his or her surrounding, ζ_i, is a function of his or her viewing angle α_i, i.e., $\zeta_i = \zeta(\alpha_i)$. Consequently, the force that actually acts on the unit, $\tilde{F}_i$, is the product of the force that might have been perceived by the driver if he or she had paid full attention, F_i, and his or her directional response ζ_i, i.e.,

$$\tilde{F}_i = F_i \times \zeta(\alpha_i) \tag{4}$$

where $\alpha_i \in [-\pi, \pi]$ is the viewing angle. For example, if one chooses $\zeta(0) = 1$ and $\zeta(\pi) = 0$, the driver responds to Fi in full when it comes from the vehicle in front (i.e., $\alpha_i = 0$) and ignores Fi when it comes from the vehicle behind (i.e., $\alpha_i = \pi$ or $-\pi$), respectively.

7.3.2.3. Postulate 3: A Driver Interacts with Others by Action at a Distance

Drivers have the ability to detect nearby vehicles and obstacles, allowing them to take evasive actions to prevent a collision. This sensing ability is thought to be a result of a field that the driver perceives as a potential collision hazard. This field can be imagined as a hill, where the higher and steeper the hill, the harder it is to navigate. The base of this hill represents a boundary beyond which the driver is not affected by the field. For example, in Figure 7.6, the dash-dotted oval labeled "Base j" shows the extent of the field perceived by driver i from unit j. This field can also be seen as the personal space of unit j, with intrusion into this space being actively discouraged. The deeper into this space driver i goes, the stronger the force repelling them. The field is represented as a curve above the x-axis on a longitudinal section.

Similarly, unit k represents another field (whose base is labeled as "Base k") which also exerts an influence on unit i. Since unit k is at the side lane of unit i, the influence is not in the longitudinal x direction but in the lateral y direction, i.e., the field results in a repelling force, F_i^k, on unit i which motivates it to shy away from unit k.

The fields and forces described above are interconnected with the personalities of drivers and the dynamics of vehicles.

An aggressive driver, for instance, tends to maintain shorter distances between cars, resulting in a smaller perceived field. Conversely, as a vehicle moves faster, it poses more risk to nearby vehicles, leading to the creation of a larger and more pronounced field.

7.3.2.4. Postulate 4: A Driver Tries to Achieve Gains and Avoid Losses

A driver's strategy for moving on roadways is to achieve mobility and safety (gains) while avoiding collisions and violations of traffic rules (losses). Such a strategy can be represented using an overall potential field U_i which consists of component fields such as those due to moving units U_i^B, roadways U_i^R, and traffic control devices U_i^C, i.e.,

$$U_i = U(U_i^B, U_i^R, U_i^C) \tag{5}$$

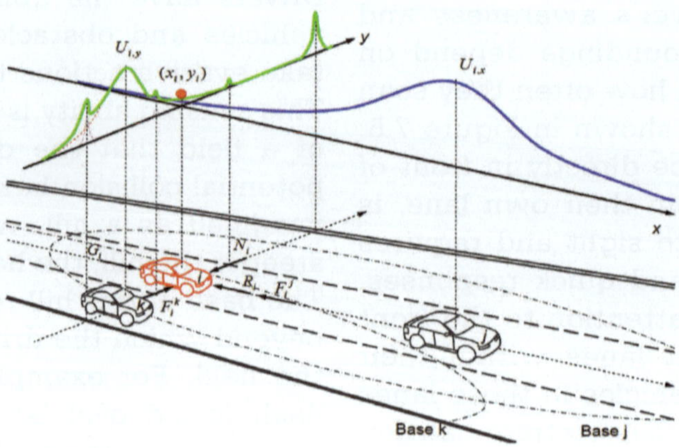

Figure 7.6. *The illustration of a field.*

Source: Daiheng Ni, A Unified Perspective on Traffic Flow Theory Part I: The Field Theory, https://doi.org/10.1061/41186(421)420.

If we imagine U_i as a mountain range where the height represents the risk of losses, the driver's goal is to move through the range following the easiest path, like traveling through a valley. For instance, in Figure 7.6, there are two sections of this kind of terrain. From the driver i's perspective, the section along the x-axis, $U_{i,x}$, is mostly influenced by unit j because it is the only nearby vehicle in the middle lane. Unit i is depicted as a ball following closely behind the curve of $U_{i,x}$, as it is within the field of unit j. Therefore, unit i is subject to a repelling force F_i^j which is derived from $U_{i,x}$ as:

$$F_i^j = -\frac{\partial U_{i,x}}{\partial x} \tag{6}$$

The effect of F_i^j is to push unit i back to keep a safe distance. By incorporating the driver's unsatisfied desire for mobility (G_i $-R_i$), the net force in the x direction can be determined as:

$$m_i\ddot{x}_i = \sum F_{i,x} = G_i - R_i - F_i^j = (m_i g_i - R_i) + \frac{\partial U_{i,x}}{\partial x} \quad (7)$$

The section of U_i in the lateral y direction, $U_{i,y}$ (the bold curve), is the union of two components: the cross-section of the field due to unit k (the dashed curve) and that due to the roadway field (the dotted curve). The former results in a repelling force F_i^k which makes unit i to shy away from unit k and the latter generates a correction force N_i should unit i depart its lane center. Therefore, the net effect can be expressed as:

$$m_i\ddot{y}_i = \sum F_{i,y} = F_i^k - N_i = -\frac{\partial U_{i,y}}{\partial y} \quad (8)$$

By incorporating time t, driver i's perception-reaction time τ_i, and driver i's directional response ζ_i, Eqns. 7 and 8 can be expressed as:

$$m_i\ddot{x}_i(t+\tau_i) = \sum \tilde{F}_{i,x}(t) = \zeta_i^0[G_i(t) - R_i(t)] + \zeta(\alpha_i^j)\frac{\partial U_{i,x}}{\partial x} \quad (9a)$$

$$m_i\ddot{y}_i(t+\tau_i) = \sum \tilde{F}_{i,y}(t) = -\zeta(\alpha_i^k)\frac{\partial U_{i,y}}{\partial y} \quad (9b)$$

The above system of equations summarizes the Field Theory in generic terms and constitutes the basic law that governs a unit's motion on a planar surface.

7.3.3. A Special Case

While the generic form of the Field Theory can provide a basic understanding of certain traffic phenomena, a more detailed and specific model is necessary for rigorous analysis of traffic flow. This section will focus on developing a specific form.

In the generic theory, the specific functions of the field Ui, roadway gravity Gi, and resistance Ri are not defined. It is clear that there are numerous potential specific forms that the theory can take, making it unfeasible to list them all. Following Occam's razor principle, it is advisable to choose the simplest form possible without unnecessary complexity. This principle states that "entities should not be multiplied unnecessarily."

Hence, the razor gives rise to the following considerations:

1. The chosen specific form should make physical sense, to which empirical observations are a good test;
2. It should take a simple functional form that involves physically meaningful parameters but not calibration coefficients; and
3. It should provide a sound microscopic basis for aggregated behavior, i.e., macroscopic equilibrium models.

With the above considerations, a specific form of the theory is formulated by focusing on longitudinal forces which receive full attention while lateral forces are only considered during a lane change.

7.3.3.1. Motion in the Longitudinal Direction

Important factors to consider are the unmet need for mobility caused by road resistance and gravity, as well as the impact of the leading vehicle on distance-related actions.

7.3.3.1.1. Unsatisfied Desire for Mobility

The term (G_i–R_i) explains a driver-vehicle unit's unsatisfied desire for mobility. Intuitively, when a unit starts from stand-still, i.e., $\dot{x}_i = 0$, its unsatisfied desire for mobility is the greatest. As the unit speeds up, (G_i – R_i) decreases accordingly but is still positive, i.e., it still accelerates the unit to higher speeds. When the unit achieves its desired speed, i.e., $\dot{x}_i = v_i$, its desire for mobility has been fully satisfied and, hence, G_i – R_i = 0 which means that the unit settles at vi if no other forces act on it. If a random perturbation brings ˙xi over v_i, the unit's desire for mobility is over-satisfied and G_i – R_i becomes negative which decelerates the unit back to v_i. With the above understanding, a specific form of the unsatisfied desire for mobility can be formulated as:

$$G_i - R_i = m_i g_i [1 - (\frac{\dot{x}_i}{v_i})^\delta] \quad (10)$$

where δ is a calibration parameter.

7.3.3.1.2. Action at a Distance

When a fast unit i catches up with a slow unit j, the former is subject to a non-contact force, F_i^j, from the latter. Such a non-contact force varies with the spacing between the two units, $s_{ij} = x_j - x_i$. For example, the force virtually has no effect on unit i when it is distant; it takes effect when unit i becomes close (e.g., within the range of its desired spacing s^*_{ij}); it increases as the spacing becomes even shorter (s_{ij} ↓); it reaches the maximum when $s_{ij} \to l_j$ where l_j represents the minimum "safety room" required by unit j. In addition, the effect of the force is also related to the speeds and relative speed of units i and j. Such an effect can be incorporated into the formulation of the driver i's desired spacing s^*_{ij}. (Figure 7.7)

Therefore, a simple way to represent the force is to use an exponential function. The general idea of this model is to set the desired spacing s^*_{ij} as a baseline, beyond which the intrusion by unit i is translated exponentially to the repelling force acting on the unit. As such, a more specific but still quite generic form of the force can be:

$$F_i^j = f(e^{s^*_{ij} - s_{ij}}) \quad (11)$$

where $s^*_{ij} - s_{ij}$ represents how far unit i intrudes into s^*_{ij}.

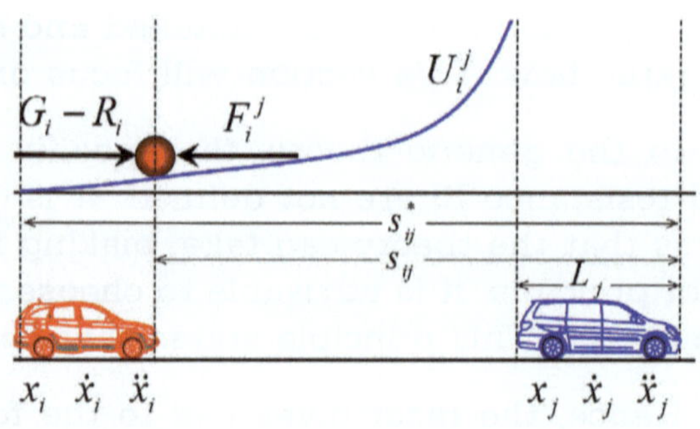

Figure 7.7. *Action at a distance.*

Source: Daiheng Ni, A Unified Perspective on Traffic Flow Theory Part I: The Field Theory, https://doi.org/10.1061/41186(421)420.

7.3.3.1.3. The Longitudinal Control Model

Combining the above, the effort that is required by driver i to control his or her vehicle in the longitudinal direction can be expressed as:

$$m_i\ddot{x}_i = m_i g_i[1 - (\frac{\dot{x}_i}{v_i})^\delta - f(e^{s^*_{ij} - s_{ij}})] \quad (12)$$

if the coefficient of term F_i^j is chosen properly. Though Eqn. (12) can be further detailed in many possible ways, the following special case is of particular interest. Putting time t and response delay τ back on and eliminating vehicle mass m at both sides, Eqn. (12) can take the following special form:

$$\ddot{x}_i(t + \tau_i) = g_i[1 - (\frac{\dot{x}_i(t)}{v_i})^\delta - e^{\frac{s_{ij}(t)^* - s_{ij}(t)}{Z}}] \quad (13)$$

If one chooses δ = 1 and Z = s_{ij} (t)*, the above equation reduces to a more specific form:

$$\ddot{x}_i(t + \tau_i) = g_i[1 - (\frac{\dot{x}_i(t)}{v_i}) - e^{\frac{s_{ij}(t)^* - s_{ij}(t)}{s_{ij}(t)^*}}] \quad (14)$$

No further motivation for this special case is provided other than the following claims: (1) it takes a simple functional form that involves physically meaningful parameters but not arbitrary coefficients, (2) it makes physical and empirical sense, and (3) it provides a sound microscopic basis to macroscopic behavior

The determination of desired spacing $s_{ij}(t)^*$ admits safety rules. An example is that one should leave a time gap of at least one perception-reaction time in front to ensure safety:

$$s^*_{ij}(t) = \dot{x}_i(t)\tau_i + l_j \quad (15)$$

Another example can be the following: the desired spacing should allow vehicle i to stop behind its leading vehicle j after a perception-reaction time τ_i and a deceleration process (at rate $b_i > 0$) should vehicle j applies an emergency brake (at rate $B_j > 0$). After some math, the desired spacing can be determined as follows:

$$s^*_{ij}(t) = \frac{\dot{x}_i^2(t)}{2b_i} + \dot{x}_i\tau_i - \frac{\dot{x}_j^2(t)}{2B_j} + l_j \quad (16)$$

7.3.3.2. Motion in the Lateral Direction

The movement of a driver-vehicle unit in the lateral y direction involves two key decisions: lane change and gap acceptance. When making a lane change decision, the driver considers switching to an adjacent lane to improve their mobility and safety. Gap acceptance comes into play when the driver actually moves the vehicle into the target lane, taking advantage of a safe opportunity (such as a gap in traffic). Figure 7.8 depicts a situation where units i and j are in the right lane, with a trailing neighbor p in the left lane.

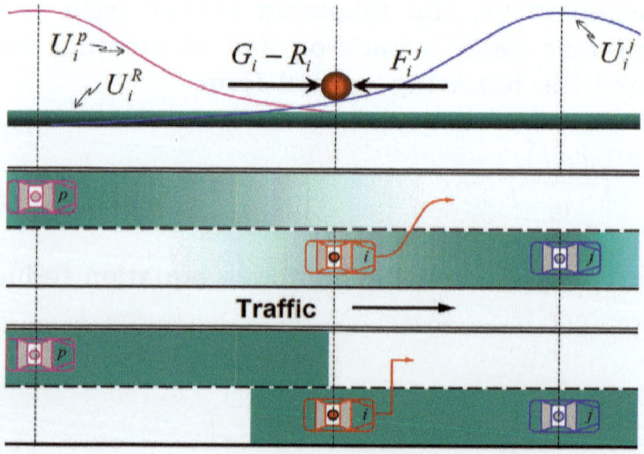

Figure 7.8. *Lane change.*

CASE STUDY

PUBLIC TRANSPORT SYSTEM IN THE GULF REGION, A CASE STUDY OF THE CITY OF RIYADH

There is no doubt that one of the most important elements of a modern city is its transportation system. If this system is efficient it is safe to say that this city is well developed, as the transportation system is the main factor affecting the infrastructure of the city. In addition, economic and social growth is dependent on the transportation system of a city. This is so because the transportation system facilitates activity for all other sectors such as agriculture, industry, mineral industry, and commerce. People benefit from a good transportation system because they can achieve their objectives easily in different areas of the city. They can also easily shop, go sightseeing, go to work, and pay visits if the system is well designed.

In many of the developed and developing countries public transport has encountered numerous difficulties that have necessitated the intervention of the governments of these countries. The interventions have ranged from ownership to operation and regulation. The aims of governmental interventions are wide and complex, but the most important of these aims are safety, transportation, efficiencies, energy conservation, and protection of the environment and better mobility for those who have specific needs. As a result of this, the public transport system had to be operated under regulated conditions and consequently, this required more effort and responsibility from both central and local government. Financial aid was thought unnecessary and inefficient in some countries because it was thought that this had reduced the efficiency of the industry while it had not helped to stop the declining trend of customer support. The shortage of supply of public transportation, lack of good access, overcrowding and high fares force passengers either to walk, use other methods of transport such as bicycles, and animals, or make the period of working hours longer which causes even longer peak hours. The shortage of public transport is mainly due to inadequate investment and it is this that led to the emergence of Para-transit operators which started to compete in the market.

The City of Riyadh

Riyadh is situated on the Najd Plateau at a height of 625 m above sea level. The topography of Riyadh is quite flat and ranges in height between 580 m in the south to over 660 m in the north. The most famous topographical feature in Riyadh is Wadi Hanifah and this watershed is the main landmark on the western side of the city. Tributaries run almost vertically to the west of the main valley. A number of them have eroded the plateau into wide steep valleys, which in some places are more than 100 m in depth. The valley floors are flat and in places are several hundred meters wide.

Historically, they were used for agriculture, mainly date plantation, but now they provide a location for extensive urban developments and some of these have had a negative impact on the environment.

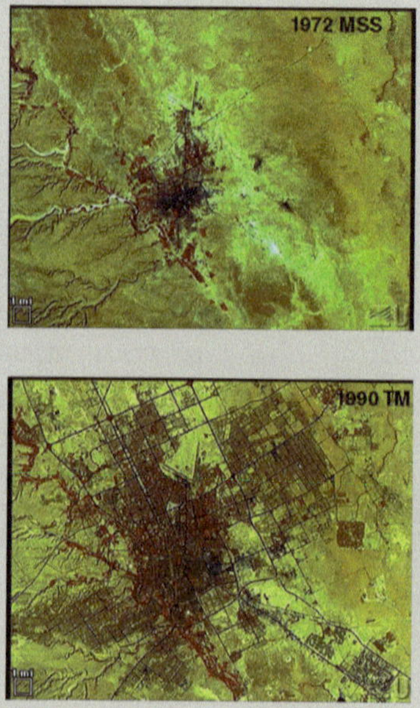

Figure 7.9. *Riyadh in 1972 and 1990.*

Source: https://www.witpress.com/Secure/elibrary/papers/UT05/UT05016FU.pdf.

The fast growth of the city of Riyadh was the main reason that led to a dramatic increase in the need for transport within the city. This increase shows the extent of the need for the improvement of the transportation system in order that these problems, as well as congestion and environmental dilemmas, can be solved adequately and efficiently.

The growth of the city is expected to continue at the same rate in the future. In the city of Riyadh, the network of roads is the most important part of the transportation system. The improvement of this network has proved to be very difficult because of many reasons such as cost, land acquisition and inconvenience during the construction periods. A survey conducted in 1999 by King Abdul Aziz City for Science and Technology found that 73% of planners and traffic engineers said that "it is very important to develop and improve" the public transport systems while the rest only said that "it is important"

From 1974 to 1999, an annual growth rate of more than 8% was recorded. This represents one of the highest rates of population growth in the region. There are many factors that drive this growth rate, which include consistently high natural birth rates of Saudi nationals, migration within the Kingdom to Riyadh and immigration of foreign nationals.

Transportation

As can be seen from Table 7.1 population growth and car ownership in Riyadh is one of the highest in the world. The average incomes are also high which is why car ownership is increasing rapidly.

Table 7.1. Average Annual Growth Rates (AAGR) of Population, Car Ownership, and Urbanization for Selected Cities

City	Population AAGR 70-80 (%)	Car Ownership AAGR 70-80 (%)	GNP/capita 1980 ($)
Ankara	4.4	14.2	1,470
Bangkok	9.1	7.9	670
Bombay	3.7	6.1	240
Cairo	3.1	17.0	580
Lima	4.2	12.1	2,050
Rio	2.4	12.1	2,050
Riyadh	10.3	37.5	11,260
Kuwait	6.3	13.8	13,200
London	- 0.9	2.6	7,920
New York	- 1.0	--	11,360
Tokyo	- 5.6	2.5	9,890

Source: https://www.witpress.com/Secure/elibrary/papers/UT05/UT05016FU.pdf.

Public Transport

The public transport system in Riyadh began in the 1960s when a number of owners of small coaches used them as a means of transport for the public. These coaches were locally known as "coasters" and the services were irregular and disorganized. The services were provided according to the demand for them and the aims of these services did not constitute a public transport system but rather they were means of increasing the income of the owners of the minibuses. These basic services continued to exist as the only means of transport until 1979 when SAPTCO was founded to provide its services within and between the cities of the Kingdom. The company was awarded a monopoly for all these services except for international transport. The company was run under the control of the Ministry of Transport (MOT). International bus services existed before the foundation of SAPTCO and are still run but are also under the control of the Ministry of Transport. After SAPTCO was created it was hoped that the services of the minibuses would be phased out but this has not happened. The number of minibuses has risen to more than 7 times the number of buses belonging to SAPTCO. This has led to a service interval of just two minutes for the minibuses whereas the service interval for SAPTCO's buses is 16 minutes. The main reason for this outcome was the increase of the ticket prices from one to two Riyals (1GBP=7.2SAR). It is relevant to mention here that SAPTCO is subsidized and regulated by the government which imposes many controls and restrictions on their buses whereas the minibuses (coasters) have much less control and restriction. This of course has led to more flexibility for the minibuses and provided them with a significant advantage over SAPTCO's buses.

Integration of Public Transport Services

"One of the big advantages of the private car is that if parking is available, it gives a door-to-door service. Public transport will never match this but with the integration of services it is possible to get closer to it than would otherwise happen."

In some cases, there might be a domination of one type of public transport which may have a negative effect on the other modes and for this reason, government's or regulator's intervention is sometimes necessary to achieve the integration of public transport.

There are means to obtain more benefits from any public transport network. These means include administrative as well as market processes, i.e., Land use patterns, traffic regulation and the level of private transport, fare level structures and systems and promoting the public transport network. Public involvement in public transport is another very important issue, i.e., people must be given the opportunity to have a say in the planning, integration and coordination of the public transport system. The reason for the need for involvement is that the public are the customers and therefore should be able to put their ideas forward and these ideas should be considered very carefully, as the people use the system and will understand the problems facing it more than others.

It is a fact that railways have a greater need for integration than public buses. One of the reasons for this is that railways are used for longer journeys and go through low-density areas whereas buses usually are not meant for that. It is usually the case that journeys start by bus before taking the train for the main journey. Another reason for the need for the integration of railways is the fact that their services are usually less frequent than those of buses. Integration is essential for the specialization of services and in suburbs of low density, it is possible to provide more frequent services for a given cost using minibuses.

In general, integration of public transport systems offers better services to passengers as the system becomes more efficient, reliable and easy to use. This will also have a positive effect on businesses as public transport plays an important role in the movement of employees and workers.

To make public transport in Riyadh more effective, integration must be implemented as this city has low-density regions in the suburbs. Buses could be used in these areas to take people to the main stations that are served by trains. The main stations should be in places where there is a large concentration of people or shopping centers.

Future Trends

According to Al-Mosaind the city of Riyadh is generating about 3.5 million vehicle trips, which corresponds to 5 million person trips daily. It is also estimated that 40 million kilometers are driven in the city every day, which amounts to a million hours of ignited automobile engines every day. Although the number of trips and the number of kilometers driven in the city may increase by 30% to 40% in the next five years, the number of hours of automobile operation may double in the same period and may increase by 4 to 5 fold in ten years time due to the anticipated increases in congestion in the city.

With current ridership levels and socioeconomic conditions in Riyadh, the city would not need a high-capacity public transport system as the expected ridership for the year 2010 is only 77,100,000 (SAPTCO Annual Report)

It has been demonstrated in many world cities, similar to Riyadh, that a good public transport system plays a major role in reducing the number of vehicle trips, thus enhancing the level of service, this, in turn, reduces the negative impact of traffic and has been exploited in many modern cities to reactivate city centers which had experienced a significant slow down in their commercial and social activities as a result of recurrent traffic jams and negative environment consequences

Conclusions

- The city of Riyadh had no real public transport system until 1979 when the government, subsidized company SAPTCO was created. In the period preceding 1979 public transport merely comprised private buses or the so–called Coasters which are minibuses that are also privately owned. All of these services were badly organized and run almost without any kind of regulation.
- The city of Riyadh was initially designed as a compact city because it was thought that it would not attract many people. This has not been the case and many people moved to the capital and the city started to grow very rapidly. The roads were designed to serve a compact city and the city grew much larger than expected and many transport problems started to arise. Planners found themselves facing major problems and dilemmas. They used traditional methods such as the expansion of roads, or building extra roads or lanes. These traditional responses were not well-thought out and many problems in transport were not properly solved.
- The city of Riyadh is in need of more integration of public transport services as the city has low-density regions in the suburbs.
- This study on the city of Riyadh will be more useful if linked with a comparison

with other of the GCC cities. Such comparison will make it possible to avoid the mistakes that have happened in Riyadh in other cities like Dubai, Kuwait, Muscat and even in smaller cities which are growing at fast rates.

- There should be an organizing body for public transport in the GCC. This body should regulate and oversee the public transport services in these countries. A one organizing body will be able to see the whole picture of the transportation problems faced by countries of the GCC.

CLASS ACTIVITY

Given the figure below, what do the values of h_1 and h_2 represent? Note that these distances must be in feet. (Hint: Think about how h_1 relates to the car in front. The box in front of h_2 represents an object in the road).

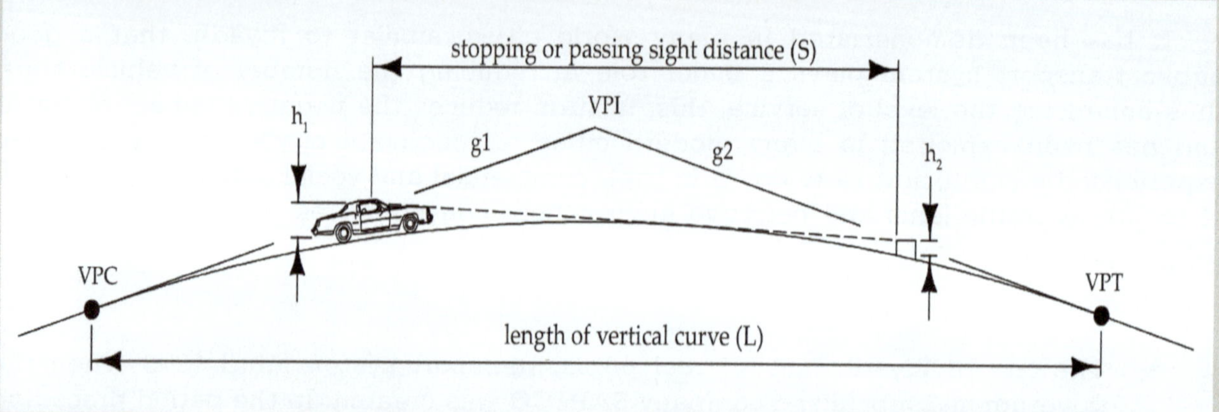

SUMMARY

- Transportation engineering is the application of technology and scientific principles to the planning, functional design, operation, and management of facilities for any mode of transportation in order to provide safe, efficient, rapid, comfortable, convenient, economical, and environmentally compatible movement of people and goods transport.
- The planning aspects of transportation engineering relate to elements of urban planning and involve technical forecasting decisions and political factors.
- A system is defined as a set of interrelated components that perform several functions in order to achieve a common goal and is therefore an entity that maintains its existence and functions as a whole through the interaction of its parts.
- The control system consists of vehicular control and flow control. The flow control system consists of the means that permit the efficient and smooth operation of streams of vehicles and the reduction of conflicts between vehicles.
- The analysis tools and models used for transportation systems include fundamental traffic analysis tools; regression techniques; basics of probability theory; queuing theory; and optimization tools.
- The geometric design of roads is the branch of highway engineering concerned with the positioning of the physical elements of the roadway according to standards and constraints.
- The profile of a road consists of road slopes, called grades, connected by parabolic vertical curves. Vertical curves are used to provide a gradual change from one road slope to another, so that vehicles may smoothly navigate grade changes as they travel.
- Sag vertical curves are those that have a tangent slope at the end of the curve that is higher than that of the beginning of the curve.
- Crest vertical curves are curves which, when viewed from the side, are convex upwards. This includes vertical curves at hill crests, but it also includes locations where an uphill grade becomes less steep, or a downhill grade becomes steeper.
- Horizontal alignment in road design consists of straight sections of road, known as tangents, connected by circular horizontal curves.
- The safety of a horizontal curve is affected by the length of the curve radius, whether spiral transition curves are used, and the superelevation of the roadway.

REVIEW QUESTIONS

1. What do you understand by transportation systems? Discuss.
2. Discuss the components of transportation systems.
3. Focus on the geography of transportation networks.
4. What is the geometric design of highways and what factors affect it?
5. Deal with Maxwell-Boltzmann distribution in traffic flow.

REFERENCES

1. Autey, J., Sayed, T., & El Esawey, M. (2013). Operational performance comparison of four unconventional intersection designs using micro-simulation. *Journal of Advanced Transportation*, *47*(5), 536–552.
2. Cai, Z., Xiong, M., Ma, D., & Wang, D. (2016). Traffic design and signal timing of staggered intersections based on a sorting strategy. *Advances in Mechanical Engineering*, *8*(4), 1–9.
3. Chen, F., Song, M., & Ma, X. (2019). Investigation on the injury severity of drivers in rear-end collisions between cars using a random parameters bivariate ordered probit model. *International Journal of Environmental Research and Public Health*, *16*(14), 2632.
4. Fraser, M. L., & Meuleners, L. B. (2021). Risk factors for unsafe events involving a motor vehicle for group riders (cyclists): A naturalistic case-crossover study. *Accident Analysis & Prevention*, *146*, 105758. doi: 10.1016/j.aap.2020.105758.
5. He, H., Guler, S. I., & Menendez, M. (2016). Adaptive control algorithm to provide bus priority with a pre-signal. *Transportation Research Part C: Emerging Technologies*, *64*, 28–44.
6. Hummer, J. E., Rasdorf, W., Findley, D. J., et al. (2010). Curve collisions: Road and collision characteristics and countermeasures. *Journal of Transportation Safety & Security*, *2*(3), 203–220.
7. Lewin, K. (1943). Defining the "Field at a Given Time." *Psychological Review*, *50*(3), 292–310.
8. Munjal, P., & Pahl, J. (1969). An analysis of the Boltzmann-type statistical models for Multi-Lane traffic flow. *Transportation Research*, *3*, 151–163.
9. Nelson, P., & Sopasakis, A. (1998). The Prigogine Herman kinetic model predicts widely scattered traffic flow data at high concentrations. *Transportation Research B*, *32*, 589–604.
10. Paveri-Fontana, S. L. (1975). On Boltzmann-like treatments for traffic flow: A critical review of the basic model and an alternative proposal for dilute traffic analysis. *Transportation Research*, *9*, 225–235.
11. Phillips, W. F. (1979). A kinetic model for traffic flow with continuum implications. *Transportation Planning and Technology*, *5*, 131–138.
12. Rubie, E., Haworth, N., Twisk, D., & Yamamoto, N. (2020). Influences on lateral passing distance when motor vehicles overtake bicycles: A systematic literature review. *Transportation Reviews*, *40*, 754–773.
13. Siam, M. R. K., Nasrin, S., Hadiuzzaman, M., Muniruzzaman, S. M., & Haque, N. (2017). VISCAL: Heuristic algorithm-based application tool to calibrate microscopic simulation parameters. *Journal of Traffic & Transportation Engineering (English Edition)*, *5*(1), 28–43.

14. Wong, C. K., & Heydecker, B. G. (2011). Optimal allocation of turns to lanes at an isolated signal-controlled junction. *Transportation Research Part B: Methodological*, *45*(4), 667–681.

15. Yan, C., Jiang, H., & Xie, S. (2014). Capacity optimization of an isolated intersection under the phase swap sorting strategy. *Transportation Research Part B: Methodological*, *60*, 85–106.

CHAPTER 8

Environmental Engineering

LEARNING OBJECTIVES

After studying this chapter, you will be able to:

- Understand water supply and distribution systems;
- Focus on wastewater collection and treatment;
- Know air pollution control measures; and
- Discuss solid waste management techniques.

KEY TERMS FROM THIS CHAPTER

Air pollution
Ecology
Geology
Hydrant flushing
Hydrology
Municipal solid-waste management
Radial system
Water distribution system
Water treatment
Dead end system
Economics
Grid iron system
Hydraulics
Microbiology
Noise pollution
Ring distribution system
Water supply network

INTRODUCTION

Environmental engineering involves creating systems and structures to manage water supply, waste disposal, and pollution control. These efforts aim to safeguard public health by preventing the spread of diseases and to maintain the quality of the environment by avoiding contamination and degradation of air, water, and land resources.

Environmental engineering is a diverse field that incorporates principles from various disciplines such as chemistry, ecology, geology, hydraulics, hydrology, microbiology, economics, and mathematics. Originally known as sanitary engineering within civil engineering, the field adopted the more precise term *environmental engineering* in the mid-1960s.

Environmental engineering projects encompass a wide range of tasks such as treating and distributing drinking water, managing wastewater, controlling air and noise pollution, handling solid and hazardous waste, cleaning up hazardous waste sites, and conducting environmental assessments. Mathematical modeling and computer analysis are commonly used to design systems for these tasks, with chemical and mechanical engineers contributing as well. Environmental engineers also engage in research, teaching, project planning, facility design and operation, equipment sales and marketing, and ensuring compliance with environmental standards and regulations.

Typically, environmental engineers receive their education through graduate-level coursework, although some schools may offer specialization or electives in environmental studies for undergraduates. Two-year associate degree programs are also available for individuals looking to become environmental technicians. Public sector employment for environmental engineers can be found at national and regional environmental agencies, local health departments, and municipal engineering and public works departments. In the private sector, environmental engineers may work for consulting engineering firms, construction contractors, water and sewerage utility companies, and various manufacturing industries.

8.1. WATER SUPPLY AND DISTRIBUTION SYSTEMS

Water for drinking comes from a range of sources. As water cycles through the environment, it falls back to Earth as precipitation. Some of it collects in lakes and rivers, while some seeps into the ground as groundwater. Groundwater can be extracted using wells, and surface water can be purified and distributed to consumers.

For water to flow out of a faucet when customers turn it on, a distribution system consisting of pipes, pumps, storage, and other components is necessary. Regardless of the type of project – commercial, residential, or agricultural – water accessibility is a key consideration in engineering. If there is not enough on-site water supply, it will need to be provided through a water distribution system.

8.1.1. Water Supply System

A water supply network, also known as a water supply system, consists of a series of intentionally designed hydrologic and hydraulic elements that distribute water. Typically, a water supply system includes the following components:

1. A drainage basin.
2. Raw water is collected from various sources such as lakes, rivers, or underground aquifers in a collection point, either above or below ground. From there, it is transported to water purification facilities through uncovered ground-level aqueducts, covered tunnels, or underground water pipes.
3. Water treatment plants process and purify water before distributing it through underground pipes.
4. Different water storage options include reservoirs, water tanks, water towers, cisterns, and pressure vessels. Tall buildings may need to store water locally in pressure vessels to ensure water reaches all floors.
5. Extra water pressurizing equipment, such as pumping stations, may need to be located at the discharge point of underground or aboveground reservoirs or cisterns if gravity flow is not possible.
6. A system of pipes used to distribute water to consumers, including private residences, industrial, commercial, and institutional establishments, as well as other points of use such as fire hydrants.
7. Connections to the sewer system, whether they be underground pipes or aboveground ditches in certain developing nations, are typically located downstream from where water consumers are situated. It is important to note that the sewer system is distinct and operates independently from the water supply system.

Public utilities within the water industry commonly manage water supply networks.

8.1.1.1. Water Abstraction and Raw Water Transfer

Untreated water, known as raw water, is sourced from either a surface water body like a lake or river, or from a groundwater source such as a well tapping into an underground aquifer within the watershed of the water supply.

Uncovered aqueducts, covered tunnels, or underground water pipes are used to transfer raw water to the water purification facilities.

8.1.1.2. Water Treatment

Almost every large system is required to treat water, which is carefully monitored and regulated by global, state, and federal agencies like the World Health Organization (WHO) and the United States Environmental Protection Agency (EPA). Water treatment is necessary before it reaches the consumer and after it is discharged. Purification usually takes place near the final delivery points to minimize pumping expenses and the risk of contamination after treatment.

Traditional surface water treatment plants typically involve three main processes: clarification, filtration, and disinfection. Clarification involves removing particles such as dirt and organic matter from the water by destabilizing their charges and causing them to settle or float out of the water stream. This is often achieved through the addition of chemicals like alum or ferric chloride. Filtration then refines the water further by removing smaller particulate matter using materials like sand, anthracite, or activated carbon filters. The final step in the process is disinfection, usually accomplished through the addition of chlorine. Chlorine effectively eliminates bacteria and viruses, ensuring the safety of the water supply as it travels through the distribution network.

8.1.1.3. Optimizing the Water Supply Network

The productivity of a system can be determined by its worth or its overall advantage. In the case of a water supply system, the real value or net benefit is a dependable water supply service that offers sufficient quantity and high-quality water. For instance, if a city's current water supply needs to be expanded to accommodate a new municipality, the introduction of the new branch must be planned carefully to meet the additional requirements while continuing to meet the needs of the existing system.

- **Single-Objective Optimization:** When designing a system, there are various criteria that must be considered, with cost being one of them. A design that minimizes cost while maintaining the same benefit leads to maximum overall benefit. However, opting for the lowest cost design often means sacrificing the capacity of a water supply network. A cost-effective model typically aims to find the solution with the lowest costs (in terms of pipe sizes) while still meeting hydraulic constraints such as required output pressures, maximum pipe flow rate, and pipe flow velocities. The cost of the system is directly influenced by the diameters of the pipes used, so the goal is to optimize the pipe sizes in order to achieve the minimum acceptable capacity at the lowest possible cost.
- Multi-Objective Optimization: However, because future demand

is unpredictable, the authors of the paper "Method for optimizing design and rehabilitation of water distribution systems" state that "the least capacity is not a desirable solution to a sustainable water supply network in the long term." It is better to have extra pipe capacity to handle unforeseen increases in demand and water shortages. The issue becomes a multi-objective optimization problem (minimizing cost and maximizing flow capacity) as opposed to a single-objective optimization problem (minimizing cost).

- Weighted Sum Method: In order to address a multi-objective optimization problem, it is essential to transform it into a single-objective optimization problem. This can be achieved by employing methods like a weighted sum of objectives or an ε-constraint method. The weighted sum technique involves assigning specific weights to each objective and combining them to create a single objective function for optimization. However, this method may not always produce the optimal solution for all original objectives as selecting the correct weights can be challenging.
- **The Constraint Method:** The constraint method, as the second approach, selects one objective function as the primary focus while treating the other objectives as limitations with restricted values. Nevertheless, the optimal solution is contingent upon the predetermined constraint boundaries.
- **Sensitivity Analysis:** Solving multiple objective optimization problems requires balancing costs and benefits to generate a range of solutions for further analysis and testing in various scenarios. It is important to note that there is no one perfect solution that addresses all objectives optimally. Since the objectives often conflict with each other, improving one may require compromising the other. In such cases, alternative methods like Pareto Analysis may be needed to determine the best possible combination of solutions.
- **Operational Constraints:** When considering the cost objective function, it must adhere to all operational constraints. Typically, the cost is primarily driven by the energy cost of pumping. Operational constraints encompass factors like customer service standards, such as ensuring minimum delivered pressure, as well as physical constraints like maintaining maximum and minimum water levels in storage tanks to prevent overflow and depletion (Martínez Fernando, Hernández Vicente, Alonso José Miguel, Rao Zhengfu, & Alvisi Stefano (2007).

To maximize efficiency in the water supply network and lower energy expenses, it is essential to forecast how altering pump and valve settings will impact network operations.

In addition to Linear and Non-linear Programming, there are alternative methods and approaches to effectively design, manage, and operate a water supply network with sustainability in mind. Examples include utilizing suitable technology paired with efficient strategies for maintenance and operation. Key components of these strategies should encompass effective management practices, technical support for customers and industries, sustainable financial mechanisms, and the establishment of reliable supply chains. These measures aim to guarantee the longevity of the system, consistent maintenance schedules, continuous operation, minimal downtime for repairs, and acceptable levels of water yield and quality.

8.1.2. Water Distribution System

The design of a water distribution system serves as a plan for constructing and managing a system that delivers safe drinking water to a community. It includes the layout of pipes, pumps, and other necessary infrastructure to transport water from its source, like a treatment plant, to the final customers. Key components of the design involve estimating water demand, choosing appropriate pipe sizes and materials, and building pump and storage systems. Additionally, the design accounts for potential future expansions and improvements to the system.

Moreover, water is essential for a variety of everyday tasks, including washing, cooking, drinking, gardening, bathing, and various commercial activities. The network of pipes and other water delivery infrastructure is typically hidden beneath the streets, mirroring the layout of the distribution system to effectively reach all areas of the region.

8.1.2.1. Components

A water distribution system includes pipelines, storage facilities, pumps, and other necessary equipment. Water mains, which are pipelines located within the public right of ways, are used to convey water throughout the distribution system. Primary feeders, which are large-diameter water mains, connect water treatment plants to service areas.

Secondary feeders are linked between primary feeders and distributors. Distributors are water mains situated near water users, providing water to individual fire hydrants. A service line is a narrow pipe connecting a water main through a small tap to a water meter at the user's location. A service valve, or curb stop, is located on the service line near the street curb to regulate the water supply to the user's location.

Distribution reservoirs, also known as storage facilities, store treated drinking water to ensure a consistent supply to meet changing demands (service reservoirs) or to stabilize operating pressure (balancing reservoirs). Additionally, they can be used temporarily for firefighting purposes during power outages. Types of distribution reservoirs include:

- **Underground Storage Reservoir or Covered Finished Water Reservoir:** A fully covered underground storage facility or large ground-excavated reservoir with impermeable lining to prevent groundwater intrusion.
- **Covered Finished Water Reservoir:** A reservoir with proper waterproof lining to prevent water runoff and intrusion, but without a top cover. This type of reservoir is not ideal as the water is vulnerable to contamination from various sources like bird waste, animal and human activities, algal bloom, and airborne particles, and is not further treated before distribution.

- **Surface Reservoir (Also Known as Ground Storage Tank and Ground Storage Reservoir):** A ground storage facility is constructed with walls made of concrete, shotcrete, asphalt, or membrane to prevent contamination. The facility is usually situated in high-elevation areas to ensure there is enough hydraulic head for distribution. If a surface reservoir at ground level does not provide adequate hydraulic head for the distribution system, booster pumps will need to be used.
- **Water Tower (Also Known as Elevated Surface Reservoir):** One option for storing water is an elevated water tank. There are several common types, including spheroid elevated storage tanks, which consist of a steel tank placed on top of a small-diameter steel column; composite elevated storage tanks, which feature a steel tank on a large-diameter concrete column; and hydropillar elevated storage tanks, which have a steel tank on a large-diameter steel column. The empty space within the large column below the water tank can also be utilized for various purposes, such as multi-story office space or storage. One potential drawback of using water towers in a water distribution system is the impact on the area's aesthetics.
- **Standpipe:** A water tank that combines features of a ground storage tank and a water tower. Unlike a traditional water tower, this tank has a standpipe that allows water storage from ground level to the top of the tank. The lower storage area is referred to as supporting storage, while the upper portion, which is comparable in height to an elevated water tower, is called useful storage.
- **Sump:** This is a reserve water storage facility that is not meant for directly distributing water. It is usually constructed underground in a circular design with a dome-shaped top above ground. The water stored in a sump will be transferred to a service reservoir as required.

Storage facilities are usually situated in the middle of service areas to minimize the distance of water mains to the service locations. This helps to minimize friction loss during water transportation over the mains.

8.1.2.2. Types of Water Distribution Systems

There are four primary types of water distribution systems:

8.1.2.2.1. Dead End System

The term "dead-end system" comes from the presence of dead ends within the pipe system, which prevents continuous water flow. This design divides the pipe network into multiple subnetworks.

Examples of these include the primary line, secondary lines, branch lines, and service connections. Initially, a major line is constructed through the center of the city or region.

Submains are established on both ends of the main line and further divided into branch lines to facilitate service connections. A shut-off valve is placed at the start of each sub-main line to control the flow for maintenance purposes.

The network diagram will have a tree-like structure; hence it is referred to as a tree system. It is often used in old cities that lack proper planning. However, this system is currently inadequate.

Advantages:

- This system is cost-effective.
- Installing pipes does not require the use of skilled labor.
- With fewer valves, it is easier to calculate the discharge amount and pressure.

Disadvantages:

- In rural places, the pressure is erratic and severely low.
- Water stagnation resulting from dead ends can lead to silt buildup. Adding more scour valves at dead ends can help to remove these sediments and improve overall efficiency.
- If there is any damage on the branch line, the entire section will need to be stopped in order to conduct repairs and prevent inconvenience to other users on the sub-main line.
- Limited discharge is provided for firefighting in this setup.

8.1.2.2.2. Radial System

The land is physically sectioned into zones using a radial technique. A main water pipeline is linked to a central distribution reservoir or storage tank. Supply pipes then extend radially from the distribution reservoir to the individual houses.

Advantages:

- This water distribution technology is well-suited for tall buildings.
- It is rare to experience water supply disruptions while maintenance or repairs are being carried out.
- The radial system efficiently delivers water at a high flow rate while minimizing head loss.
- This system delivers quick service.

Disadvantages:

- The project's cost has risen because of the growing number of individual distribution reservoirs.

8.1.2.2.3. Ring Distribution System

The water distribution system is encompassed by a radial or rectangular main pipeline, as shown in the diagram. The sub-main pipeline serves the smaller sections. If there is a system failure, only a small area will experience the impact.

Water can be supplied to the area in front of the damaged section from other points within the system. The ring distribution system requires a larger quantity of valves. In this setup, water can be obtained from two different directions.

Advantages:

- The discharge rate is significant when compared to other distribution systems.
- It is capable of conducting maintenance and repairs without disrupting the flow of water.
- There is minimal head loss due to the reduced number of links.
- There is little to no water stagnation due to the absence of dead ends.

Disadvantages:

- Longer and wider pipes are needed. Additional shutdown valves are required. Skilled workers are vital for pipeline installation.

8.1.2.2.4. Grid Iron System

The primary pipeline, the subaqueous

pipeline, and the secondary pipelines are linked together in a grid formation known as a grid-iron system. Gridiron water distribution systems are also referred to as interlaced and reticulated systems. To ensure consistent water pressure, the total length of the pipeline needs to be increased as the number of connections grows.

A gridiron water distribution system is perfect for contemporary, well-organized cities as it features main pipelines and branches that are rectangular in shape.

Advantages:

- Water flows constantly through pipes because there are no dead ends.
- It is capable of conducting maintenance and repairs without disrupting the flow of water.
- This water distribution system meets the requirements for fire suppression discharge regulations.
- The linking of pipes minimizes head loss as much as possible.

Disadvantages:

- Due to the flow recirculation coming from all directions, the pipes required for this system must be of substantial diameter and extended length.
- Since we are unable to determine the pipe's discharge, velocity, or pressure, the design could potentially be simplified.
- Professional laborers will do pipe laying, which will be more expensive.
- This system requires additional shutdown valves to be installed.

8.1.2.3. The Integrity of the Systems

The systems' integrity is categorized into physical, hydraulic, and water quality aspects. Physical integrity focuses on the barriers' ability to prevent outside contaminants from entering water distribution systems, which can deteriorate due to either physical or chemical factors.

Hydraulic integrity refers to the capacity to uphold sufficient water pressure within pipes across distribution systems, as well as the flow and duration of water movement within the system. These factors can affect the effectiveness of disinfectants used in water treatment.

Maintaining the integrity of water quality involves monitoring and preventing degradation as water moves through distribution systems. Factors such as physical damage or hydraulic issues can impact water quality. Degradations may occur within distribution systems, leading to issues like microorganism growth, nitrification, and internal corrosion of pipes.

8.1.2.3.1. Network Analysis and Optimization

Analyzes are conducted to aid in the design, operation, maintenance, and optimization of water distribution systems. Two primary types of analyzes are performed: hydraulic analyzes, which focus on the flow of water through the system, and water quality analyzes, which examine the behavior of water as it moves through the distribution system. [Mala-Jetmarova, Helena; Barton, Andrew; Bagirov, Adil (2015 Designing water distribution networks is a complicated process that involves optimizing various factors. Many methods have been suggested for this task, with a focus on metaheuristics (De Corte Annelies; & Sörensen & Kenneth,

2013). Using mathematical optimization techniques can result in significant cost savings in the construction of these types of infrastructures (Aragones David, Calvo Gabriel, & Galan Alvaro, 2021).

8.1.2.4. Methods of Water Distribution System

In order to have a functioning distribution system, it is important that different locations have sufficient water pressure. The distribution system receives water through various methods, depending on the source's elevation, topography, and other local conditions.

- **Gravity Supply:** Gravity is used to move water from its origin to the area where it is distributed in this water distribution method. The source of water, which could be a reservoir or lake, needs to be located at a higher elevation than the distribution area for this method to work effectively. Since no pumps are needed, this technique is both energy-efficient and easy to maintain. Additionally, there are no sudden changes in pressure within the system, ensuring a consistent flow of water.
- **Pumped Supply:** Pumps are used to push water into the system when the source is at a lower elevation than the distribution zone. This method is necessary when the source cannot provide enough pressure for the distribution system to work properly. However, this approach can be costlier due to the electricity consumption and regular maintenance required for the pumps.
- **Combined Supply:** This technique utilizes both gravity and pump force by using pumps and storage reservoirs to distribute water to the designated area. It is suitable for situations where there are two water sources available, when an elevated tank is connected to the system, or when the water source is located lower than the consumer area. While this method offers a more stable and dependable water supply, it is also more costly and requires greater maintenance.

8.1.2.5. Equipment Used in Water Distribution System

8.1.2.5.1. Infrastructure

In this discussion, we will briefly cover the different components of a drinking water distribution system, such as pipes, pumps, meters, storage tanks, valves, reservoirs, couplings, and other hydraulic elements. These elements connect treatment facilities or well-supplied to consumers' faucets, forming the infrastructure of the distribution system. We will also touch on the characteristics, typical maintenance needs, and important qualities of these fundamental components.

8.1.2.5.2. Industrial Valves

There are two common types of valves used in water distribution systems: isolation valves (or stop/shutdown valves) and control valves. Isolation valves, such as gate or butterfly valves, are utilized to isolate specific sections for maintenance and repairs. These valves are strategically placed to minimize disruption to surrounding service areas (Figure 8.1).

Valve maintenance is a key responsibility for utilities, with many implementing regular valve-turning programs to ensure the proper functioning of their systems. It is recommended to turn each system valve at least once a year to ensure that water can be easily shut off or redirected when needed, especially during emergencies, and to prevent accidental valve closures.

Figure 8.1. *Industrial valves.*

Source: unsplash, https://www.dombor.com/designing-a-water-distribution-system/.

Control valves are utilized to regulate both flow and pressure within a distribution system. These valves are typically chosen based on the intended maximum and minimum flow rates, as well as the differences in pressure upstream and downstream, and the velocity of the flow. Control valves consist of pressure-reducing, pressure-sustaining, and pressure-relief valves, flow-control valves, throttling valves, float valves, and check valves. The majority of industrial valves are typically constructed of steel or cast iron, with the exception of premise plumbing valves that are often made of brass for easy shut-off during repairs.

Unlike the smaller pipes, the transmission mains in the distribution system are spread further apart. Additionally, blow-off valves and air-release/vacuum valves are important components used for flushing water mains and releasing entrained air in a water system.

8.1.2.5.3. Pipes

Pipes that transport water from a source like a treatment facility to the customer are typically categorized as transmission or trunk mains, distribution mains, service lines, and premise plumbing. Transmission or trunk mains are designed to carry large volumes of water over extensive distances, moving water from a treatment facility to a storage tank in a distribution system. Transmission mains are typically larger in diameter compared to distribution mains, which commonly run along city streets. Service lines deliver water from the distribution main to individual buildings or properties.

Service lines are designed according to the utility's specifications to ensure the desired pressure is maintained at the customer's property for anticipated water usage. These lines can vary in size based on the customer's water needs. Premise plumbing, on the other hand, refers to the internal plumbing within a building that distributes water to various points of use. The pipe sizes used in premise plumbing are generally smaller, which increases the surface-to-volume ratio compared to traditional distribution system pipes.

Today, the most commonly used materials for pipes are steel, ductile iron, pre-stressed concrete, polyvinyl chloride (PVC), and reinforced plastic, as they are essential for meeting the three requirements of delivering the correct amount of water, withstanding external and internal forces, and ensuring durability and longevity. This study is unique in that it considers service lines and premise plumbing as integral components of the distribution system, examining their impact on the quality of drinking water. When accounting for premise plumbing, the total length of the distribution system increases significantly, from approximately 1 million miles to over 6 million miles.

Residential plumbing and service lines experience longer residence times, lower flow conditions, more stagnation, and higher temperatures compared to the main distribution system. Due to the changing ownership of these lines, determining responsibility for their maintenance can be complicated. Therefore, it is not common for premise plumbing and service lines to be included in a public water supply distribution system. Most drinking water utilities and regulatory bodies are only accountable for water delivered to the curb stop, which may not cover the entire service line. The building owner is responsible for the portion of the service line that is not controlled by the utility, as well as the entire plumbing system on the property.

8.1.2.5.4. Pipe-Network Configurations

Water distribution systems are typically divided into branches and grids/loops. In a branched system, smaller pipes branch off from larger pipes to distribute water in a single route from the source to the consumer, similar to the branches of a tree. This technology is commonly used in rural areas. In large municipal regions, the most common configuration is a looped/grid system, with interconnected pipes forming loops throughout the area to supply water. This type of system allows water to follow multiple pathways to reach the consumer.

Looped systems offer a high level of reliability in the event of a line failure because the issue can be isolated without affecting users further than the immediate surrounding area. Additionally, looping helps prevent water stagnation problems, such as reactions with pipe walls and enhances firefighting capabilities. However, loops can also create dead ends, particularly in suburban areas like cul-de-sacs, leading to water quality concerns. Most systems typically have a combination of looped and branched sections.

8.1.2.5.5. Pumps

Pumps are equipment that increases the energy of water to raise its height or pressure. Typically constructed of steel or cast iron, pumps in distribution systems are typically centrifugal, with water entering through an intake pipe and being expelled outward through vanes and output pipes. The electricity cost for pumping is often one of the highest expenses in maintaining a water supply.

8.1.3. Design a Water Distribution System

Water distribution systems are crucial infrastructures that play a vital role in providing water to consumers. The main purpose of water distribution design is to ensure that a suitable amount of water is delivered at the right pressure. When designing, constructing, and maintaining water delivery systems, factors like environmental preservation, natural resource conservation, and cost-effectiveness should also be considered.

- **Water Distribution Systems Based on Gravity and Pressure:** There are two types of water distribution systems: gravity and pressure. Gravity systems distribute water from a reservoir at a higher elevation to the distribution area, while pressure systems use pumps to create the necessary water pressure. Typically, water systems include both gravity and pressure components. Gravity systems are generally more affordable to maintain.

- **Water Distribution Network Topology:** In the structure of the water delivery system, we categorize networks as either branching or annular. A branch network distributes water through one or more distribution lines that end at designated locations, while an annular network allows water to be delivered from multiple routes to a single destination. Water supply networks may include both annular and branching elements. One challenge of a branched network is the reduced flow rate of water at the extremities of the branches.
- **Design Objective:** In addition to meeting the water needs of industry and households, water distribution systems must also conform to fire prevention regulations. These requirements can result in a substantial increase in the cost of constructing and upkeeping water distribution networks. The design of water delivery systems follows the same principles and project categories as other types of construction projects. Project documentation typically includes a conceptual solution, conceptual design, main project, or implementation project. Each project type includes details on the concept, content, and specifics of the design.
- **Variables Influencing Water Supply System Design:** The design of water distribution systems is greatly impacted by spatial variables. This includes how geography affects pressure distribution within the pipeline network. Additionally, construction costs are influenced by factors such as land usage and ownership, while excavation costs are influenced by soil spatial features. The installation of water supply lines along existing or proposed highways may also present challenges with existing communal infrastructures.

8.1.4. Maintenance

8.1.4.1. Internal Corrosion Control

The degradation of water quality is caused by corrosion of metal pipe surfaces and connections within distribution systems. This corrosion can manifest in the water as changes in color, taste, and odor, all of which may pose health risks. The presence of trace metals like lead, copper, or cadmium in the water can lead to various health concerns. For instance, exposure to lead can result in developmental delays in children. Long-term exposure to copper may lead to liver and kidney damage, while high or prolonged exposure to cadmium can cause harm to multiple organs. The corrosion of iron pipes can result in rusty or red water, while corrosion of zinc and iron pipes may introduce a metallic taste to the water.

There are several methods available to prevent internal corrosion. These include adjusting the pH levels, controlling carbonate and calcium levels to create a calcium carbonate coating on the pipe surface, and using corrosion inhibitors. One example of a corrosion inhibitor is phosphate products that form films on the pipe surface, reducing the risk of trace metals leaching into the water.

8.1.4.2. Hydrant Flushing

Hydrant flushing involves releasing water from fire hydrants or special flushing hydrants to remove mineral deposits from the water main. This process also tests the pressure of water supplied to fire hydrants for fire fighting. Consumers may observe a rust color in their water during flushing due to stirred-up iron and mineral deposits.

8.1.4.3. Water Main Renewals

As water mains age, they experience structural, water quality, and hydraulic degradation. Structural decay can result from various factors, such as internal and external corrosion in metal pipes leading to thinning or degradation of the pipe walls, potentially causing leaks or bursts. Cement pipes are also susceptible to deterioration of the cement matrix and reinforced steel. Additionally, joint failures can occur in all types of pipes.

Water quality degradation can manifest in various ways, including scaling, sedimentation, and the formation of biofilms. Scaling occurs when hard deposits accumulate on the inner walls of pipes, often caused by pipe corrosion and the presence of calcium in the water, known as tuberculation. Sedimentation refers to the settling of solids in the pipes, typically at spots where scaling has built up. Disruption in water flow velocity, such as sudden usage of a fire hydrant, can disturb the settled solids and lead to discoloration of the water. Biofilms can develop in pipes with rough surfaces and heavy scaling, providing a breeding ground for bacteria to thrive. The rougher the interior wall of the pipe, the more challenging it becomes for disinfectants to effectively eliminate bacteria. Additionally, hydraulic deterioration, which impacts water pressures and flows, can occur as a result of other forms of deterioration that impede the flow of water.

When selecting the method for water main renewal, there are several options to consider. These include open-trench replacement as well as various pipeline rehabilitation methods such as pipe bursting, sliplining, and pipe lining.

Using an in-situ rehabilitation method offers the advantage of lower costs since there is no need to excavate the entire water main pipeline. Instead, only small pits are dug to access the existing water main. However, this method requires the construction of a temporary water bypass system to supply water to the affected area during the rehabilitation process. This temporary bypass system, also known as temporary bypass piping, must be carefully designed to ensure a continuous water supply to customers in the project area. Water is drawn from a feed hydrant into the temporary pipe, which is covered when it crosses driveways or roads to allow for vehicular traffic. Temporary service connections can be made by connecting a garden hose to the temporary pipe, and fire hydrants for fire protection should also be included in the temporary system.

Since water main work has the potential to disrupt lead service lines, leading to increased lead levels in drinking water, it is advised that water utilities coordinate with property owners to replace these lines during water main renewal projects.

To create an efficient, reliable, and safe water distribution system, engineers must carefully consider various factors including water pressure, flow rate, system structure, and regulatory compliance. Understanding the fundamental elements of design is essential for ensuring the success of the system.

8.2. WASTEWATER COLLECTION AND TREATMENT

Various industries produce wastewater streams that contain organic compounds. These streams typically undergo a series of processes- including collection, contaminant treatment, and storage- before being released into either a body of water or a municipal treatment plant. In some cases, the wastewater is exposed to the environment, allowing volatile organic compounds (VOC) to be emitted into the atmosphere.

8.2.1. Collection Systems

Wastewater collection systems are specifically engineered to gather and transport wastewater from residential, commercial, and industrial sources. This water contains both dissolved waste and solid particles. To effectively transport the wastewater, the collection system must be designed to maintain a velocity of around 2.5 feet per second. Excessive or inadequate velocities can hinder the system's functionality and upkeep. The collected wastewater and solids are then transported to a treatment facility where pollutants are removed prior to discharge into a body of water. Typically, the wastewater collection system relies on gravity to transport wastewater, following the natural slope of the land. Water quality treatment facilities are usually situated in low-lying areas near water bodies to facilitate this process. However, when the terrain does not allow for wastewater to flow to the treatment plant by gravity alone, the use of lift stations or pumps becomes necessary. These stations lift the wastewater from lower points to higher points, allowing it to continue its gravity flow journey. In areas with hilly or rolling landscapes, multiple lift stations may be needed to transport wastewater to the treatment plant. Furthermore, the farther the distance the wastewater needs to travel to reach the treatment plant, the higher the likelihood of requiring multiple lift stations for the journey.

8.2.2. Wastewater Treatment and Disposal

The process of wastewater treatment involves eliminating contaminants from sewage or wastewater before it is released into aquifers or natural water sources like rivers, lakes, estuaries, and oceans. Because truly pure water does not exist in nature, determining whether water is clean or polluted depends on the presence and levels of impurities, as well as the intended purpose of the water.

Water is considered polluted when it has enough impurities to make it unsuitable for specific purposes like drinking, swimming, or fishing. While natural conditions can impact water quality, pollution typically stems from human activities. The primary cause of water pollution is the release of contaminated wastewater into surface water or groundwater, making wastewater treatment a key component of pollution control efforts.

The size and capacity of wastewater treatment systems are based on the amount of sewage generated by residential, commercial, and industrial properties connected to sewer systems, as well as the expected inflows and infiltration (I&I). The type of treatment plant – whether on-lot, clustered, or centralized – is determined by factors like the number of customers served, geographical considerations, site limitations, sewer connections, flow rates, influent characteristics, regulatory requirements, feasibility of technology, energy usage, and operation and maintenance expenses.

In large cities and towns, the most common way of disposing of wastewater is by releasing it into a surface water body. Suburban and rural areas, on the other hand, tend to rely more on underground disposal methods. In both cases, wastewater needs to undergo some form of purification or treatment in order to safeguard public health and maintain water quality. This includes removing suspended particles and biodegradable organic materials to different degrees, as well as destroying harmful bacteria. Additionally, it may be necessary to eliminate nitrates and phosphates (which serve as plant nutrients), as well as neutralize or eliminate industrial waste and toxic chemicals. The treatment of wastewater is determined by local environmental conditions and government regulations. There are two main types of standards that dictate the level of treatment required: stream standards and effluent standards. Stream standards focus on maintaining water quality by setting limits on specific pollutants in streams, rivers, and lakes based on their intended use. Parameters regulated by stream standards include dissolved oxygen, coliforms, turbidity, acidity, and toxic substances. Effluent standards, on the other hand, concentrate on the quality of treated wastewater discharged from sewage treatment plants, controlling factors such as biochemical oxygen demand (BOD), suspended solids, acidity, and coliforms.

There are three stages of wastewater treatment: primary, secondary, and tertiary (or advanced). Primary treatment eliminates approximately 60% of total suspended solids and about 35% of BOD; dissolved impurities are not eliminated. It is generally utilized as the initial step before secondary treatment. Secondary treatment eliminates over 85% of both suspended solids and BOD. A basic level of secondary treatment is typically mandatory in the United States and other developed nations.

Tertiary treatment methods are employed when the removal of over 85% of total solids and BOD, or the reduction of dissolved nitrate and phosphate levels is necessary. These processes can eliminate over 99% of impurities from sewage, resulting in an effluent that is nearly of drinking water quality. However, tertiary treatment can be costly, often doubling the expenses of secondary treatment, and is therefore only utilized in specific situations.

Before sewage effluent can be discharged into a body of surface water, disinfection is necessary to eliminate any remaining pathogens and ensure public health is protected. Typically, this is achieved by mixing the effluent with chlorine gas or liquid hypochlorite chemicals in a contact tank for a minimum of 15 minutes. However, to prevent any adverse effects on aquatic life, a dechlorination chemical may need to be added. Ultraviolet radiation is also emerging as a competitive alternative to chlorine for wastewater disinfection as it can effectively disinfect without leaving any residual in the effluent.

8.2.2.1. Primary Treatment

The initial stage of treatment eliminates materials that are capable of floating or settling out naturally through gravity. This process involves screening, grinding, removing grit, and allowing sedimentation to occur. Screens consist of thin, closely arranged metal bars that prevent large debris like wood and rags from causing blockages in pipes or pumps. Nowadays, screens are cleaned automatically, and the collected material is promptly buried on the plant premises. A comminutor may be utilized to break down debris that passes through the screens, and the shredded material is later eliminated through sedimentation or flotation methods. Grit chambers are elongated tanks created to decrease the speed of water flow, allowing heavy solids like sand, coffee grounds, and eggshells to sink to the bottom. Grit can damage pumps and other equipment in wastewater treatment plants, making its removal crucial. This process is especially important in urban areas with combined sewer systems that often carry large amounts of silt, sand, and gravel washed off streets or land during storms.

Solid materials that are not caught by screens and grit chambers are separated from sewage in sedimentation tanks, also known as primary clarifiers. Gravity allows for settling to occur over a period of about two hours as the sewage slowly passes through these tanks. During this process, the solids settle to the bottom of the tank. The accumulated solids, referred to as raw or primary sludge, are then pushed along the tank bottom by mechanical scrapers. The sludge is collected in a hopper and pumped out for disposal. Additionally, mechanical surface-skimming devices are used to remove floating materials such as grease from the sewage.

8.2.2.2. Secondary Treatment

The secondary treatment process eliminates the soluble organic material that was not captured in the primary treatment, as well as a higher proportion of suspended solids. This is typically achieved through biological processes where microorganisms consume the organic pollutants, turning them into carbon dioxide, water, and energy for their growth and reproduction.

The sewage treatment plant creates an appropriate setting, though it may be made of steel and concrete, for the natural biological process to take place. By removing soluble organic matter at the treatment plant, it helps maintain the dissolved oxygen balance of a stream, river, or lake where the treated water is released. The three main biological treatment methods used are the trickling filter, activated sludge process, and oxidation pond. Another less frequently used method is the rotating biological contactor.

8.2.2.2.1. Trickling Filter

A trickling filter consists of a tank filled with a deep bed of stones, where settled sewage is continuously sprayed over the top. The sewage trickles down to the bottom of the tank, where it is collected for additional treatment. Bacteria gather and multiply on the stones as the wastewater flows down, helping to lower the biochemical oxygen demand (BOD) of the sewage by absorbing dissolved organics. Circulating air among the stones provides the necessary oxygen for metabolic processes to occur. After passing through the trickling filters, the wastewater flows into settling tanks known as secondary clarifiers. These tanks are designed to remove microbes that are dislodged from the rocks during the treatment process. To improve treatment efficiencies, multiple trickling filters can be connected in series and sewage may be recirculated.

8.2.2.2.2. Activated Sludge

The activated sludge treatment system includes an aeration tank and a secondary clarifier. Sewage that has settled, along with fresh sludge from the secondary clarifier, is added to the aeration tank. Compressed air is then pumped into the mixture using porous diffusers at the tank's bottom. This creates bubbles that rise to the surface, providing oxygen and mixing the contents quickly. Additionally, mechanical mixers resembling propellers at the tank's surface can churn to add more air.

In an oxygen-rich environment, microorganisms flourish, creating a lively mixture of biological solids known as activated sludge, primarily composed of bacteria. The aeration tank allows the microbes to absorb dissolved organics from the sewage for approximately six hours, which helps reduce the BOD. The activated sludge then moves into the secondary clarifier, where it settles out by gravity.

The clear water is collected from the top of the clarifier, treated, and released as secondary effluent. The sludge is removed from the bottom of the tank and about 30% of it is mixed back into the aeration tank with the primary effluent. This recycling process is crucial in the activated sludge method as the microbes are accustomed to the sewage environment and efficiently break down organic materials. The other 70% of the sludge needs to be properly treated and disposed of.

Different types of the activated sludge process include extended aeration, contact stabilization, and high-purity oxygen aeration. Extended aeration and contact stabilization systems do not require the primary settling step, making them ideal for treating smaller sewage flows from motels, schools, and other isolated wastewater sources. These treatments are typically found in prefabricated steel tanks known as package plants. Oxygen aeration systems combine pure oxygen with activated sludge, resulting in a higher concentration of oxygen that can reduce aeration time from six hours to two hours and decrease the needed tank volume.

8.2.2.2.3. Oxidation Pond

Oxidation ponds, also known as lagoons or stabilization ponds, are large, shallow bodies of water that are specifically designed for the treatment of wastewater. The treatment process involves the interaction of sunlight, bacteria, and algae. Algae use energy from the sun along with carbon dioxide and inorganic compounds released by bacteria in the water to grow. Through the process of photosynthesis, algae release oxygen which is essential for aerobic bacteria. In some cases, mechanical aerators are used to increase the oxygen levels in the pond and therefore reduce the required size of the pond. Eventually, the sludge that accumulates in the pond must be removed through dredging. Any remaining algae in the pond effluent can be eliminated through filtration or a combination of chemical treatment and settling.

8.2.2.2.4. Rotating Biological Contactor

In this treatment system, a horizontal shaft is fitted with a series of large plastic disks that are partially submerged in primary effluent. As the shaft turns, the disks are alternately exposed to air and wastewater, creating an environment for bacteria to grow and break down the organics in the wastewater.

8.2.2.3. Tertiary Treatment

If the receiving water is particularly sensitive to pollution, additional treatment of secondary effluent may be necessary using various tertiary processes.

8.2.2.3.1. Effluent Polishing

Effluent polishing is a proven method for removing extra suspended solids and BOD from secondary effluent. This process typically involves granular media filters similar to those used in purifying drinking water. Prefabricated polishing filters are commonly used, with tanks positioned above the filters to hold backwash water. Alternatively, microstrainers like those used in municipal water treatment can also be utilized for effluent polishing in wastewater treatment.

8.2.2.3.2. Removal of Plant Nutrients

When treatment standards mandate the extraction of plant nutrients from sewage, it is typically carried out during the tertiary stage. Phosphorus in wastewater is commonly found in organic compounds and phosphates, which can be effectively removed through chemical precipitation. However, this process results in a higher volume and weight of sludge. Nitrogen, another essential plant nutrient, exists in sewage as ammonia and nitrates. Ammonia is harmful to fish and contributes to oxygen depletion in water bodies when converted to nitrates. Nitrates, like phosphates, fuel algae growth and the eutrophication of lakes. Nitrification-denitrification is a method used to eliminate nitrates, involving a two-step biological process where ammonia nitrogen is first converted to nitrates by microorganisms, which are then metabolized by another type of bacteria to produce nitrogen gas that is released into the atmosphere. This method necessitates the installation of additional aeration and settling tanks, substantially raising treatment costs.

One method to remove ammonia from sewage is ammonia stripping, a physicochemical process. Ammonium ions are converted into ammonia gas by adding chemicals. The sewage is then cascaded through a tower, allowing the gas to escape from the solution and into the air. While stripping is more cost-effective than nitrification-denitrification, it is not as efficient in cold temperatures.

8.2.2.3.3. Land Treatment

In certain areas, secondary effluent can be directly applied to the ground to produce polished effluent through natural processes as it passes over vegetation and seeps through the soil. Land treatment can be categorized into three types: slow rate, rapid infiltration, and overland flow.

Effluent is applied onto the land using the slow-rate method, where it is spread out in ridge-and-furrow patterns or by sprinkler systems. The water and nutrients are absorbed by the roots of vegetation. In the rapid infiltration method, wastewater is stored in recharge basins and mostly percolates into the groundwater, with little being absorbed by vegetation. Highly permeable soils are necessary for this method to be effective. In the overland flow method, wastewater is sprayed onto a sloped vegetated terrace and flows slowly to a collection ditch. Purification is achieved through physical, chemical, and biological processes before the water is discharged into a nearby stream.

Using land for sewage treatment can help supply moisture and nutrients for plants like corn or grains used in animal feed. It can also help refresh groundwater sources. Overall, land treatment allows sewage to be reused in a positive way. However, this method requires a lot of space and may not be as practical in certain areas due to soil type and climate.

8.2.2.3.4. Clustered Wastewater Treatment Systems

In situations where connecting individual homes or units to public sewer systems is not possible, communities may choose to use a clustered wastewater treatment system. These facilities are smaller than centralized treatment plants and only serve a small number of connections. The technology used for clustered wastewater treatment can be similar to that used in centralized systems or individual on-site systems, depending on the specific needs and level of treatment required. Once treated, effluent from clustered wastewater systems can be disposed of through surface or subsurface methods.

8.2.2.3.5. On-site Septic Tanks and Leaching Fields

In suburban or rural areas with low population density, it is often not cost-effective to construct sewage collection systems and a centralized treatment facility. As a result, individual treatment and disposal systems are typically installed for each household. These on-site systems offer efficient and affordable long-term solutions for wastewater disposal, as long as they are designed, installed, and maintained correctly. Approximately one-third of private homes in the United States utilize on-site subsurface disposal systems.

The most common on-site system consists of a buried, watertight septic tank and a subsurface absorption field, also known as a drain field or leaching field. The septic tank acts as a primary sedimentation and sludge storage chamber, removing most of the settleable and floating material from the incoming wastewater. Though the sludge decomposes anaerobically, it eventually accumulates at the bottom of the tank and needs to be pumped out every two to four years. A baffle at the tank outlet traps floating solids and grease, allowing settled sewage to flow into the absorption field. From there, the wastewater percolates downward into the ground, slowly passing through layers of soil. During this process, the settled wastewater undergoes further treatment and purification through physical and biological processes before reaching the water table.

An absorption field consists of multiple perforated pipelines that are positioned within long, shallow trenches containing gravel. These pipes evenly distribute the wastewater across a large area, allowing it to filter through the gravel and into the layers of soil beneath. In cases where there is limited space for a traditional leaching field, deeper seepage pits can be utilized in place of shallow trenches. Seepage pits require less land space than leaching fields. It is important that both leaching field trenches and seepage pits are situated above seasonal high groundwater levels.

In order for on-site wastewater disposal to be successful, the soil must have an appropriate level of permeability, also known as hydraulic conductivity. If the soil is not permeable enough, the effluent may not flow effectively and could leak onto the surface, posing a risk to public health. Conversely, if the permeability is too high, the effluent may not be adequately purified before reaching the water table, potentially contaminating the groundwater. The ability of the soil to absorb wastewater is determined primarily by its texture, including the proportions of gravel, sand, silt, and clay. Permeability can be assessed by inspecting the soil in excavated test pits and by conducting a percolation test, also known as a "perc test." This test measures how quickly water infiltrates the soil in small holes dug on the disposal site. The perc rate obtained from the test can help determine the size of the required absorption field or the number of seepage pits needed.

In cases where unfavorable conditions prevent the use of absorption fields and seepage pits for on-site sewage disposal, mound systems can be utilized instead. Mound systems involve building an absorption field above the natural ground surface to allow for proper percolation and to keep the drain field separated from the water table. The septic tank effluent is pumped periodically from a chamber and distributed onto the mound. Other options for on-site sewage disposal include intermittent sand filters or small, prefabricated aerobic treatment units. If the effluent is discharged into a nearby stream, disinfection (typically through chlorination) is necessary to meet safety regulations.

8.2.2.3.6. Wastewater Reuse

As urban populations grow and fresh water becomes scarce, wastewater can offer a valuable solution in cities or towns. Recycling wastewater not only helps alleviate the pressure on limited freshwater sources but also benefits the environment by decreasing the amount of harmful discharge into streams and lakes. Wastewater can be treated and reused for activities like irrigating crops, replenishing groundwater, or even for recreational purposes. While using reclaimed water for drinking is feasible, it often encounters resistance from the public.

There are two main methods of wastewater reuse: direct and indirect. Direct reuse involves sending treated wastewater directly into a water system without diluting it in a natural body of water first. An example of this is using treated wastewater from a municipal treatment plant to irrigate a golf course. Indirect reuse, on the other hand, involves mixing reclaimed wastewater with another body of water before using it. For instance, any community that uses surface water downstream from another community's treatment plant discharge pipe is indirectly reusing wastewater. Indirect reuse can also take place by injecting reclaimed wastewater into a groundwater aquifer and withdrawing it later for use. This process, known as artificial recharge, can be done through deep-well injection or shallow surface spreading.

As the likelihood of direct human contact and ingestion of reclaimed wastewater increases, the quality and treatment standards become more strict. The specific impurities that need to be eliminated depend on the intended use of the water. For instance, if the water is going to be used for landscape irrigation, removing phosphates or nitrates may not be necessary. However, if the goal is to use the water as a potable supply, a tertiary treatment process with multiple barriers against contaminants is mandatory. This could involve secondary treatment, granular media filtration, ultraviolet radiation, granular activated carbon adsorption, reverse osmosis, air stripping, ozonation, and chlorination. The implementation of gray-water recycling systems in newly constructed commercial buildings provides a way to conserve water and decrease overall sewage levels. These systems effectively filter and treat water from bathtubs and sinks, then repurpose it for non-drinking uses, such as flushing toilets and urinals. To prevent confusion, recycled water may be colored with a blue dye to indicate that it is not safe for consumption.

8.2.3. Sludge Treatment and Disposal

The remaining material found in sewage treatment plants, known as sludge or biosolids, is a mixture of solid, semisolid, or slurry residual substances that are created during the wastewater treatment process. This residue is typically categorized as either primary or secondary sludge. Primary sludge is the result of chemical precipitation, sedimentation, and other initial processes, while secondary sludge consists of activated waste biomass from biological treatments. In some cases, sewage plants may also receive septage or solids from septic tanks used in household wastewater treatment systems. Often, these different types of sludges are combined for further processing and disposal.

The treatment and disposal of sewage sludge are important considerations in the planning and functioning of wastewater treatment facilities. Two main objectives of treating sludge prior to disposal are decreasing its volume and stabilizing the organic materials. By stabilizing the sludge, unpleasant odors are eliminated, making it easier to manage without posing a health or environmental risk. Additionally, reducing the volume of sludge lowers the expenses related to pumping and storage.

8.2.3.1. Treatment Methods

The process of treating sewage sludge may involve a mix of thickening, digesting, and dewatering methods.

8.2.3.1.1. Thickening

The initial step in sludge treatment is typically thickening, as thin sludge is difficult to manage. Thickening is typically carried out in a gravity thickener tank, which can decrease the sludge volume to less than half of its original size. Another option for thickening is dissolved-air flotation, where air bubbles transport the solids to the surface, forming a layer of thickened sludge.

8.2.3.1.2. Digestion

The process of sludge digestion involves breaking down organic solids into stable substances through biological means. This reduces the mass of solids, eliminates harmful pathogens, and simplifies the process of dewatering or drying the sludge. After digestion, the sludge becomes odorless and resembles nutrient-rich soil suitable for gardening.

Most large sewage treatment plants utilize a two-stage digestion system in which bacteria metabolize organics anaerobically, without oxygen. In the initial stage, sludge with a dry solids content of approximately 5% is heated and mixed in a sealed tank for a few days. Acid-forming bacteria break down large molecules like proteins and lipids into smaller, water-soluble molecules, which are then fermented into various fatty acids. The sludge is then transferred to a second tank, where other bacteria convert the dissolved matter into biogas, consisting of carbon dioxide and methane. The methane can be used as a fuel to heat the first digestion tank and produce electricity for the plant.

Anaerobic digestion is highly influenced by temperature, acidity, and various other factors. It necessitates precise supervision and regulation. In some instances, additional hydrolytic enzymes are added to the sludge at the start of the initial digestion phase to enhance the bacteria's activity. Research has shown that this enzymatic treatment is capable of eliminating more harmful pathogens in the sludge and can also lead to increased production of biogas during the second phase of digestion.

An additional improvement to the traditional two-stage anaerobic digestion process involves thermal hydrolysis, where large molecules are broken down using heat before digestion begins. In this process, sludge with a dewatered DS content of around 15% is mixed with steam in a pulper. This mixture is then heated and pressurized at approximately 165 °C (330 °F) for about 30 minutes in a reactor to complete hydrolytic reactions. After this, some steam is released, and the sludge is suddenly transferred to a "flash tank" under pressure, causing the cell walls of solid matter to burst. The hydrolyzed sludge is then cooled, slightly diluted with water, and sent directly to the second stage of anaerobic digestion.

Sludge digestion can also occur aerobically, meaning in the presence of oxygen. The sludge is aerated vigorously in an open tank for approximately 20 days. Methane gas is not produced during this method. While aerobic systems are simpler to manage than anaerobic systems, they typically require higher operating costs due to the energy needed for aeration. Aerobic digestion is commonly integrated with small extended aeration or contact stabilization systems.

Both aerobic and traditional anaerobic digestion processes can convert approximately 50% of organic sludge solids into liquids and gases. However, by implementing thermal hydrolysis before anaerobic digestion, it is possible to convert 60–70% of the solid matter into liquids and gases. This method not only reduces the volume of solids generated compared to conventional digestion but also increases biogas production. This excess biogas can potentially enable some wastewater treatment plants to generate enough energy to become self-sufficient.

8.2.3.1.3. Dewatering

Digested sewage sludge is typically dehydrated before being disposed of. Even though dehydrated sludge still contains a significant amount of water, usually around 70%, it no longer behaves like a liquid and can be handled as a solid material. The simplest method of dehydrating sludge is through sludge-drying beds. The sludge is spread on a bed of sand and left to dry through a combination of evaporation and gravity drainage. A network of pipes underneath the sand collects the water, which is then pumped back to the plant. After approximately six weeks of drying, the resulting sludge cake, with a solids content of around 40%, can be easily removed with a pitchfork or front-end loader. To speed up the drying process in wet or cold weather, a glass enclosure can be built over the sand beds. Because drying beds require a large amount of land, this method is commonly used in rural or suburban areas rather than densely populated cities.

Instead of using sludge-drying beds, other options such as the rotary drum vacuum filter, centrifuge, and belt filter press can be used. These mechanical systems take up less space and provide more operational control compared to sludge-drying beds. However, they typically require sludge conditioning beforehand, where chemicals are added to the liquid sludge to help solidify particles and enhance drainage.

8.2.3.2. Disposal

Treated sewage sludge is typically disposed of on land as its final destination. Dewatered sludge can either be buried underground in a sanitary landfill or applied on agricultural land to enhance soil quality and serve as a fertilizer. However, due to the potential presence of harmful industrial chemicals, sludge is never spread on land used for growing crops intended for human consumption.

In urban areas where suitable land disposal sites are not available, sludge can be incinerated as an alternative. Incineration removes moisture and transforms organic solids into inert ash, which must then be properly disposed of. Although incineration reduces the volume of waste, air pollution control is crucial. To mitigate pollution, scrubbers and filters must be utilized during the incineration process.

Previously, dumping sludge in the ocean was a cost-effective way for coastal communities to dispose of waste. However, this practice is now banned in the United States and other coastal nations as it is no longer seen as a sustainable option.

8.3. AIR POLLUTION CONTROL MEASURES

Methods employed to reduce or eliminate harmful emissions into the atmosphere are known as air pollution control. This is a key component of overall pollution control efforts, alongside wastewater treatment, solid waste management, and hazardous waste management.

Polluted air is defined by the presence of certain substances at levels that can cause harm or negative effects over extended periods of time. These effects can impact human health, property, and visibility in the atmosphere. Pollution can come from both natural sources, like volcanic eruptions and forest fires, as well as human activities. However, only pollution from human activities, such as industry and transportation, can be regulated and reduced.

Most air pollutants stem from combustion processes. In the Middle Ages, the burning of coal for fuel caused recurring air pollution issues in London and other major European cities. The Industrial Revolution in the 19th century led to a greater reliance on fossil fuels, exacerbating the severity and frequency of air pollution events. The rise of gasoline-powered vehicles in the 20th century further worsened air quality in urban areas. It wasn't until the mid-20th century that significant efforts were made to regulate emissions from stationary and mobile sources to improve air quality on regional and local levels.

Stringent emission regulations have been put in place to address harmful pollutants, including greenhouse gases. The extensive and lasting impacts of these gases on atmospheric composition and climate have been acknowledged, leading to collaborative global initiatives to mitigate their effects. Examples of greenhouse gases include carbon dioxide, CFCs, methane, nitrous oxide, and ozone. In 2009, the U.S. Environmental Protection Agency declared that greenhouse gases were harmful to human health and may be regulated as air pollutants.

One effective method for preserving air quality is by decreasing the release of pollutants through the adoption of cleaner fuels and processes. Any pollutants that cannot be eliminated should be captured or contained by the use of suitable air purification equipment before they are able to be released into the environment. The following are descriptions of these devices.

8.3.1. Control of Particulates

There are several methods to extract airborne particles from a contaminated airstream. Equipment like cyclones, scrubbers, electrostatic precipitators, and baghouse filters are commonly used to gather fine particulates. After collection, the particulates stick together, creating agglomerates that can be easily removed from the equipment and properly disposed of, often in a landfill.

Due to the uniqueness of each air pollution control project, it is typically impossible to determine beforehand the most suitable particle-collection device (or combination of devices); control systems must be individually designed. Critical particulate factors that impact the selection of collection devices include corrosiveness, reactivity, shape, density, and particularly size and size distribution (the variety of particle sizes present in the airstream).

Additional design considerations include the characteristics of the airstream, such as pressure, temperature, and viscosity, as well as the flow rate, required removal efficiency, and permissible resistance to airflow. Cyclone collectors are commonly employed for controlling industrial dust emissions and serving as pre-cleaners for other collection devices. Wet scrubbers are typically utilized for handling flammable or explosive dust or mists in environments like industrial and chemical processing facilities and hazardous waste incinerators, where they can manage hot airstreams and adhesive particles. Electrostatic precipitators and fabric-filter baghouses are frequently found in power plant applications.

8.3.1.1. Cyclones

A cyclone uses a spiral path inside a cylindrical chamber to remove particulates from dirty air. The dirty air enters the chamber from the outer wall in a tangential direction, creating a vortex. Larger particulates, due to their greater inertia, move outward and are pushed against the chamber wall. Slowed down by friction with the wall, they then slide into a dust hopper at the bottom of the cyclone. The cleaned air flows upward in a narrower spiral through an inner cylinder and comes out from an outlet at the top. The collected dust in the hopper is removed regularly for disposal.

Cyclones excels at removing large particulates and can reach efficiencies of up to 90% for particles over 20 micrometers in size. However, cyclones alone are not enough to meet strict air quality regulations. They are commonly used as a first step in cleaning air and are often paired with more effective equipment like electrostatic precipitators and baghouses.

8.3.1.2. Scrubbers

Wet scrubbers are devices that capture airborne particles by coming into direct contact with a water or liquid spray. The scrubber effectively removes the particles from the polluted air by causing them to collide with and be carried away by the numerous small droplets in the spray.

Different types of wet scrubbers are currently being utilized. One example is the spray-tower scrubber, where an upward-flowing airstream is cleansed by water sprayed downward from a set of nozzles. The water is reused once it has been effectively purified to avoid nozzle blockage. Spray-tower scrubbers have the ability to eliminate around 90% of particulates that are larger than approximately 8 μm.

In both orifice scrubbers and wet-impingement scrubbers, the air and droplet mixture comes into contact with a solid surface, causing the droplets to break into smaller particles. This results in an increased surface area for contact. These scrubbers require less water recirculation and can effectively remove particles larger than 2 µm with an efficiency of approximately 90%.

Venturi scrubbers are known to be the most effective wet collectors, with the ability to achieve efficiencies greater than 98% for particles larger than 0.5 µm in size. The efficiency of a scrubber is determined by the speed difference between droplets and particulates. Venturi scrubbers are able to create high relative velocities by introducing water into the narrow throat of a venturi channel, which is a constriction in the airflow where air containing particles is moving at a high velocity.

8.3.1.3. Electrostatic Precipitators

Electrostatic precipitation is a commonly used technique for removing fine particulates from air streams. In this method, particles in the air are electrically charged upon entry into the unit and then captured by an electric field. The unit includes airflow baffles, discharge and collection electrodes, a dust cleanup system, and collection hoppers. High voltage direct current (DC) up to 100,000 volts is used to charge the particles, which are then drawn to oppositely charged collection electrodes where they are trapped.

In a usual setup, the collection electrodes consist of numerous large rectangular metal plates suspended vertically and aligned in parallel inside a box-shaped enclosure. There can be hundreds of plates with a total surface area of tens of thousands of square meters. Discharge electrode wires are positioned between the collection plates in rows, with the wires negatively charged and the plates grounded, resulting in a positive charge. Particles trapped on the collection plates are periodically removed by shaking or "rapping," a mechanical technique that dislodges the particles from the plates. These plates typically accumulate a 6-mm (0.2-inch) layer of dust. Rapping can be done using either impulse (single-blow) or vibrating rappers. The dislodged particles are collected in a hopper at the bottom of the unit for disposal. Electrostatic precipitators are capable of removing particles as small as 1 µm with an efficiency exceeding 99%. The high effectiveness of electrostatic precipitators in removing fly ash from the combustion gases of fossil-fuel furnaces explains their frequent use in power stations.

8.3.1.4. Baghouse Filters

A baghouse, made up of a collection of fabric-filter bags, is considered one of the best tools for eliminating suspended particulates. These bags, each approximately 25 cm (10 inches) in diameter, hang upside down in a spacious enclosure. Fans blow dust-laden air upwards through the bottom of the enclosure, where the particulates are captured within the filter bags. Clean air then filters through the fabric, exiting at the top of the baghouse.

A fabric-filter dust collector is capable of removing almost 100% of particles as small as 1 µm and a significant portion of particles as small as 0.01 µm. However, fabric filters create a high airflow resistance, resulting in a substantial amount of energy needed for the fan system. To extend the lifespan of the filter fabric, the air must be cooled to below 300°C (570°F) before passing through the unit, requiring additional energy for cooling coils. Certain filter fabrics, such as ceramic or mineral materials, can operate at higher temperatures.

In many multi-compartment baghouse installations, filter bags are utilized for optimal filtration. This setup enables individual compartments to be cleaned independently, while the others continue to operate. The cleaning process involves removing the accumulated surface dust layer from the bags, which can be achieved through mechanical shaking, temporarily reversing the airflow to collapse the bags, or sending a quick burst of air down through the bags to briefly expand them.

Once the dust is cleared from the filters, it drops into a receptacle underneath where it can be gathered for disposal or potential reuse. It is important to be cautious when cleaning the filters to avoid removing too much of the accumulated surface dust, known as "dust cake." For many filter types, the filter serves as a base for the dust cake to form and capture most of the particles. Filters with a membrane coating like polytetrafluoroethylene (Teflon) do not rely on dust cake for optimal performance.

8.3.2. Control of Gases

Three main techniques – absorption, adsorption, and incineration (or combustion) – are commonly used to control gaseous criteria pollutants, volatile organic compounds (VOCs), and other gaseous air toxics. These techniques can be used individually or in combination, and are also effective in reducing major greenhouse gases. Additionally, a fourth technique, carbon sequestration, is being developed as a way to control carbon dioxide levels.

8.3.2.1. Absorption

When it comes to controlling air pollution, absorption is the process of moving a gas pollutant from the air to a liquid, like water. The liquid needs to either dissolve the pollutant or chemically react with it to capture it.

8.3.2.2. Wet Scrubbers and Packed Scrubbers

Wet scrubbers, like those mentioned previously for managing suspended particles, can also be utilized for gas absorption. Additionally, gas absorption can be conducted in packed scrubbers or towers, where the liquid is coated on a wet surface rather than being in droplet form in the air. The countercurrent tower is a popular type of packed scrubber.

Once inside the tower base, the contaminated air moves upwards through a wet column filled with light packing material that is chemically inert. The liquid absorbent trickles down and spreads evenly throughout the packing, maximizing contact between the gas and liquid. Countercurrent scrubber towers typically use thermoplastic materials as the packing material, achieving gas removal efficiencies of 90–95%.

Packed scrubbers for gas absorption can be designed in either cocurrent or cross-flow configurations. In the cocurrent design, both gas and liquid flow downwards through the scrubber in the same direction. While not as efficient as countercurrent designs, cocurrent devices are able to handle higher liquid flow rates and prevent packing from becoming blocked when the gas stream contains a high level of particulates. On the other hand, cross-flow designs involve horizontal gas flow through the packing with vertical liquid flow. This design is beneficial when dealing with high particulate levels, as it allows for lower airflow resistance. Both designs offer advantages such as reduced tower size and improved airflow resistance in certain conditions.

Scrubbers are commonly utilized in fertilizer production facilities to eliminate ammonia from the airstream, in glass production plants to remove hydrogen fluoride, in chemical plants to get rid of water-soluble solvents such as acetone and methyl alcohol, and in rendering plants to manage odors.

8.3.2.3. Flue Gas Desulfurization

Flue gas desulfurization (FGD) is a process that helps control sulfur dioxide emissions from fossil-fuel power plants. There are two types of FGD systems: wet scrubbing and dry scrubbing. In wet FGD systems, flue gases come into contact with an absorbent, which can be a liquid or a slurry, allowing the sulfur dioxide to dissolve or react with the absorbent. Dry FGD systems use dry pulverized lime or limestone as the absorbent, with the solid particles being removed by baghouse filters after absorption. Dry FGD systems are more cost-effective and energy-efficient than wet systems, but they require more chemicals and can only be used with low-sulfur coal combustion flue gases.

FGD systems are divided into categories of regenerable and nonregenerable, based on whether the removed sulfur is reused or disposed of. In the United States, nonregenerable systems are more common due to their lower costs. However, regenerable systems are widely used in Japan and mandated by law in Germany. Nonregenerable FGD systems create a sulfur-containing sludge that must be properly disposed of. On the other hand, regenerable systems require extra steps to convert sulfur dioxide into beneficial by-products such as sulfuric acid.

There are various FGD methods available, with differences primarily based on the choice of chemicals used. FGD techniques utilizing lime or limestone slurries as reactants are commonly used. In the limestone scrubbing process, sulfur dioxide reacts with limestone particles in the slurry, producing calcium sulfite and carbon dioxide. In the lime scrubbing process, sulfur dioxide reacts with slaked lime, resulting in calcium sulfite and water.

Depending on the levels of sulfur dioxide and the oxidation conditions, calcium sulfite may react with water to form calcium sulfate, also known as gypsum. Both calcium sulfite and calcium sulfate have low solubility in water and can be separated from the solution as a slurry through gravity settling. This thick slurry, referred to as FGD sludge, poses a significant challenge for disposal. Flue gas desulfurization is used to reduce sulfur dioxide levels in the environment and address the issue of acid rain. However, this process is costly and the expense is passed on to consumers through higher electricity rates. Despite these efforts, millions of tons of FGD sludge are produced annually.

8.3.2.4. Adsorption

Gas adsorption is a process where gas molecules are attracted to and held on the surface of a solid. Unlike absorption, which involves the gas being taken into the solid material, adsorption occurs on the surface. This method is commonly used for odor control in chemical-manufacturing and food-processing facilities, as well as for recovering volatile solvents like benzene. It is also used to control volatile organic compounds (VOCs) at industrial sites.

Activated carbon, also known as heated charcoal, is a widely used adsorbent material due to its high porosity and a high surface area-to-volume ratio. It is especially effective for cleaning air containing volatile organic compounds, as well as for recovering solvents and controlling odors. A well-designed activated carbon adsorption unit can remove gases with an efficiency of over 95%.

Adsorption systems can be set up as either stationary bed units or moving bed units. In stationary bed adsorbers, the contaminated air enters from the top, flows through a layer of activated carbon, and exits at the bottom. Moving bed adsorbers involve the activated carbon moving downwards through channels by gravity while the air in need of cleaning passes through in a cross-flow direction.

8.3.2.5. Incineration

The process known as incineration or combustion involves the rapid oxidation of VOCs and other gaseous hydrocarbon pollutants to produce carbon dioxide and water. This process is typically carried out in a specialized incinerator called an afterburner. For successful combustion, the afterburner must have the right amount of turbulence and burning time, while maintaining a high temperature. Effective turbulence plays a crucial role in reducing the necessary burning time and temperature. In cases where the waste gas is already a combustible mixture and does not require additional air or fuel, direct flame incineration can be utilized.

An afterburner is usually constructed with a steel shell that is lined with refractory materials like firebrick. The refractory lining not only protects the shell but also acts as a thermal insulator. When exposed to sufficient time and high temperatures, gaseous organic pollutants can be nearly completely oxidized, resulting in an incineration efficiency close to 100%. Additionally, certain substances like platinum can help facilitate the combustion reaction as catalysts, enabling complete oxidation of combustible gases at lower temperatures.

Afterburners are utilized to manage odors, eliminate harmful compounds, or lower the quantity of photochemically reactive substances discharged into the atmosphere. They are utilized in a range of industrial settings where volatile organic compound (VOC) vapors are released from burning procedures or solvent evaporation, such as petroleum refineries, paint-drying facilities, and paper mills.

8.3.2.6. Carbon Sequestration

One of the most effective ways to lower carbon dioxide levels in the atmosphere is by improving energy efficiency and shifting away from the use of fossil fuels towards alternative energy sources like nuclear, wind, tidal, and solar power. Another method is carbon sequestration, which involves storing carbon dioxide underground or in forests and oceans to help reduce greenhouse gas emissions.

Natural processes like forest growth and ocean absorption play a key role in carbon sequestration. However, deforestation for agriculture and other activities, as well as ocean pollution, reduces the effectiveness of these processes. To address this issue, there is the ongoing development of technologies like geosequestration or carbon capture and storage, which involve storing carbon dioxide underground by injecting it into geological "reservoirs." This process requires separating carbon dioxide from power plant emissions, which can be expensive.

8.4. SOLID WASTE MANAGEMENT TECHNIQUES

Solid waste management involves the collection, treatment, and disposal of solid materials that are no longer needed. If municipal solid waste is not disposed of properly, it can lead to unsanitary conditions, environmental pollution, and outbreaks of diseases spread by rodents and insects. The management of solid waste presents technical challenges as well as administrative, economic, and social issues that must be addressed.

8.4.1. Solid-waste Collection

8.4.1.1. Collecting and Transporting

Proper solid waste collection plays a crucial role in safeguarding public health, safety, and the environment. This task is labor-intensive and constitutes a significant portion of the overall cost of managing solid waste. While public employees are commonly responsible for waste collection, it may be more cost-effective for private companies to handle this service through municipal contracts or be paid directly by homeowners. Typically, each collection vehicle is operated by a driver and one or two loaders using enclosed, compacting trucks with capacities of up to 30 cubic meters (40 cubic yards). Trash can be loaded from the front, rear, or side of the vehicle, with compaction reducing the volume of waste to less than half its initial size.

Choosing the best collection route can be a challenging task, particularly in big cities with high population density. An ideal route maximizes the use of labor and equipment efficiently, which requires advanced computer analyzes to consider various design factors in a large and intricate network. Factors such as collection frequency, distance of haulage, type of service, and climate must all be taken into account. Collection of waste in rural areas poses its own set of challenges due to low population densities, resulting in increased unit costs.

Garbage collection typically happens weekly due to food waste decomposing quickly. The use of garbage grinders or disposals can help reduce the amount of waste in individual homes. Although ground garbage can strain sewer systems, most are able to handle it. Many communities now have programs for source separation and recycling, where recyclable items are separated from garbage and placed in separate containers for pick up. Some communities also have drop-off centers for residents to bring recyclables.

8.4.1.2. Transfer Stations

If the final destination of the waste is not close to the community where it is produced, one or more transfer stations may be required. These stations serve as a central location where waste from multiple collection vehicles is consolidated into larger vehicles for transportation, such as tractor-trailer units. Open-top trailers can hold up to 76 cubic meters (100 cubic yards) of uncrushed waste, which is then taken to a regional processing or disposal site. Closed compactor-type trailers are also an option, but they need to have ejector mechanisms installed. There are two types of transfer stations: direct discharge, where collection trucks unload directly into the transport vehicle, and storage discharge, where the waste is first deposited into a storage area and then pushed or lifted into the transport vehicle using machinery. Some large transfer stations are capable of handling over 500 tons of waste per day.

8.4.2. Solid-Waste Treatment and Disposal

After being gathered, municipal solid waste can undergo treatment to decrease the overall size and weight of materials that need to be disposed of. Treatment alters the waste's structure, making it more manageable. This process can also help recover specific materials and heat energy for recycling or reusing.

8.4.2.1. Incineration

8.4.2.1.1. Furnace Operation

Burning is an efficient way to reduce the size and weight of solid waste, but it does release greenhouse gases. Modern incinerators burn waste in a controlled furnace, where the combustible parts combine with oxygen to produce carbon dioxide, water vapor, and heat. This process can decrease the volume of waste by over 90%, leaving behind ash, glass, metal, and other materials known as bottom ash. The gaseous by-products and fine particles, called fly ash, are removed by emission control devices in the incinerator. Bottom ash and fly ash are typically disposed of in landfills, but if toxic metals are present, they must be treated as hazardous waste.

Municipal solid-waste incinerators are specifically designed to receive and burn a continuous stream of refuse. A deep refuse storage pit, also known as a tipping area, is allocated to store approximately one day's worth of waste. The refuse is lifted out of the pit using a crane equipped with a bucket or grapple device, and then transported to a hopper and chute located above the furnace. Once inside, the waste is released onto a charging grate or stoker, where it is shaken and moved through the furnace to allow proper circulation of air around the burning materials. Most modern incinerators are built with a rectangular furnace design, although rotary kiln furnaces and vertical circular furnaces are also options. These furnaces are constructed using refractory bricks that are able to withstand the high temperatures generated during combustion.

Furnace combustion happens in primary and secondary stages. During primary combustion, moisture is removed, and waste is ignited and volatilized. Secondary combustion oxidizes remaining unburned gases and particulates, ridding odors and reducing fly ash in the exhaust. If the refuse is very moist, auxiliary gas or fuel oil may be burned to kickstart primary combustion. To ensure sufficient oxygen for primary and secondary combustion, it is important to thoroughly mix air with the burning refuse. Air can be supplied from below the grates or through openings above the area. The plant operator must determine the proper balance of underfire air and overfire air to achieve optimal combustion efficiency. An uninterrupted flow of air can be achieved through either a natural draft in a tall chimney or by using mechanical forced-draft fans.

8.4.2.1.2. Energy Recovery

Refuse can potentially have an energy value equal to about one-third of that of coal, depending on its paper content. By using a refractory-lined furnace connected to a boiler, the heat generated during incineration can be captured and converted into steam or hot water. This process allows for the recycling of the energy content of the refuse. Waste-to-energy plants are incinerators that recycle heat energy in this manner. Additionally, a water-tube wall furnace can be utilized for energy recovery instead of a separate furnace and boiler. This type of furnace is lined with closely spaced vertical steel tubes that form continuous sections of wall. The walls are insulated to minimize heat loss. Water circulating through the tubes absorbs heat to produce steam and helps regulate combustion temperatures, reducing the need for excessive air and subsequently lowering air pollution control costs.

Waste-to-energy plants can operate as either mass burn or refuse-derived fuel systems. In a mass burn system, all the refuse is burned without any prior treatment. On the other hand, a refuse-derived fuel system separates combustible wastes from noncombustibles like glass and metal before burning. If a turbine is installed at the plant, both steam and electricity can be produced through cogeneration. While waste-to-energy systems are more costly to build and operate compared to plain incinerators due to the need for special equipment, skilled personnel, and auxiliary fuel systems, the sale of generated steam or electricity helps offset these additional expenses. Recovering heat energy from refuse is seen as a practical solid waste management solution from both an engineering and economic perspective. In the United States, approximately 80% of municipal refuse incinerators are waste-to-energy facilities.

8.4.2.2. Composting

Composting is another approach to managing municipal solid waste, where the organic part of the waste decomposes with the help of microbes in controlled conditions. This process can reduce the waste volume by up to 50%, resulting in a stable product known as compost or humus. Compost has a similar texture and smell to potting soil, and can be utilized as a soil conditioner or mulch. Composting provides a way to process and recycle both garbage and sewage sludge together in one step. With stricter environmental regulations and limitations on incineration and landfills, composting is expected to become more common. The process involves sorting, separating, reducing size, and breaking down the waste.

8.4.2.2.1. Sorting and Shredding

The decomposable materials in refuse are separated from glass, metal, and other inorganic items through mechanical sorting and separating operations. These operations utilize differences in physical characteristics such as size, density, and magnetic properties. Shredding or pulverizing is then used to reduce the size of the waste articles, resulting in a uniform mass of material. This process is typically carried out with hammer mills and rotary shredders.

8.4.2.2.2. Digesting and Processing

Crushed waste is prepared for composting using either the open windrow technique or an enclosed mechanical facility. Windrows are extended piles of refuse that are regularly aerated by turning or mixing every few days to support microbial breakdown of organics. Complete digestion of the waste can take anywhere from five to eight weeks, depending on moisture levels. The high metabolic activity of aerobic bacteria in an active compost pile raises temperatures to around 65°C (150°F), effectively eliminating any harmful pathogens present in the waste material.

Windrow composting requires significant land space, while enclosed mechanical composting facilities can reduce land needs by up to 85%. Mechanical composting systems use closed tanks with rotating vanes to mix and aerate waste, allowing for complete digestion in about a week.

Before digested compost can be utilized as mulch or soil conditioner, it must undergo processing. This involves steps such as drying, screening, and granulating or pelletizing. Through this process, the market value of the compost is enhanced, which is a key factor in the success of composting as a waste management solution. Agricultural demand for digested compost is typically limited due to the expensive transportation costs and competition with synthetic chemical fertilizers.

8.4.2.3. Sanitary Landfill

Municipal solid waste is commonly managed through land disposal, where refuse is disposed of in a sanitary landfill that is carefully designed to protect the environment and public health. It is crucial that the waste does not come into contact with surface water or groundwater. Landfills must meet engineering design requirements, including a minimum distance from the groundwater table and the use of impermeable liners and monitoring systems. Upon completion, landfills must be capped with an impermeable cover to prevent precipitation or runoff from reaching the buried waste. Liners at the bottom and cap can be made of plastic membranes, clay soil layers, or a combination of both.

8.4.2.3.1. Constructing the Landfill

The main component of a sanitary landfill is the refuse cell, which is a designated area where waste is spread and compacted in thin layers. Multiple layers can be compacted on top of each other, reaching a maximum depth of around 3 meters (10 feet). The compacted waste takes up about a quarter of its original volume. Each day, the waste is covered with a layer of soil to prevent litter, odors, and pest issues. A refuse cell contains the daily amount of compacted waste and soil cover. Multiple adjacent refuse cells form a lift, and a landfill can consist of multiple lifts stacked on top of each other. The final cap of a completed landfill may be covered with topsoil to support vegetation growth.

Each day, soil cover can either be found on-site or brought in and stored from external sources. Different types of heavy machinery, such as crawler tractors or rubber-tired dozers, are utilized to distribute and compress the waste and soil. In addition, heavy steel-wheeled compactors may be used to achieve a high level of compaction for the waste.

The new landfill is carefully planned and prepared, with the area and depth determined before the construction of any necessary liner and leachate-collection system. When using a plastic liner, a layer of sand is evenly spread over it to protect it from damage by landfill vehicles. Depending on the site conditions, either the trench method or area method may be followed for construction. If excavation below ground is possible, the trench method is used; otherwise, the area method is employed, resulting in a mound or hill above the original ground level. Soil may need to be transported to the site if excavation is not possible. Different variations of the area method may be used for landfill sites on sloping ground, in valleys, or ravines. Ultimately, the completed landfill will blend in with the surrounding landscape.

8.4.2.3.2. Controlling By-Products

When organic material is buried in a landfill, it breaks down through anaerobic microbial processes. This decomposition process can take more than 20 years to complete. One of the by-products of this decomposition is methane gas, which is both toxic and flammable when mixed with air. Methane is also a potent greenhouse gas, contributing to climate change. It has the ability to travel long distances through soil layers and can accumulate in enclosed spaces, posing a danger to people. To prevent these risks, modern landfills utilize impermeable barriers and gas-venting systems to control the movement of methane. Some landfills even capture and use the methane gas as a fuel source, either directly or as part of biogas.

Leachate, a highly contaminated liquid formed during decomposition in sanitary landfills, is a major concern due to its environmental impact. It is mainly created when runoff from decomposing garbage penetrates refuse cells. If not properly managed, leachate can pollute groundwater and surface water, endangering drinking-water supplies. Techniques to control leachate include intercepting surface water to avoid landfill infiltration and installing impermeable barriers to protect groundwater. New landfills should have monitoring wells and systems to collect and treat leachate to prevent pollution.

8.4.2.3.3. Importance in Waste Management

In communities where suitable sites are available, sanitary landfills are typically the most cost-effective option for disposing of nonrecyclable waste. However, finding sites with enough capacity, accessibility, and favorable environmental conditions is becoming increasingly challenging. Despite these challenges, landfills will continue to play a crucial role in waste management. Not all solid waste can be recycled, and there will always be leftover residues from treatment processes that need to be buried. Additionally, properly managed landfills can actually help improve low-quality land by being converted into recreational areas like parks, playgrounds, or golf courses in some communities.

8.4.2.4. Recycling

Recycling is the process of separating, recovering, and reusing components of solid waste that still have economic value. One form of recycling is the recovery and reuse of heat energy, which is covered in the discussion of incineration. Composting is another type of recycling that involves reclaiming organic materials from solid waste for use as mulch or soil conditioner. Additionally, paper, metal, glass, plastic, and rubber are examples of waste materials that can be recycled and their recovery is addressed in this discussion.

8.4.2.4.1. Separation

In order to recycle any material, it needs

to first be separated from the raw waste and organized. This separation can be done either at the location where the waste is generated or at a central facility. When done at the source, also known as curbside separation, individual citizens gather newspapers, bottles, cans, and trash separately and place them at the curb for collection. Some communities permit the mixing of non-paper recyclables (such as glass, metal, and plastic). However, municipal collection of source-separated waste is typically more costly than regular waste collection.

Instead of relying on centralized mechanical processing plants to separate recyclable materials from garbage, it is now understood that source separation by citizens yields higher-quality recyclables. Citizens are encouraged to separate refuse into specific categories such as newspapers, magazines, other paper wastes, commingled metals, glass, plastics, and garbage/non-recyclables. The separated materials are then collected and processed at a centralized material recycling facility (MRF), where modern facilities can handle up to 300 tons of recyclable waste per day. This method minimizes contamination with moist garbage and broken glass, resulting in better quality recyclables being recovered.

8.4.2.4.2. Reuse

Crushed broken glass can be utilized in asphalt pavement, while color-sorted glass is sold to glass manufacturers as a cullet for glassmaking. Steel cans are compressed into bales and sent to steel mills as scrap, while aluminum is baled or compacted for reuse by smelters. Although aluminum is a minor component of municipal solid waste, it is the most valuable recyclable material. Plastic recycling is challenging due to the variety of polymeric materials used in production. Mixed thermoplastics are only suitable for creating lower-quality products like "plastic lumber."

Recycled rubber is often obtained from solid waste, shredded, reformed, and remolded through a process known as revulcanization. However, it is generally not as durable as the original material. Shredded rubber can serve as an additive in asphalt pavements and synthetic turf, as well as being sold directly as outdoor mulch. Old tires can be repurposed as swings and other recreational structures in "tire playgrounds" for children to enjoy.

Typically, the biggest challenge with recycling solid waste materials is identifying appropriate uses and markets for them. Simply recycling alone is not enough to address the increasing issue of managing and disposing of solid waste. There will always be some waste that remains unusable and without any value, needing to be disposed of properly.

CASE STUDY

ENVIRONMENTAL PROTECTION AND ENERGY FOR QUEEN MARY II

Background

The Queen Mary II is one of the largest and fastest passenger ships ever built. The vessel is equipped with cutting-edge technology from the mast to the keel. This luxury liner produces enough electricity for a large city and has its own on-board wastewater treatment facility.

Task and Target

New legislation for environmental standards in relation to waste treatment on board vessels meant that an upgrade of their facilities was necessary.

The new process required the installation of a sludge cake handling system to be fitted within a limited space which was capable of metering approximately 20%ds dewatered sludge from a centrifuge into a sludge dryer process to increase the dry solids content prior to being incinerated to maximize energy production.

Key Facts

- Range BTI
- Conveying of sewage sludge
- Dry solids content: 20%

Solution

SEEPEX pumps were installed in the ship's new, modernized facilities which process waste in an environmentally friendly fashion, turning it into energy.

Special BTI range pumps which have the ability to handle highly viscous non-flowable products, as well as being able to accurately meter the product, were fitted with variable speed drives and installed on space-saving easy-maintenance pump skids.

Centrifuges dewater the sludge to approx. 20% dry solids, which are then metered with SEEPEX pumps into a thermal dryer. To optimize the dryer efficiency, it is important that the correct feeding technology is selected to ensure that the right amount of dry solids is fed into the dryer for optimal efficiency.

Key Specifications

- Flexible dewatered sludge conveying system

- Enclosed pipework for transporting dewatered sludge
- System that meters dewatered sludge into a sludge dryer

Results

The technology installed within the new wastewater treatment process enabled compliance with the new environmental standards by burning the accumulated waste in a high-temperature incinerator and then reusing the heat generated within the ship's energy cycle.

The technical solution delivered by SEEPEX results in cost savings for the customer due to reduced disposal costs and assists their energy production. SEEPEX experts were aware of the limited space on board and created a compact design for the sludge cake handling system that fitted perfectly in the available space.

Benefits

- Reduced disposal costs
- Lower energy costs

CLASS ACTIVITY

HOW DO WE DECREASE AIR POLLUTION?

Students brainstorm a list of what can be done to reduce the amount of air pollution. Hold a class discussion in which students share their ideas. Discuss which ideas are easy to implement and which would take more effort.

Discuss, one way to decrease air pollution is to decrease our use of fossil fuels. Fossil fuels are used for many energy purposes, such as transportation, electricity and manufacturing. Alternative Energy sources can be used to replace fossil fuels as renewable energy sources.

Renewable sources of energy include;

- Solar Energy
- Photovoltaic Cell (Solar Cell)
- Wind Power
- Wind Turbines
- Waterpower
- Germany: Eifel Region
- Geothermal Energy

Create a list of ways students can individually decrease air pollution/pollution. Challenge the students to take action and commit to one of the suggestions below.

- Use public transport or car sharing, cycle or walk.
- Switch to a hybrid or electric vehicle and request electric taxis.
- Turn off the car engine when stationary.
- Reduce your consumption of meat and dairy to help cut methane emissions.
- Compost organic food items and recycle non-organic trash.
- Switch to high-efficiency home heating systems and equipment.
- Save energy: turn off lights and electronics when not in use.
- Choose non-toxic paints and furnishings.
- Audit your family's habits – ask questions like; How often do people have a shower? Do you throw clothes in the wash when they could be worn again? Which appliances do not need to be plugged in all the time? Where has your food come from, locally or imported?

SUMMARY

- Drinking water comes from a variety of different sources. As water makes its way through the hydrologic cycle, it comes back to land in the form of precipitation (rain, sleet, snow, etc.). Some of it is captured in lakes and rivers, while some of it percolates into the earth's surface and becomes groundwater.
- A water supply network or water supply system is a system of engineered hydrologic and hydraulic components that provide water supply.
- A water distribution system design is a blueprint for building and operating a water distribution system that provides drinkable water to a community. The arrangement of pipes, pumps, and other infrastructure required to carry water from a source, such as a treatment plant, to end customers is included in the design.
- Water distribution systems are among the most important types of infrastructure. The Water distribution design's principal role is to provide water to all consumers. A sufficient amount of water must be supplied at an adequate pressure.
- Wastewater collection systems are designed and built to collect and convey wastewater from homes, businesses, and industries. Water carries waste as dissolved material and suspended solids.
- Wastewater treatment, the removal of impurities from wastewater, or sewage, before it reaches aquifers or natural bodies of water such as rivers, lakes, estuaries, and oceans. Since pure water is not found in nature (i.e., outside chemical laboratories), any distinction between clean water and polluted water depends on the type and concentration of impurities found in the water as well as on its intended use.
- The residue that accumulates in sewage treatment plants is called sludge (or biosolids). Sewage sludge is the solid, semisolid, or slurry residual material that is produced as a by-product of wastewater treatment processes. This residue is commonly classified as primary and secondary sludge.
- Air pollution control, the techniques employed to reduce or eliminate the emission into the atmosphere of substances that can harm the environment or human health. The control of air pollution is one of the principal areas of pollution control, along with wastewater treatment, solid-waste management, and hazardous-waste management.
- Solid-waste management, is the collecting, treating, and disposing of solid material that is discarded because it has served its purpose or is no longer useful. Improper disposal of municipal solid waste can create unsanitary conditions, and these conditions in turn can lead to pollution of the environment and to outbreaks of vector-borne disease—that is, diseases spread by rodents and insects.

REVIEW QUESTIONS

1. What are water distribution systems and why do they matter?
2. What are the types of water supply systems?
3. Write the process of a water supply system.
4. How do water distribution systems work?
5. What variables influence water supply system design?
6. What are the methods of collection systems for wastewater?
7. What are the measures to control air pollution caused by industry?
8. solid-waste management

REFERENCES

1. Aragones, D. G., Calvo, G. F., & Galan, A. (2021). A heuristic algorithm for optimal cost design of gravity-fed water distribution networks: A real case study. *Applied Mathematical Modelling, 95*, 379–395. https://doi.org/10.1016/j.apm.2021.02.015.
2. De Corte, A., & Sörensen, K. (2013). Optimization of gravity-fed water distribution network design: A critical review. *European Journal of Operational Research, 228*(1), 1–10. https://doi.org/10.1016/j.ejor.2012.11.046.
3. Ferguson, J. (1999). Environmental engineering. *Water Environment Research, 71*(6), 1139. https://doi.org/10.1002/j.1554-7531.1999.tb00198.x.
4. Lincoln, T. (1997). Environmental engineering. *Nature, 388*(6637), 27. https://doi.org/10.1038/40285.
5. Mala-Jetmarova, H., Barton, A., & Bagirov, A. (2015). A history of water distribution systems and their optimization. *Water Science and Technology: Water Supply, 15*(2), 224–235. https://doi.org/10.2166/ws.2014.115.
6. Martínez, F., Hernández, V., Alonso, J. M., Rao, Z., & Alvisi, S. (2007). Optimizing the operation of the Valencia water-distribution network. *Journal of Hydroinformatics, 9*(1), 65–78. https://doi.org/10.2166/hydro.2006.018.
7. Maxwell, S. (1997). Environmental advisor: Environmental engineering firms. *Journal of Management in Engineering, 13*(2), 16–17. https://doi.org/10.1061/(asce)0742-597x(1997)13:2(16).
8. Pawłowski, L. (1990). Environmental engineering. *Science of The Total Environment, 92*, 297. https://doi.org/10.1016/0048-9697(90)90356-y.
9. Safferman, S. I., Utgikar, V. P., & Sandhu, S. S. (1996). Environmental engineering forum: Undergraduate environmental engineering education. *Journal of Environmental Engineering, 122*(9), 779–784. https://doi.org/10.1061/(asce)0733-9372(1996)122:9(779).
10. Takakura, T. (1997). Bio-environmental engineering. *Trends In the Sciences, 2*(3), 60–61. https://doi.org/10.5363/tits.2.3_60.
11. Walski, T. M. (1993). Environmental engineering forum. *Journal of Environmental Engineering, 119*(2), 21.

CHAPTER 9

Hydraulic Engineering

LEARNING OBJECTIVES

After studying this chapter, you will be able to:

- Examine the fundamentals of the hydraulics system; and
- Evaluate the flow through orifices and mouthpieces in hydraulics.

KEY TERMS FROM THIS CHAPTER

Canals	Civil engineering
Coastal management	Environmental Health
Geographic information systems (GIS)	Hydraulic systems
Pipelines	Pneumatic control systems
Remote sensing	Reservoirs
Sewage systems	Sustainable management
Water resources	

INTRODUCTION

The flow and conveyance of fluids, mainly water and sewage, is the focus of the specialized field of hydraulic engineering within civil engineering. The field of water resource management involves the planning, building, and upkeep of various structures and systems. For agriculture, industry, human life, and environmental health, water must be managed sustainably and with sufficient quantity and quality. This discipline is essential to achieving this goal. Fundamentally, the study of hydraulic engineering involves applying the concepts of fluid mechanics to the solution of intricate issues involving water in all of its forms and states. Water treatment plants, sewage systems, canals, reservoirs, pipelines, dams, levees, and other engineering projects are among the many projects that engineers in this field work on. They design and carry out irrigation plans, drainage systems, and flood control strategies as well. In order to minimize the effects of water-related disasters and maintain the ecological health of water bodies, it is important to efficiently and safely utilize available water resources. Hydraulic structure design constitutes a major field of hydraulic engineering. Flood control, hydroelectric power generation, and water storage are just a few of the uses for dams. Hydrological conditions, geological conditions, environmental impact, and the socio-economic consequences of their designs are just a few of the variables that hydraulic engineers have to take into account. Similarly, in order to safeguard populated areas from floodwaters, levees and floodwalls are built to withstand extreme conditions through rigorous construction methods and precise calculations.

Urban water management is a significant field in hydraulic engineering. The need for efficient sewage treatment and clean water is growing as the population of cities increases. In order to avoid contaminating natural water bodies, engineers must create systems that can consistently provide drinkable water and treat wastewater. This entails using cutting-edge technologies for sewage treatment, water recycling, and water purification. Stormwater management systems, which lessen runoff pollution and prevent flooding, are another aspect of sustainable urban water management. In addition, hydraulic engineering is essential to the preservation and repair of the environment. Initiatives like river engineering, coastal management, and wetland restoration seek to protect and improve natural ecosystems while also benefiting people. Engineers strive to create wildlife habitats, stabilize eroding coastlines, and restore rivers to their natural flow. In addition to preserving biodiversity, these initiatives help make communities more resilient to the effects of climate change. Advanced technologies and interdisciplinary approaches have been increasingly incorporated into hydraulic engineering in recent years. Hydraulic system design can be optimized, hazards can be evaluated, and water flow patterns can be predicted by engineers using computer modeling and simulation. Remote sensing technologies and Geographic Information Systems (GIS) offer useful data for large-scale water resource management. Hydraulic engineering projects are made sustainable and in line with larger societal objectives through collaboration with environmental scientists, urban planners, and legislators.

9.1. FUNDAMENTALS OF HYDRAULICS SYSTEM

Similar to electricity and pneumatics, "hydraulics" refers to a drive system used to operate machinery and equipment. The first commercially available practical hydraulic applications appeared in the early 1900s. The practical uses of fluids, especially liquids, in motion, are the focus of the scientific field of hydraulics. It is associated with fluid mechanics, which serves as its theoretical basis to a great extent. Liquid flow in pipes, rivers, and channels, as well as how those flows are constrained by tanks and dams, are all aspects of hydraulics.

Certain principles are also applicable to gases, typically in situations where density variations are not very significant. Thus, the field of hydraulics encompasses pneumatic control systems as well as mechanical devices like fans and gas turbines. Before the laws of modern hydraulic power technology were developed by Swiss physicist Daniel Bernoulli and French scientist-philosopher Blaise Pascal, liquids in motion or under pressure performed beneficial tasks for humans for many centuries. About 1650, Pascal developed Pascal's law, which states that pressure in a liquid is transmitted equally in all directions; i.e., For example, applying pressure to any point within a closed container of water will cause the pressure to be distributed to all sides of the container. Pascal's law increases the force in the hydraulic press by applying a small force to a small piston in a small cylinder, which is then transferred through a tube to a large cylinder, where it presses equally against the large piston, and all of the cylinder's sides.

Although each hydraulic system has a different purpose, they all operate on the same fundamental idea. In short, hydraulic systems work and carry out tasks by means of a pressurized fluid. To put it another way, everything functions because of the pressurized fluid. Because liquid fuel has such a large power in hydraulics, heavy machinery frequently uses hydraulics. Pressure applied to a contained fluid at any point in a hydraulic system is transmitted without diminishing. Every part of the containing vessel is affected by the pressurized fluid, which produces force or power. Operators can effortlessly perform precise repetitive tasks and lift heavy loads thanks to the application of this force, which varies in intensity. Heavy machinery depends heavily on hydraulic systems to function. When designing hydraulic power train systems, automatic transmissions, power assisted steering, brake systems, steering systems, and hydraulic implement systems, hydraulic principles are applied. Before moving on to machine systems, it is necessary to grasp the fundamentals of hydraulics. Equipment used in mining, construction, agriculture, and material handling heavily relies on hydraulics. Tools for pushing, lifting, and moving materials are operated by hydraulics. The widespread application of hydraulics on earthmoving machinery did not occur until the 1950s. Since then, using this kind of power to run machinery has become commonplace. The liquid's applied forces are transferred to a mechanical mechanism in hydraulic systems. The fundamentals of hydraulics must be understood in order to comprehend how hydraulic systems work (Butterworth, 1981)

9.1.1. Why Are Hydraulic Systems Used?

There are many reasons. Some of these are that hydraulic systems are versatile, efficient, and simple for the transmission of power. This is the hydraulic system's job, as it changes power from one form to another.

The science of hydraulics can be divided into two sciences:

1. Hydrodynamics
2. Hydrostatics.

1. **Hydrodynamics:** This describes the science of moving liquids.

   Applications of hydrodynamics:

   - water wheel or turbine; the energy that is used is that created by the water's motion
   - torque converter.

2. **Hydrostatics:** This describes the science of liquids under pressure.

   Applications of hydrostatics:

   - hydraulic jack or hydraulic press
   - hydraulic cylinder actuation.

In hydrostatic devices, pushing on a liquid that is trapped (confined) transfers power. If the liquid moves or flows in a system then movement in that system will happen.

9.1.2. Hydraulic Principles

There are several advantages to using a liquid:

- Liquids conform to the shape of the container.
- Liquids are practically incompressible.
- Liquids apply pressure in all directions.

1. **Liquids Conform to Shape:** Liquids adapt to fit into any shape container. Additionally, liquids can pass through hoses and lines of any size or shape and in any direction. All of them were joined and had the same amount of liquid inside. The liquid has taken on the form of the receptacles.
2. **A Liquid is Practically Incompressible:** When hydraulic oil is compressed to 3,000 psi (20,685 kPa), it compresses between 1 and 1½ percent. Hydraulic oil is the best choice for machine hydraulic applications because it is completely non-compressible. A substance requires less space when it is compressed. Even under pressure, a liquid has the same volume or amount of space. Because gas compresses and takes up less space, it would not be appropriate for use in hydraulic systems.
3. **Liquids Apply Pressure in All Directions:** A liquid has an equal distribution of pressure. Anywhere pressure is measured in a hydraulic cylinder or line, the results will always be the same. A change in volume in one cylinder will transfer the same volume to the other when two cylinders of the same size are connected by a pipe. "Displacement" refers to the volume or space occupied by any substance. Power can be transferred through pipes, over short or long distances, up and down, and around corners with the help of liquids. The force exerted at one end of a pipe will instantly transfer to the other end with the same force. Since oil cannot be compressed and lubricated the system, it is used in the majority of hydraulic systems.

Water would be unsuitable because:

- It freezes at cold temperatures and boils at 100°C
- It causes corrosion and rusting and furnishes little lubrication.

9.1.3. Pressure

Whether it is stored in a tank or is running through a pipe, water or wastewater exerts force and pressure against the walls of its container. Despite their close relationship, force and pressure are not the same thing. In particular, force per unit area is the definition of pressure. Within the U. S. pressure is commonly measured in pounds per square inch, or psi. Newtons per square meter (N/m^2) is the unit of measurement for pressure in SI units. The unit N/m^2 is referred to as a Pascal (Pa) for convenience. In most real-world hydraulics (Andreasen, J-O. 1979) applications, the term kilopascal (kPa) will be used because 1 Pa is a relatively small pressure (1 Pa = 0.000145 psi). One kPa is equal to 1,000 Pa, or 0 psi.

In this equation, pressure can be expressed as:

$P = F/A$

Where P is the pressure; F is the force; and A is the area over which the force is distributed

9.1.3.1. Hydrostatic Pressure

The pressure water exerts is called hydrostatic pressure. These principles always apply to hydrostatic pressure:

1. Pressure depends only on the depth of water above the point in question (not on the water surface area).
2. Pressure increases in direct proportion to the depth of water.
3. Pressure in a continuous volume of water is the same at all points that are at the same depth or elevation.
4. Pressure at any point in the water acts in all directions at the same magnitude.

Consider the two tanks connected by a horizontal pipe shown in Figure 9.1. The water surfaces in both tanks are at the same elevation. Pressure at the water's surface equals zero. Because of the weight of the air above, pressure is also present at the free surface. We refer to this pressure as barometric or atmospheric pressure. At sea level, the atmospheric pressure is roughly 101 kPa, or 14.7 psi. However, atmospheric pressure is regarded as a zero reference or starting point in practical applications and is not used for hydraulic calculations. Rather than absolute pressure, this application is called gauge pressure. Two storage tanks with varying volumes that are side by side and the same height are depicted in Figure 9.1. This situation may have arisen in many water systems when they needed to upgrade and discovered that the older tank might have been larger than the newer one.

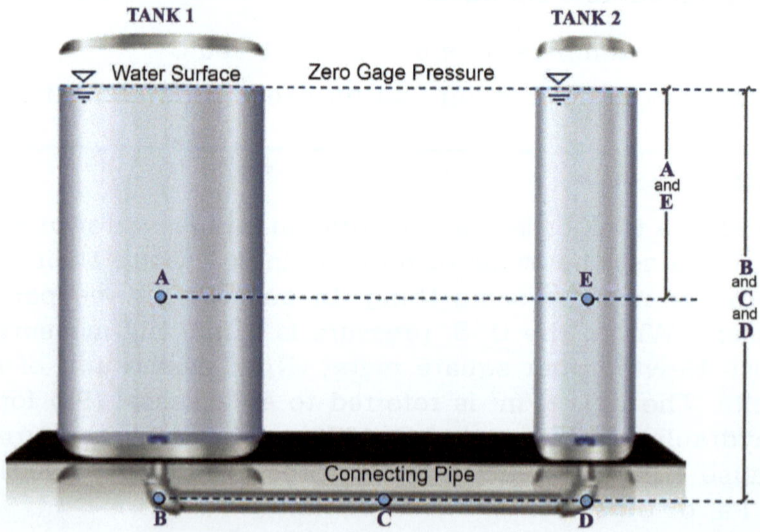

Figure 9.1. *The pressure at point A equals the pressure at point E, since these points are at the same depth in the water. Likewise, the hydrostatic pressure at points B, C, D are equivalent.*

Source: https://scontent.fdel3-2.fna.fbcdn.net/v/t1.18169-9/13015117_944327942283310_6693601076852577764_n.png?_nc_cat=107&ccb=1-7&_nc_sid=5f2048&_nc_ohc=xuxTPCms3XsQ7kNvgGbAzHY&_nc_ht=scontent.fdel3-2.fna&oh=00_AYD-Y_7Mvtus9jmLINRlIAxmJFO5ASvAgmuqfDpXTE_T2A&oe=6697450A.

This figure illustrates how pressure remains constant at the same elevations, moving from a storage tank with a larger diameter (more water storage) to one with a smaller diameter. Point B, which is close to the tank's bottom, has a higher pressure than Point A. Because pressure varies in direct proportion to depth, in this application, the gauge pressure at point B would be exactly twice that of point A if point B were exactly twice as deep as point A. Once more, take note of point E in Tank 2 in Figure 9.1. The pressure at point E is the same as the pressure at point A because they are both located at the same depth below the water's surface. Keep in mind that the fact that Tank 2 is smaller than Tank 1 is irrelevant. Hydrostatic pressure is independent of the water's surface area and volume, only relying on the height of the water above the points. The connecting pipe's point C has the same pressure as points B and D even though there isn't any water directly above it.

- **Calculating Pressure:** In customary U. S. units, 1 ft3 of water weighs 62.4 lbs. So, a 1-ft depth of water on an area of 1 ft2, or 144 in^2, the pressure at the bottom would be P = 62.4 lbs/144 in^2 = 0.43 psi per foot of water (lbs/in^2)/ft). This is usually written as:

  P = 0.43 x h

  Where P = hydrostatic pressure in psi (lbs/in^2) 0.43 psi/ft or (lbs/in^2)/ft

  h = water depth from surface in ft

  In SI-metric units, 1 m^3 oxf water weighs 9.8 kN. So, a 1-meter (m) depth of water on an area of 1 m^2 would have a pressure of 9.8 kN/m^2 usually written as 9.8 kPa. This can be written as:

P = 9.8 x h

Where P = hydrostatic pressure in kPa or kN/m^2 9.8 kPa or $(kN/m^2)/m$

h = water depth from surface in meters (m)

9.1.3.2. Atmospheric Pressure

The force per unit area that the weight of the air above a surface exerts against it is known as atmospheric pressure. As the weight of the air above it increases, so does the pressure at point X. The same holds true for decreasing pressure, where a drop in air weight above point X causes a corresponding drop in pressure at that location. When considering air molecules, an increase in the number of molecules above a surface means that there are more molecules to press against it, which raises pressure.

Conversely, a decrease in the quantity of air molecules above a surface will likewise cause a drop in pressure. Because atmospheric pressure is measured using a device known as a barometer, the term barometric pressure is also used to describe atmospheric pressure. Meteorologists use millibars (mb), the unit of pressure shown on weather maps, while aviation and television weather reports use inches of mercury (Hg). Take one square inch as an example of a unit of measurement. The air above this unit area would weigh, on average, 14 points 7 pounds at sea level. This means that the air would apply pressure to the unit area of 14 points 7 pounds per square inch. The average pressure at sea level is 1013.25 millibars, the metric unit of pressure used by meteorologists.

9.1.3.3. Gauge Pressure

Gauge pressure is a more practical unit of measurement than absolute pressure for the majority of applications, especially those that involve pressure measurements. Gauge pressure, which is the absolute pressure less the atmospheric pressure, is a relative pressure measurement that assesses pressure in relation to atmospheric pressure. Instead of providing an absolute pressure reading, the majority of pressure measuring apparatuses provide a gauge pressure reading for a system. For instance, blood pressure and tire pressure are conventional gauge pressures, but altimeter, deep vacuum, and atmospheric pressures need to be absolute. In closed systems with most working fluids, gauge pressure measurement is the most common method. The system's pressure instruments will display pressures in relation to the ambient pressure at that moment. When measuring extremely high vacuum pressures, the circumstances change; absolute pressures are usually utilized instead. The atmospheric pressure must then be added to the gauge pressure in order to determine the system's absolute pressure. Although gauge pressure is very helpful for practical pressure measurements, gauge pressure must be converted to absolute pressure because most pressure calculations, including the ideal gas law, require pressure values to be expressed in absolute terms.

9.1.3.4. Absolute Pressure

The pressure of a perfect vacuum, or the absence of all matter within a space, is known as absolute pressure. This absolute zero serves as the reference point for all pressure measurements. The barometric pressure measurement is the best illustration of an absolute referenced pressure. The manufacturer will create an absolute pressure sensor by creating a high vacuum behind the sensing diaphragm. An absolute pressure transmitter will thus read the real barometric pressure if you hold open the process pressure connection to the air.

9.1.3.5. Vacuum Pressure

When talking about pressure measurements and selecting pressure transducers, the term "vacuum" is frequently misinterpreted and used incorrectly as a result. A vacuum is, by definition, an area that has been artificially (with an air pump, for example) partially (to the maximum extent feasible) exhausted. In this definition, a "hard" or "high" vacuum is meant. The connection between gauge and absolute pressure, where 0 PSIA corresponds to a hard or high vacuum.

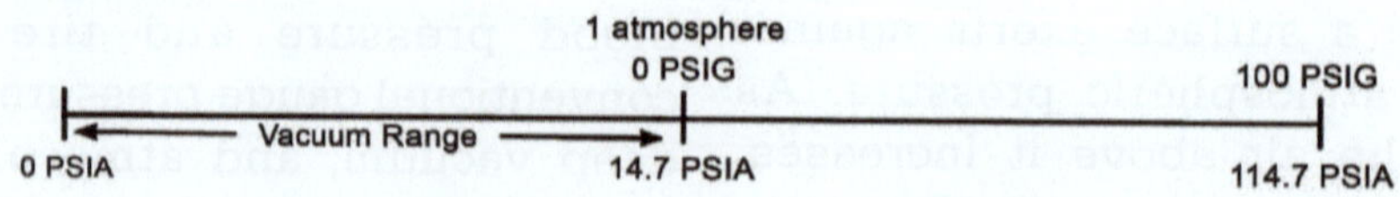

Figure 9.2. *Relationship of absolute and gauge pressure with 0 PSIA.*

Source: https://www.setra.com/hs-fs/hub/211498/file-2786334973-jpg/blog-files/vacuum-measurements-range-1024x147.jpg?width=600&height=86&name=vacuum-measurements-range-1024x147.jpg.

Recall that gauge pressure is the measurement of pressure in relation to 14.7 PSIA, or the ambient atmospheric pressure. PSIG, or pounds per square inch (gauge), is the term used to describe it. Gauge pressure transducers have two electrical outputs: full-scale output (usually 5 VDC) at full-scale pressure (in PSIG) and 0 VDC at 0 PSIG (14.7 PSIA). When measuring absolute pressure, a high vacuum (0 PSIA) is used as a reference. PSIA, or pounds per square inch (absolute), is the term used to describe it. An absolute pressure transducer's electrical output is 0 VDC at 0 PSIA and full-scale output, or 5 VDC, at full-scale pressure (in PSIA). The connection between vacuum and absolute pressures (Figure 9.3).

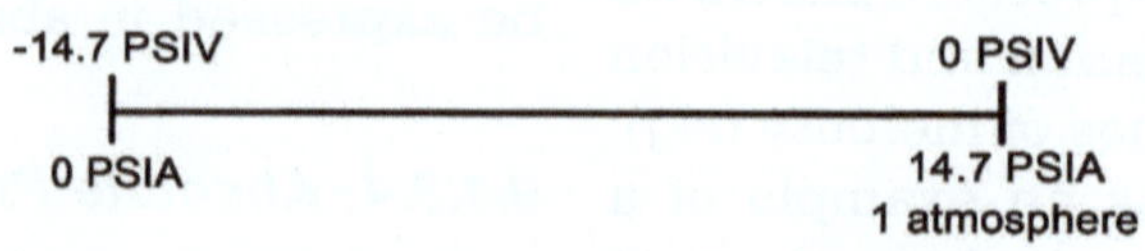

Figure 9.3. *Measured relative to absolute pressure.*

Source: https://www.setra.com/hs-fs/hub/211498/file-2786334983-jpg/blog-files/relationship-of-absolute-and-vacuum-pressures-1024x202.jpg?width=378&height=75&name=relationship-of-absolute-and-vacuum-pressures-1024x202.jpg.

The measurement of vacuum pressure is done in relation to the outside air pressure. PSIV, or pounds per square inch (vacuum), is the term used to describe it. At 0 PSIV

(14.7 PSIA), a vacuum pressure transducer's electrical output is 0 VDC; at full scale vacuum, 14.7 (0 PSIA), it is full scale output, usually 5 VDC. A higher positive voltage output is produced by the vacuum pressure transducer in proportion to a decrease in pressure (increasing vacuum). A higher positive voltage is produced by the absolute pressure transducer in direct proportion to rising pressure (falling vacuum).

9.1.3.6. Measurement of Pressure

There are two types of pressure measurements: static and dynamic. Static pressure is the pressure in cases where there is no motion. The pressure of the water in a basin or the air inside a balloon are two instances of static pressure. A fluid's motion frequently modifies the force exerted on its surroundings. Consider a hose where the nozzle is closed and the water pressure is 40 pounds per square inch (force per unit area).

When you open the nozzle and release water, the pressure decreases. Pressure can be influenced by numerous elements, such as flow, the fluid's compressibility, and outside forces. While a wide variety of pressure measuring instruments are available for various applications, they range from basic low-accuracy gauges to intricate high-accuracy gauges; some are designed to measure very high pressures, while others use fluids; mechanical, electrical, and even optical gauges are among the options.

9.1.3.6.1. Manometer

Manometers are devices that measure pressure by means of liquid columns in either vertical or inclined tubes. The most popular is the water-filled u-tube manometer, which is used to gauge the pressure differential in pitot orifices in the airflow of ventilation or air handling systems.

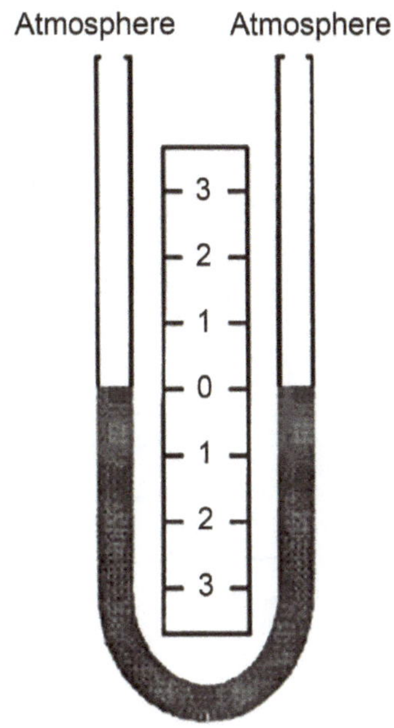

Figure 9.4. *With both legs of a U-tube manometer open to the atmosphere and subjected to the same pressure, the liquid maintains the same level in each leg, establishing a zero reference.*

Source: https://qtxasset.com/files/sensorsmag/nodes/2001/1073/fig1.gif.

Due to its inherent accuracy and ease of use, one of the first pressure-measuring instruments is still in widespread use today. It's a U-shaped glass tube that has liquid partially inside of it; it's called a U-tube manometer. There are no moving parts on this manometer, so calibration is not necessary. The U-tube manometer meets NIST standards for accuracy because its measurements are dependent on the density of the liquid and gravity.

As shown in Figure 9.4, with each leg of a U-tube manometer exposed to the atmosphere, the height of liquid in the columns is equal. Using this point as a reference and connecting each leg to an unknown pressure, the difference in column heights indicates the difference in pressures (see Figure 9.5).

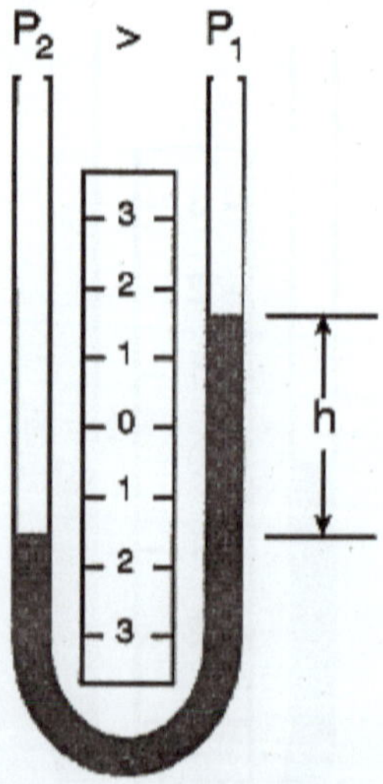

Figure 9.5. *With a greater pressure applied to the left side of a U-tube manometer, the liquid lowers in the left leg and rises in the right leg. The liquid moves until the unit weight of the liquid, as indicated by h, exactly balances the pressure.*

Source: https://qtxasset.com/files/sensorsmag/nodes/2001/1073/fig2.gif.

The fundamental relationship for pressure expressed by a liquid column is:

$$\Delta p = P_2 - P_1 = ⊠gh$$

Where; Δp is the differential pressure; P_1 is the pressure at the low-pressure connection; P_2 is the pressure at the high-pressure connection; ⊠ is the density of the indicating fluid (at a specific temperature); g is the acceleration of gravity (at a specific latitude and elevation); and h is the difference in column heights.

With pounds per square inch (psi) or newtons per square meter (pascals) as the units, the resulting pressure is the difference between the forces applied per unit of surface area of the liquid columns. Since the manometer is so frequently used to measure pressure, the variation in column heights is also a widely used unit. This can be converted using a conversion table from standard units of pressure to inches or centimeters of water or mercury at a given temperature. Every pressure measurement is differential. A complete vacuum or zero absolute pressure, atmospheric pressure (also known as the barometric pressure), or any other pressure can be used as the reference. The measured pressure is that which is greater than the atmospheric pressure, which is 14.7 psi, 101.3 kPa, or 76 cmHg at sea level when one leg of a manometer is exposed to the atmosphere (see Figure 9.6A).

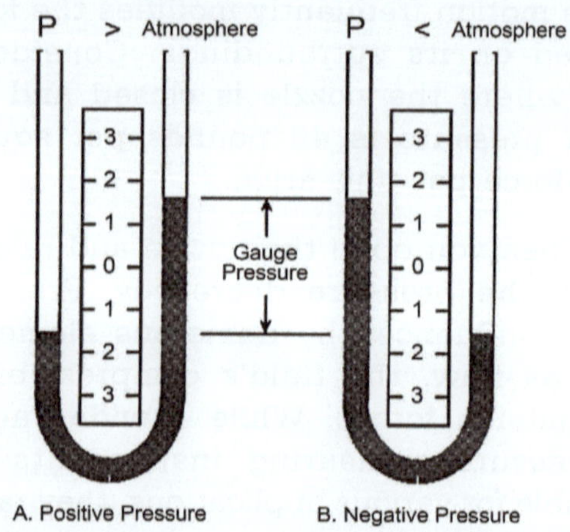

Figure 9.6. *Gauge pressure is a measurement relative to atmospheric pressure and it varies with the barometric reading. A gauge pressure measurement is positive when the unknown pressure exceeds atmospheric pressure (A), and is negative when the unknown pressure is less than atmospheric pressure (B).*

Source: https://qtxasset.com/files/sensorsmag/nodes/2001/1073/fig3.gif.

Gauge pressure is the name given to this measurement. The relationship for a positive pressure is as follows: absolute pressure = atmospheric pressure + positive gauge pressure. The column heights reverse for a negative pressure (vacuum) measurement (see Figure 9.6B), and the relationship is as follows: absolute pressure = atmospheric pressure + negative gauge pressure. Figure 9.7 displays these pressure relationships.

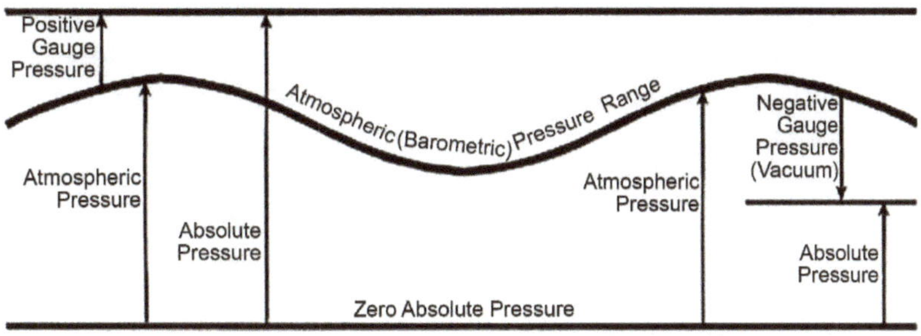

Figure 9.7. *A graphical representation of positive and negative gauge pressure shows the differential aspect of all pressure measurements, where gauge pressure is the difference between absolute pressure and atmospheric pressure.*

Source: https://qtxasset.com/files/sensorsmag/nodes/2001/1073/fig4.gif.

Absolute pressure can be measured directly with a manometer. In a sealed leg above a mercury column, the manometer in Figure 9.8 measures the pressure in relation to zero absolute pressure. The traditional mercury barometer, which is used to measure atmospheric pressure, is the most widely used type of manometer. This configuration allows the measurement of pressures above and below atmospheric pressure with a single connection.

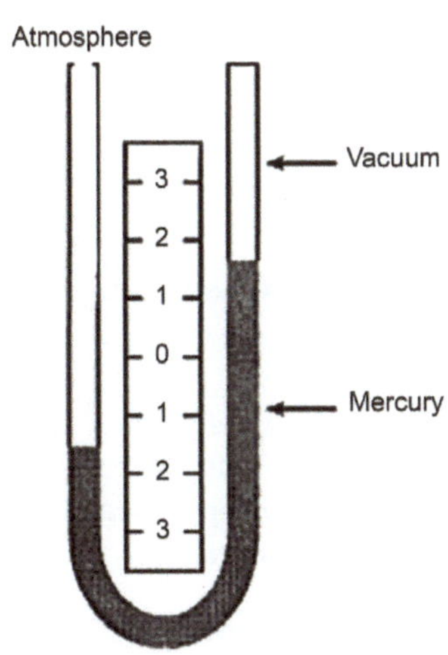

Figure 9.8. *In a sealed-tube manometer, the pressure reference is a vacuum, or zero absolute pressure. The most common form of a sealed-tube manometer is the conventional mercury barometer used to measure atmospheric pressure.*

Source: https://qtxasset.com/files/sensorsmag/nodes/2001/1073/fig5.gif.

9.1.3.6.2. Variations on the U-Tube Manometer

Regardless of the tubes' dimensions, the differential pressure is always equal to the difference in column heights. The indicating fluids are at the same level and both manometers' legs are exposed to the atmosphere, as seen in Figure 9.9A. Each manometer's level decreases when the same pressure is applied to its left leg. The fluid in each column travels a different distance due to the variation in volume in the manometer legs. The fluid levels in the two manometers differ by the same amount, though (see Figure 9.9B).

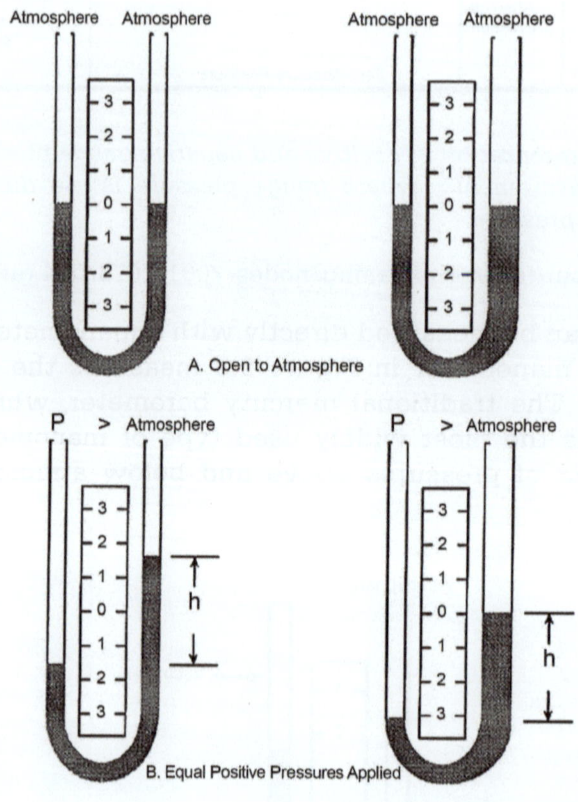

Figure 9.9. *The pressure reading is always the difference between fluid heights, regardless of the tube sizes. With both manometer legs open to the atmosphere, the fluid levels are the same (A). With an equal positive pressure applied to one leg of each manometer, the fluid levels differ, but the distance between the fluid heights is the same.*

Source: https://qtxasset.com/files/sensorsmag/nodes/2001/1073/fig6.gif.

The well-type (also known as reservoir) manometer (refer to Figure 9.10) carries this range of tube sizes even farther. In comparison to the level rise in the column, the well's level slightly drops as pressure is applied. Direct differential pressure readings can be obtained by correcting the column's scale graduations for the well drop. In contrast to the U-tube style, well-type manometers have the following connection guidelines:

- Connect pressures higher than atmospheric to the well; connect pressures lower than atmospheric to the tube.
- For differential measurements, connect the higher pressure to the well.
- For raised-well manometers, the well connection can be used for gauge and vacuum measurements.

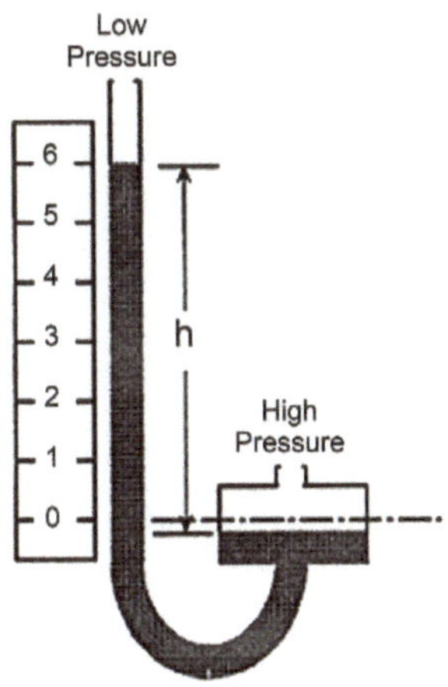

Figure 9.10. *In a well-type manometer, the cross-sectional area of one leg (the well) is much larger than the other leg. When pressure is applied to the well, the fluid lowers only slightly compared to the fluid rise in the other leg.*

Source: https://qtxasset.com/files/sensorsmag/nodes/2001/1073/fig7.gif.

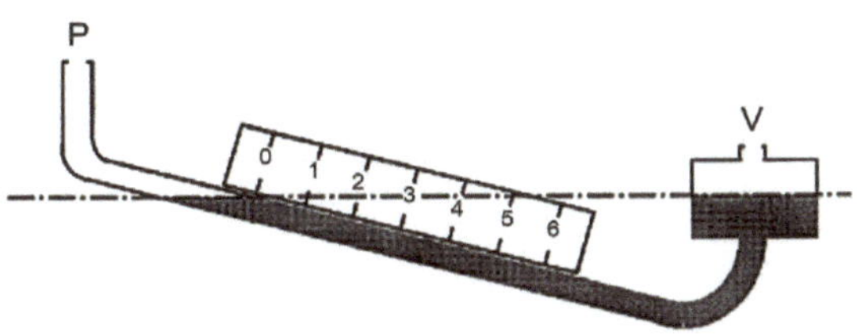

Figure 9.11. *Low pressure and low differentials are better handled with an inclined-tube manometer, where 1 in. of vertical liquid height can be stretched to 12 in. of scale length.*

Source: https://qtxasset.com/files/sensorsmag/nodes/2001/1073/fig8.gif.

The inclined-tube (or draft gauge) manometer shown in Figure 9.11 is a variant of the well-type manometer. With an inclined indicating tube, 1 in. stretches across several inches of scale length in a vertical rise. For low pressures, the inclined-tube manometer offers superior sensitivity and resolution.

9.1.3.6.3. Indicating Fluids

By balancing a liquid's weight between two pressures, liquid manometers detect differential pressure. Small pressure variations can be measured with light liquids like water; large pressure differences require the use of heavy liquids like mercury. The pressure measurement range increases threefold for an indicating fluid three times heavier than water, but the resolution decreases. Colored water, oil, benzenes, bromides, and pure mercury are examples of indicator fluids. Examine the specifications for specific gravity, operating temperature range, vapor pressure, and flash point when choosing an indicating fluid. Other factors to take into account are toxicity, solubility, and corrosive qualities.

9.1.3.7. Digital Manometers

The use of a liquid manometer is limited. Glass tubing, requirements for level mounting, and indicating fluids are better suited for a laboratory setting than for the field. It is also incapable of being interfaced with a PLC or computer. Digital manometers can be used to get around these restrictions. These microprocessor-based instruments come in panel or stand-alone mounting styles, with outputs for process control or measurement data transfer, and in handy, portable sizes for field use. Liquid manometer pressure measurements require manual correction for deviations from standard density and gravity conditions. This is made simpler by the fact that digital manometers allow for the software to account for some of the correction factors for liquid manometers while ignoring others. The dual ports enable the user to switch between gauge, absolute, and differential pressure readings simply by swapping out the sensors.

Other common features of digital manometers include:

- Onboard memory for data logging or storing min./max. readings
- Averaging a number of readings to dampen pressure pulses

Higher-accuracy digital manometers are used to calibrate pressure transmitters and other pressure instrumentation in the field. Digital calibrators are faster and simpler as they require no boxes, gas cylinders, regulators, or weights to set up and have no special platforms or critical leveling requirements.

9.2. FLOW THROUGH ORIFICES AND MOUTHPIECES IN HYDRAULICS

An orifice is a closed-perimeter opening drilled into the side or bottom of a tank that allows fluid to be released. A mouthpiece is a small tube that is attached to a circular opening of the same size in a tank so that liquid can be released through it. The mouthpiece and orifice are used to gauge the liquid flow rate. The device's purpose is to gauge the mouthpiece's and orifice's discharge coefficient. The device is made up of a supply tank with a universal fixture for mounting a mouthpiece or orifice attached to its side. Water is supplied to the supply tank by a centrifugal pump. A bypass valve at the pump discharge regulates the head over the orifice/mouthpiece. To measure the discharge, a measuring tank is offered. A tiny opening in any cross-section, whether it be rectangular, triangular, circular, etc., is called an orifice. An orifice is typically located at the bottom or side of a tank through which fluid is passing. A mouthpiece is a short pipe that is installed in a tank or other fluid-containing vessel and is two to three times longer than its diameter. Mouthpieces and orifices are both used to gauge the fluid flow rate.

9.2.1. Classification of Orifices

The orifices are classified on the basis of their size, shape, nature of discharge, and shape of the upstream edge. The following are the important classifications:

1. Depending on the size and liquid head from the orifice's center, the orifices are categorized as small or large orifices. A small orifice is one where the head of liquid at the center of the orifice is greater than five times the depth of the orifice. It is referred to as a large orifice if the head of liquids is less than five times the depth of the orifice.
2. The orifices are classified as
   - Circular orifice,
   - Triangular orifice,
   - Rectangular orifice and

- Square orifice depending upon their cross-sectional areas.

3. The orifices are classified as
   - Sharp-edged orifice and
   - Bell-mouthed orifice depending upon the shape of the upstream edge of the orifices.
4. The orifices are classified as
   - Free discharging orifices and

- Drowned or sub-merged orifices depending upon the nature of the discharge.

The submerged orifices are further classified as:

- Fully sub-merged orifices and
- Partially sub-merged orifices.

9.2.2. Vena Contracta

Imagine a tank installed in an orifice. The liquid particles travel in all directions in the direction of the orifice in order to flow out through it. A small portion of the particles descend initially, then veer to fit into the opening and eventually pass through it (Figure 9.12).

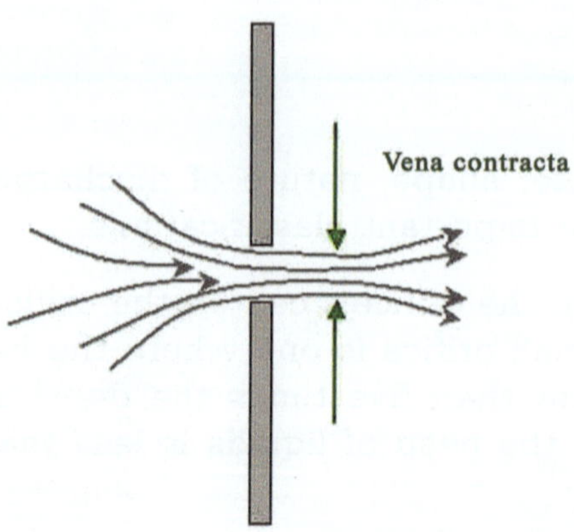

Figure 9.12. *Vena contracta.*

Source: https://www.codecogs.com/users/23547/image001.jpg.

It should be noted that when the liquid particles turn to enter the orifice, some of their energy is lost. It is evident that the jet contracts after exiting the orifice. The jet is roughly horizontal at the point where the maximum contraction occurs, which is a little way downstream of the orifice. Vena-contracta is the name given to such a section.

9.2.3. Hydraulic Coefficients

The following four coefficients are known as hydraulic coefficients or orifice coefficients.

- Coefficient of contraction
- Coefficient of velocity
- Coefficient of discharge
- Coefficient of resistance

1. **Coefficient of Contraction:** The ratio of the area of the jet, at vena-contracta, to the area of the orifice is known as the *coefficient of contraction*. Mathematically coefficient of contraction;

$$C_c = \frac{\text{Area of jet at Vena contracta}}{\text{Area of the orifice}}$$

The value of Coefficient of contraction varies slightly with the available head of the liquid, size and shape of the orifice. The average value of C_c is 0.64.

2. **Coefficient of Velocity:** The ratio of actual velocity of the jet, at vena-contracta, to the theoretical velocity is known as coefficient of velocity.

The theoretical velocity of jet at vena-contracta is given by the relation, $v=\sqrt{2gh}$, where h is the head of water at vena-contracta.

Mathematically coefficient of velocity;

$$C_v = \frac{Actual\ velocity\ of\ the\ jet\ at\ Vena\ contracta}{Theorotical\ velocity\ of\ the\ jet}$$

The orifice's friction is the cause of the velocity differential. Because the orifice's edges have different shapes, the coefficient of velocity's value varies slightly. This value is negligible for orifices with sharp edges. The value of C_v rises with the head of water in a sharp-edged orifice.

3. **Coefficient of Discharge:** The ratio of an actual discharge through an orifice to the theoretical discharge is known as the *coefficient of discharge*. Mathematically coefficient of discharge,

$$C_d = \frac{Actual\ discharge}{Theorotical\ discharge}$$

$$= \frac{Actual\ velocity \times Actual\ are}{Theorotical\ velocity \times Theorotical\ area}$$

$$= C_v \times C_c$$

Thus, the value of the coefficient of discharge varies with the values of C_v and C_c. The average coefficient of discharge varies from 0.60 to 0.64.

4. **Coefficient of Resistance:** The ratio of loss of head in the orifice to the head of water available at the exit of the orifice is known as the *coefficient of resistance*.

$$C_r = \frac{Loss\ of\ head\ in\ the\ orifice}{Head\ of\ water}$$

The loss of head in the orifice takes place because the walls of the orifice offer some resistance to the liquid as it comes out. The coefficient of resistance is generally neglected while solving numerical problems (Bathurst et al., 1981).

5. Experimental Determination of Hydraulic Co-Efficient:

- **Determination of C_d:** The water is allowed to flow through an orifice fitted to a tank under a constant head, H as shown in Figure 9.13. The water is collected in a measuring tank for a known time, t. The height of water in the measuring tank is noted down.

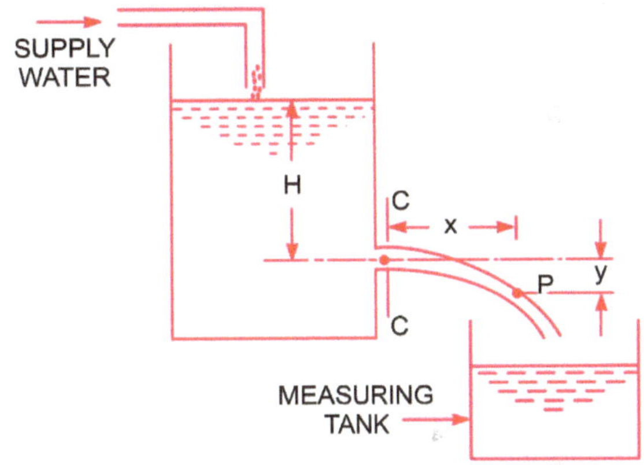

Figure 9.13. *Value of C_d.*

Source: https://www.hkdivedi.com/ezoimgfmt/blogger.googleusercontent.com/img/b/R29vZ2xl/AVvXsEhHaZBHkvljT1dfotWN2h9uks1wE3tUfvaZDCbDOmUmixizIsjIAA8wcaSe9ENrmfxNFs-YLCV1-mYL6bqfQQLm_hdnVET9NV6y5EgQtyw1QgHWVVTEJ4tSrqI37dK18XVNv-dKmwj_RAjd/s1600/1.jpg?ezimgfmt=rs:295x235/rscb2/ng:webp/ngcb2.

The actual discharge through an orifice,

$$Q = \frac{Area\ of\ measuring\ tank \times Height\ of\ water\ in\ measuring\ tank}{Time(t)}$$

$$and\ theoretical\ discharge = Area\ of\ orifice \times \sqrt{2gh}$$

$$Therefore,\ C_d = \frac{Q}{a \times \sqrt{2gh}}$$

- **Determination of Co-Efficient of Velocity (C_v):** Let C-C represent the vena-contracta of a jet of water coming out from an orifice under constant head H as shown in figure 9.13. Consider a liquid particle which is at vena-contracta at any time and takes the position at P along the jet in time 't.'

  Let x = horizontal distance traveled by the particle in time 't'

  y = vertical distance between P and C-C

  V = actual velocity of jet at vena-contracta.

  Then horizontal distance, x = V × t (1)

  and vertical distance, $y=\frac{1}{2}gt^2$ (2)

  For Eqn. (1) $t=\frac{x}{V}$

  Substituting this value of 't' in (2), we get

  $$y=\frac{1}{2}g\times\frac{x^2}{V^2}$$

  $$V^2=\frac{gx^2}{2y}$$

  Therefore, $V=\sqrt{\frac{gx^2}{2y}}$

But theoretical velocity,

$$V_{th}=\sqrt{2gh}$$

Therefore, Co-efficient of velocity,

$$C_v=\frac{V}{V_{th}}=\sqrt{\frac{gx^2}{2y}}\times\frac{1}{\sqrt{2gh}}=\sqrt{\frac{x^2}{4yH}}=\frac{x}{\sqrt{4yH}}$$

- **Determination of Co-efficient of Contraction (C_c):**

  The co-efficient of contraction is determined from the equation of the co-efficient of discharge as;

  $$C_d=C_v\times C_c$$

  Therefore, $C_c=\frac{C_d}{C_v}$

9.2.4. Flow Through an Orifice

As illustrated in Figure 9.14, imagine a tank with a circular orifice drilled into one of its sides. Let H represent the liquid's head above the orifice's center. A jet of liquid with a cross-sectional area smaller than the orifice is created by the liquid passing through the orifice. The fluid jet's area continues to decrease, reaching its minimum at section CC. This section is situated roughly half the orifice's diameter away. The streamlines are perpendicular to the orifice plane, straight, and parallel to one another at this point. The Vena-contracta is the name of this section. Once past this point, the jet diverges and becomes drawn downward by gravity.

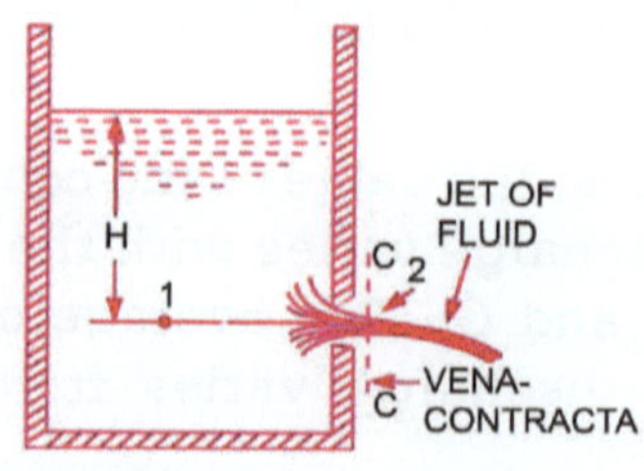

Figure 9.14. *Tank with an orifice.*

Source: https://blogger.googleusercontent.com/img/b/R29vZ2xl/AVvXsEjoR1Y7r9vuqt1W37uY4X4BNV4Rq6mhMBnHu68t4TE4YDtTVz-09t1_fOFh7X7bEVOLqwNEh9gmrJrzpPY0CBfxkaLNuVk24TldueXDzarZ-jwWNfxycZo8c6zLEMGeTiEHejAs4hXnUcGe/s640/Capture1.JPG.

Consider two points 1 and 2 as shown in Figure 9.14. Point 1 is inside the tank and point 2 is at the vena-contracta. Let the flow be steady and at a constant head H. Applying Bernoulli's equation at points 1 and 2.

$$\frac{p_1}{\rho g}+\frac{v_1^2}{2g}+z_1=\frac{p_2}{\rho g}+\frac{v_2^2}{2g}+z_2$$

But $z_1 = z_2$

Therefore, $$\frac{p_1}{\rho g}+\frac{v_1^2}{2g}=\frac{p_2}{\rho g}+\frac{v_2^2}{2g}$$

Now, $\frac{p_1}{\rho g}=H$

and $\frac{p_2}{\rho g}=0\ (atmospheric\ pressure)$

v_1 is very small in comparison to v_2 as the area of the tank is very large as compared to the area of the jet of liquid.

Therefore, $H+0=0+\frac{v_2^2}{2g}$

Therefore, $v_2=\sqrt{2gh}$

This is theoretical velocity. Actual velocity will be less than this value.

9.2.5. Flow Through Large Orifices

If the head of liquid is less than 5 times the depth of the orifice, the orifice is called a large orifice. In the case of the small orifice, the velocity in the entire cross-section of the jet is considered to be constant and discharge can be calculated by $Q=C_d\times a\times\sqrt{2gh}$. But in the case of a large orifice, the velocity is not constant over the entire cross-section of the jet and hence Q cannot be calculated by $Q=C_d\times a\times\sqrt{2gh}$.

- **Discharge Through a Large Rectangular Orifice:** Consider a large rectangular orifice on one side of the tank discharging freely into the atmosphere under a constant head, H as shown in Figure 9.15.

  Let – H_1 = height of liquid above the top edge of the orifice

  H_2 = height of liquid above the bottom edge of the orifice

  b = breadth of orifice

  d = depth of orifice = $H_2 - H_1$

  C_d = Co-efficient of discharge.

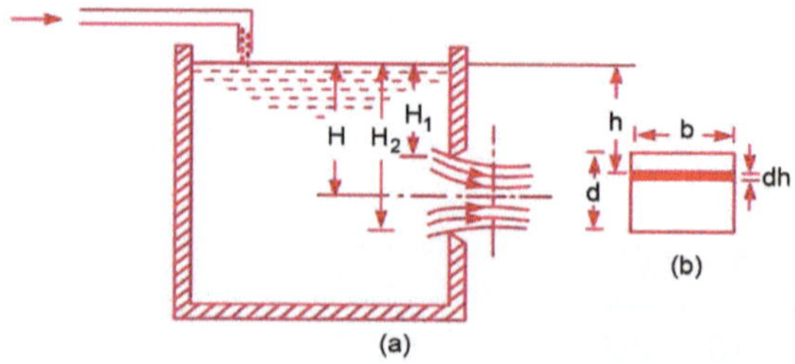

Figure 9.15. *Large rectangular orifice.*

Source: https://www.hkdivedi.com/ezoimgfmt/blogger.googleusercontent.com/img/b/R29vZ2xl/AVvXsEhFv3buCmRWr6Hpmz4xmawJgzdzNcMyIFPX03xnuDgJjhqB_yBGJuS17tkpLIVQrX8A3FwD6MpxfI5eXHzGP2vHAFlFXIBJraeVnZ_vvuj59Oj_HkoSjeNUleoAblu9XDNQOqZLuWCcnKB3/s1600/12.jpg?ezimgfmt=rs:338x191/rscb2/ng:webp/ngcb2.

Consider an elementary horizontal strip of depth 'dh' at a depth of `h' below the free surface of the liquid in the tank as shown in Figure 9.15(b).

Therefore, Area of strip = b × dh

and theoretical velocity of water through strip = $\sqrt{2gh}$

Therefore, Discharge through the elementary strip is given

$$dQ = C_d \times Area\ of\ strip \times velocity$$
$$= C_d \times dh \times \sqrt{2gh}$$
$$= C_d b \times \sqrt{2gh}\, dh$$

By integrating the above equation between the limits H_1 and H_2, the total discharge through the whole orifice is obtained

Therefore;

$$Q = \int_{H_1}^{H_2} C_d \times b \times \sqrt{2gh}\, dh$$
$$= C_d \times b \times \sqrt{2g} \int_{H_1}^{H_2} \sqrt{h}\, dh = C_d \times b \times \sqrt{2g}\left[\frac{h^{3/2}}{3/2}\right]_{H_1}^{H_2}$$
$$= \frac{2}{3} C_d \times b \times \sqrt{2g}\left[H_2^{3/2} - H_1^{3/2}\right]$$

9.2.6. Discharge Through Fully Sub-Merged Orifice

A fully submerged orifice is one that releases a jet of liquid into another liquid of the same kind with the entirety of its outlet side submerged under liquid.

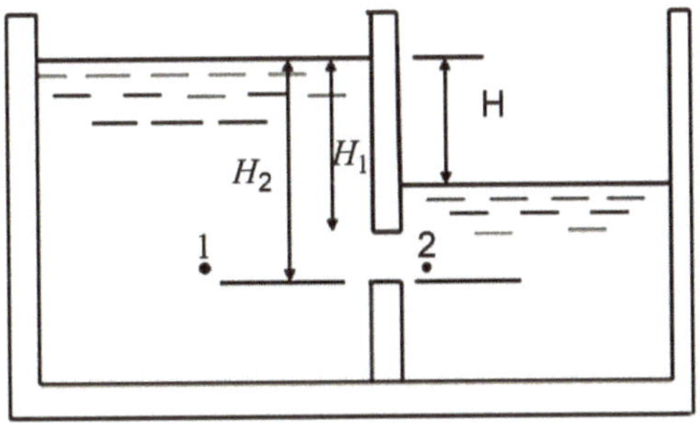

Figure 9.16. *Fully submerged orifice.*

Source: https://blogger.googleusercontent.com/img/b/R29vZ2xl/AVvXsEhUHJQufE8j9R3kJsNZGGbSaMPNy420vkVex3I2EZrKJZ2Ch0Ayhv2Tr7M6diwEegjWih2LQSUh7kV8W22GlNPsDkMZvJC4jAv0FlyCfMXL-2CIrC9fv9tHO10cI18oJBisOEAFqGQT3ubt/s320/B1.JPG.

Another name for it is completely drowned orifice. The fully submerged orifice is depicted in Figure 9.16. Think about points (1) and (2), which are represented in figure 9.16. Point 1 is in the reservoir on the upstream side of the orifice, and point 2 is at the vena-contracta.

Let H_1 = Height of water above the top of the orifice on the upstream side.

H_2 = Height of water above the bottom of the orifice

H = Difference in water level

b = Width of orifice

C_d = Co-efficient of discharge.

Height of water above the center of the orifice on the upstream side

$$=H_1+\frac{H_2-H_1}{2}=\frac{H_1+H_2}{2} \quad (3)$$

Height of water above the center of the orifice on the downstream side

$$=\frac{H_1+H_2}{2}-H \quad (4)$$

Applying Bernoulli's Eqn. at (3) and (4), we get

$$\frac{p_1}{\rho g}+\frac{V_1^2}{2g}=\frac{p_2}{\rho g}+\frac{V_2^2}{2g}$$

Now,

$$\frac{p_1}{\rho g}=\frac{H_1+H_2}{2}, \frac{p_2}{\rho g}=\frac{H_1+H_2}{2}-H \text{ and } V_1 \text{ is negligible}$$

$$\therefore \frac{H_1+H_2}{2}+0=\frac{H_1+H_2}{2}-H+\frac{V_2^2}{2g}$$

$$\therefore \frac{V_2^2}{2g}=H$$

$$\therefore V_2=\sqrt{2gh}$$

Area of orifice = b × (H_2 − H_1)

Therefore, Discharge through orifice = C_d × Area × Velocity = C_d × b (H_2 − H_1) × $\sqrt{2gh}$

Therefore, Q = C_d × b (H_2 − H_1) × $\sqrt{2gh}$

9.2.7. Discharge Through Partially Sub-Merged

An orifice that has part of its outlet side partially submerged in a liquid is said to be partially submerged, as Figure 9.17 illustrates. Another name for it is a partially drowned orifice. There are two parts to the partially submerged orifice as a result.

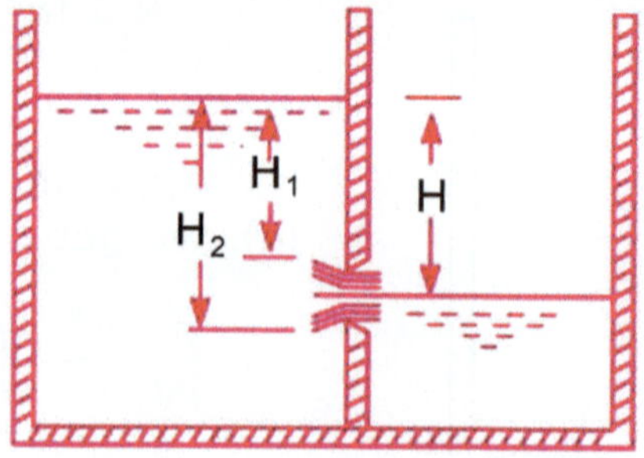

Figure 9.17. *Partially submerged orifice.*

Source: https://www.hkdivedi.com/ezoimgfmt/blogger.googleusercontent.com/img/b/R29vZ2xl/AVvXsEjnAlCpUxypUZVfb9vW5zEALDiC OBbnQyQWOVO9kR1Dlji_fEODtOnEE06COsRGKQwJhMiYbAkaMmZ7-qTynDqYANG4xfxS4LPbR3byyB5nOcW7HIiiBZZ-f-YpH9XT2sDkwqkcAVsuWsMr/s1600/51.jpg?ezimgfmt=rs:165x158/rscb2/ng:webp/ngcb2.

While the lower portion functions as a submerged orifice, the upper portion functions as a free-draining orifice. The only orifice that functions as a partially submerged orifice is a large one. The sum of the discharges through the free and submerged sections determines the total discharge Q through the partially submerged orifice. The discharge through the part that is submerged is provided by.

$$Q_1 = C_d \times b\,(H_2 - H) \times \sqrt{2gh}$$

Discharge through the free portion is given as

$$Q_2 = \frac{2}{3} C_d \times b \times \sqrt{2g}\left[H_2^{3/2} - H_1^{3/2}\right]$$

$$\therefore \textit{Total discharge}, Q = Q_1 + Q_2$$

$$= C_d \times b \times (H_2 - H) \times \sqrt{2gH} + \frac{2}{3} C_d \times b \times \sqrt{2g}\left[H_2^{3/2} - H^{3/2}\right]$$

ROLE MODEL

SUNSHU AO: A CHINESE HYDRAULIC ENGINEER AND POLITICIAN.

Biography

Sunshu Ao (孫叔敖) was an ancient Chinese court minister serving the administration of Duke Zhuang of Chu during the reign of King Ding of Zhou (606 BC-586 BC), during the Eastern Zhou Dynasty. During his governmental career, Sunshu Ao was given notice by Duke Zhuang, who had him promoted to the rank of Prime Minister in the State of Chu. Sunshu Ao was entrusted with many endeavors of the state, and because of a large dam, reservoir, and irrigation project he had established, he is also credited as the first known hydraulic engineer of China. Needham, Volume 4, Part 3, 271.] Dams, reservoirs, and irrigation.

Both the ancient historian Sima Qian (in his "Shiji") and the author of the "Huai Nan Zi" wrote of Sunshu Ao and his works. Their records state that Duke Zhuang had given Sunshu Ao the responsibility and oversight of the construction of a large river dam that would create an enormous planned reservoir for means of agricultural irrigation. The erection of this dam flooded a flat valley in modern-day northern Anhui province (south of Shouxian City), and the reservoir created spanning a circumference of 62 miles. Since Sunshu's time, this ancient reservoir accumulated tons of north-flowing water that came from the mountains north of the Yangtze River and supplied an irrigated area of some 6 million acres (24000 km^2). In ancient times, the Chinese had called this reservoir the "Si-Si Bei," as well as the "Shao Bei" (Peony Dam). Today, the large reservoir created thousands of years ago by Sunshu Ao still exists, known in modern times as the Anfeng Tang (Anfeng Reservoir).

Subsequent Chinese authorities following the Zhou Dynasty repaired and maintained the dam that flooded the reservoir created by Sunshu Ao, such as the Chinese states of the Han Dynasty (202 BC-220 AD) era and the Tang Dynasty (618–960 AD) era.

- **Class Activity:** "Design and Analysis of a Simple Hydraulic System"

- **Objective:** To understand the principles of hydraulic engineering by designing a basic hydraulic system and analyzing its performance.
- **Materials Needed:** Clear plastic tubing, water, small water pumps, pressure gauges, flow meters, measuring tools, personal protective equipment, etc.
- **Assessment**: Each group designs a simple hydraulic system to lift a small weight. The system should include a reservoir, pump, tubing, and a lifting mechanism. Groups draw a schematic of their hydraulic system, labeling all components and indicating the flow direction. Groups analyze their test results, comparing them with theoretical predictions based on hydraulic principles.

SUMMARY

- Hydraulic engineering is a specialized field within civil engineering that focuses on the flow and conveyance of fluids, primarily water and sewage. It encompasses the design, construction, and maintenance of structures and systems that control and manage water resources.
- Hydraulic engineers must consider various factors such as the hydrology of the region, geological conditions, environmental impact, and the socio-economic implications of their designs.
- Hydraulics deals with such matters as the flow of liquids in pipes, rivers, and channels and their confinement by dams and tanks. Some of its principles apply also to gases, usually in cases in which variations in density are relatively small.
- The power of liquid fuel in hydraulics is significant and as a result, hydraulics are commonly used in heavy equipment. In a hydraulic system, pressure, applied to a contained fluid at any point, is transmitted undiminished.
- Liquids are useful for transmitting power through pipes, for small or large distances, and around corners and up and down. The force applied at one end of a pipe will immediately be transferred with the same force to the other end of the pipe.
- Atmospheric pressure is defined as the force per unit area exerted against a surface by the weight of the air above that surface.
- The term vacuum is often misunderstood and consequently misapplied when discussing pressure measurements and choosing pressure transducers. By definition, a vacuum is a space that is partially exhausted (as to the highest degree possible) by artificial means (such as an air pump).
- Vacuum pressure is measured relative to ambient atmospheric pressure. It is referred to as pounds per square inch (vacuum) or PSIV. The electrical output of a vacuum pressure transducer is 0 VDC at 0 PSIV (14.7 PSIA) and full-scale output (typically 5 VDC) at full-scale vacuum, 14.7 (0 PSIA).

REVIEW QUESTIONS

1. What are the primary reasons for the widespread use of hydraulic systems in various engineering applications?
2. How do the fundamental principles of hydraulics underpin the design and operation of hydraulic machinery?
3. In what ways does atmospheric pressure influence hydraulic systems and fluid mechanics?
4. How do variations in the U-tube manometer affect the accuracy and reliability of pressure measurements in hydraulic systems?
5. What is the significance of the vena contracta in fluid flow through orifices, and how does it impact hydraulic efficiency?

6. How do hydraulic coefficients contribute to the analysis and optimization of fluid flow in hydraulic engineering applications?

REFERENCES

1. Andreasen, J. O. (1979). Grödens hydralike virkning. Laboratory for Physical Geography, Geological Institute, Århus University, Report No. 3, 64pp.
2. ASCE. (1963). Frictional factors in open channels. *Journal of the Hydraulics Division, Proceedings of the American Society of Civil Engineers, 89*, 97–143. (Discussion: 1963, 89 (July), 283; 1963, 89 (Sept), 169; 1964, 90 (July), 223.)
3. Autos.com. (2015). 7 things to know about brake fluid. Retrieved from: http://www.autos.com/auto-repair/7-things-to-know-about-brake-fluid (accessed on 15 July 2024).
4. Bathurst, J. C., Li, R.-M., & Simons, D. B. (1981). Resistance equation for large scale roughness. *Journal of the Hydraulics Division, Proceedings of the American Society of Civil Engineers, 107*, 1593–1613.
5. Bruk, S., & Oef, Z. (1975). Determination of roughness coefficients for very irregular rivers with large flood plains. In *Proceedings of the 12th Congress of the International Association for Hydraulic Research, Paper A.12*.
6. Butterworth, B. (1981). Biofilm growth and hydraulic performance. *Journal of the Hydraulics Division, Proceedings of the American Society of Civil Engineers, 107*, 629–649.
7. Chen, C. L. (1976). Flow resistance in broad shallow grassed channels. *Journal of the Hydraulics Division, Proceedings of the American Society of Civil Engineers, 102*, 307–322. (Discussion: 1977, 103, 90–91; 1977, 103, 1356–1357.)
8. Chow, V. T. (1981). Hydraulic exponents. *Journal of the Hydraulics Division, Proceedings of the American Society of Civil Engineers, 107*, 1489–1499.
9. Chow, V. T. (1981). *Open-Channel Hydraulics*. McGraw-Hill Book Co. Inc. New York, 680pp. (First edition 1959.)
10. Cook, H. L. (1938). Spartanburg Outdoor Hydraulic Laboratory. *Journal of the Civil Engineering Division, Proceedings of the American Society of Civil Engineers, 8*, 653–655.
11. Cornish, B. A., Yong, K. C., & Stone, D. M. (1967). Hydraulic characteristics of low-cost surfaces for farm dam by wash spillways. University of New South Wales, Water Research Laboratory, Report No. 93.
12. Dictionary.com. (2015). The definition of fluid. Retrieved from: http://dictionary.reference.com/browse/fluid (accessed on 15 July 2024).
13. Fenzl, R. N. (1962). Hydraulic resistance of broad shallow vegetated channels. University of California, Davis. PhD Thesis.

14. Horsforth School. (2015). A question of pressures. Retrieved from: http://www.sciwebhop.net/sci_web/science/ks3/worksheets/word/9L%5C./9Lb3_files/image009.jpg (accessed on 15 July 2024).
15. Mechanicalmania.blogspot.com. (2015). Hydraulic brake ~ mechanical engineering. Retrieved from: http://mechanicalmania.blogspot.com/2011/07/hydraulic-brake.html (accessed on 15 July 2024).

CHAPTER

10 Project Management in Civil Engineering

LEARNING OBJECTIVES

After studying this chapter, you will be able to:

- Discuss the project planning and scheduling
- Explain the construction project management
- Focus on cost estimation and budgeting
- Describe the risk management in construction projects

KEY TERMS FROM THIS CHAPTER

Civil engineering
Crucial skills
Project efficiency
Risk assessment
Construction manager
Mitigation
Project planning
Risk management

INTRODUCTION

In the field of civil engineering, project management is essential to the successful completion of major infrastructure projects, such as the construction of skyscrapers, bridges, and highways. The discipline includes a range of methods and procedures designed to maintain safety and quality standards while allocating resources, controlling expenses, and completing projects on schedule. Effective planning is one of the most important components of project management in civil engineering. This includes preliminary feasibility studies, in-depth design creation, and thorough scheduling. Project managers and engineers collaborate to set project goals, assign materials, labor, and equipment, and define the project's scope. Effective planning lays the groundwork for a project's successful completion and reduces risks throughout its whole lifecycle.

One more essential element is risk management. Numerous risks, such as those related to the environment, changing regulations, and unforeseen technical difficulties, are frequently present in civil engineering projects. Early risk identification, assessment of likelihood and impact, and development of mitigation strategies are all done by project managers in order to reduce potential effects. This proactive approach ensures the safety and integrity of the infrastructure being built while also helping to keep projects within budget and schedule constraints.

Budget management and cost control are essential to the accomplishment of civil engineering projects. Project managers must monitor costs during the course of the project, estimate costs precisely from the beginning, and make adjustments as necessary. Good cost control guarantees that financial resources are used effectively and helps avoid budget overruns.

In civil engineering, quality control and assurance are critical to delivering infrastructure that satisfies client expectations and engineering standards. Project managers oversee the execution of quality management plans, carry out inspections, and ensure adherence to guidelines and standards. To keep everyone in agreement throughout the project, they also help stakeholders—engineers, contractors, clients, and regulatory agencies—communicate with one another.

Finally, successful project collaboration and communication are critical. Projects involving civil engineering entail multidisciplinary groups and stakeholders with a range of specialties and objectives. In order to maintain project momentum and accomplish intended results, project managers establish and maintain open lines of communication, mediate disputes, and promote teamwork. Civil engineering project management is a complex field that calls for careful planning, strict risk management, careful cost control, strict quality assurance, and effective communication. Project managers can guarantee the effective implementation of infrastructure projects that improve communities and advance society development by incorporating these components.

10.1. PROJECT PLANNING AND SCHEDULING

The efficient and successful completion of projects depends on the fundamental project management processes of project planning and scheduling. In civil engineering, careful planning and exact scheduling are essential because complex infrastructure projects take a long time to complete and involve many stakeholders.

10.1.1. Project Planning

In civil engineering, project planning starts with a thorough analysis of the goals and requirements of the project. Determining the project's scope entails determining its deliverables, objectives, and limitations. In close coordination with stakeholders, engineers, and project managers set specific, attainable goals that meet regulatory requirements and client expectations.

Creating a thorough project plan is the next stage after defining the scope. This plan outlines the tasks, deadlines, milestones, and resources needed, acting as a roadmap. It consists of a task breakdown, task sequencing, and dependency analysis between various project components. Better organization and control are made possible by the systematic breakdown of projects into manageable sections using tools like Work Breakdown Structures (WBS).

Risk management and assessment are essential components of planning. Early risk identification enables project teams to create plans for risk mitigation and contingency. In civil engineering projects, risks can include everything from technical difficulties and stakeholder conflicts to environmental factors and regulatory changes. Proactively managing these risks reduces interruptions and guarantees more seamless project execution.

Project planning also entails allocating resources, such as supplies, machinery, and labor. Effective resource management throughout the project lifecycle is ensured by efficient allocation, which helps with cost containment and timeline adherence. Budgeting and financial management are also included in planning; resource allocation and expenditure tracking are guided by comprehensive cost estimates and financial forecasts.

10.1.2. Project Scheduling

The project plan is translated into a timeline of tasks and deadlines through project scheduling. It entails giving each task a defined start and end date and taking critical path analysis, dependencies, and resource availability into account. In civil engineering, schedules that maximize project duration and resource utilization are typically developed using the critical path method (CPM) and program evaluation and review technique (PERT).

Setting reasonable deadlines and timetables is essential to project scheduling. Project timelines can be affected by a number of factors, including lead times for procurement, weather, and regulatory approvals. These factors must be factored into schedules. Schedules are designed with flexibility in mind, allowing for last-minute changes or delays without compromising the overall goals of the project.

Throughout the scheduling process, it is crucial to track and manage progress. Project managers use tools like Gantt charts and project management software to monitor actual progress against scheduled milestones and timelines. This makes it possible to quickly identify schedule deviations and take appropriate corrective action. Throughout the project, transparency and alignment are maintained through regular stakeholder meetings and progress reports.

In order to facilitate coordination among team members, contractors, suppliers, and clients, communication is essential to project scheduling. Open lines of communication promote cooperation and reduce miscommunication by guaranteeing that all parties involved in the project are informed of schedule updates, modifications, and milestones.

The success of civil engineering projects is largely dependent on the efficient planning and scheduling of said projects. Project managers can effectively manage risks, allocate resources efficiently, set clear goals, create realistic schedules, and navigate the complexities of infrastructure projects to produce results that satisfy stakeholder's and society's demands. Proactive communication, flexibility in response to changing circumstances, and ongoing monitoring are essential for guaranteeing that projects are finished on schedule, within budget, and to the appropriate standard of quality.

10.1.3. What Does a Civil Engineer Project Manager Do?

Large-scale construction projects are managed by professionals known as civil engineering project managers. They frequently work on several projects at once, and in order to move building supplies around the site, they must be physically strong and active. You can decide if this is the right career path for you by learning about the duties and skill requirements of a civil engineering project manager.

It's a good idea to familiarize yourself with the roles and responsibilities of civil engineering project managers if you're trying to figure out what they do. Often referred to as civil project managers, civil engineering project managers supervise the construction of major projects, such as roads, bridges, and commercial buildings. These experts handle every aspect of the project, including pre-planning, budgeting, resource allocation, contract negotiation, employee coordination, and goal implementation. In addition, civil project managers oversee subcontractors and other workers on the construction site, participate in weekly project meetings, and keep an eye on the observance of health and safety regulations. These specialists can operate independently as private practitioners or full-time as employees of government organizations, engineering firms, or construction companies. The need for civil project managers is rising due to advancements in technology and the complexity of construction procedures. These managers may have to deal with time and money constraints in addition to their daily responsibilities because most projects have tight deadlines that require their undivided attention.

10.1.4. Duties of a Civil Engineer Project Manager

Some typical duties of a civil engineer project manager include:

- **Project Planning:** As civil project managers frequently oversee several projects at once, one of their most important responsibilities is to establish the appropriate plan and procedures to be followed during the building process. They begin with the pre-planning phase, during which they investigate potential locations and assess the suitability of the land and sites. These experts may also evaluate the budget, estimated completion date, and project estimations. In the process of developing design plans that act as a visual reference point for specific activities, they also look at existing structures that are pertinent to these projects. Reviewing possible safety concerns and creating backup plans for the future are additional tasks included in project planning.
- **Oversee Construction Personnel:** Supervising the work performance of subcontractors and other construction workers on the job site is a crucial aspect of a civil project manager's duties. In addition to negotiating contracts with subcontractors and keeping in touch with all pertinent parties, including inspectors and architects, they oversee the composition and distribution of tasks among the site crew. In order to address their needs and make sure the work meets their expectations, they also schedule meetings with construction managers and their clients. In addition to managing staff, civil engineer project managers arrange with vendors and suppliers and ensure that materials are delivered on schedule.
- **Prepare Progress Reports:** Through progress reports, civil project managers evaluate team performance and graphically depict the objectives they still haven't met. This enables them to guarantee that the team achieves their objectives and to clarify their vision to everyone. Construction staff can regularly prepare progress reports to provide upper management with an understanding of the tasks completed by each professional by tracking their professional development. In order to specify each employee's role in the finished project, these project reports also include project costs.

10.1.5. Preparing Construction Schedules

Creating construction schedules pertinent to the overall project completion is a significant portion of a civil project manager's work. They arrange the tasks for the project according to the weather, construction crew availability, and financial resources. They frequently rearrange a lot of their assignments to accommodate unanticipated events. These experts may plan their work and oversee the team's daily output using a variety of software programs.

- **Reviewing Compliance:** Project managers who are civil engineers create project plans and assist in coming up with ways to carry them out. In addition, they verify that their project or site location complies with all applicable state and federal regulations. When there are discrepancies, they address the problem and take the necessary precautions to keep it from happening again in future projects. In addition, these experts make sure that the project conforms to safety standards and laws and that their workers have a healthy workplace while construction is underway. In addition to following safety protocols, they guarantee that their projects meet environmental standards.

10.1.6. Essential Skills for a Civil Engineer Project Manager

Some of the crucial skills essential for a civil engineer project manager include:

- **Communication Skills:** There are several parties involved in any civil engineering construction project, including the client, subcontractors, external suppliers, and other construction workers. In order to guarantee seamless coordination between all stakeholders, civil project managers keep in regular contact with these experts. Assuring efficiency in back-office operations and expediting the processing of paperwork, including contracts, are two examples of how important communication is to a civil project manager's job. Clear communication creates clarity and increases productivity within the construction team, which can result in outstanding outcomes.
- **Organizational Skills:** A civil project manager frequently works with multiple deadlines at once and multitasks. They might be working on multiple assignments at once, which frequently calls for the application of strong organizational abilities in order to effectively manage and prioritize these tasks. Working with this understanding is necessary for these managers, and their ability to stay organized can enable them to quickly adjust when unforeseen tasks arise and effectively readjust their schedules.
- **Risk Management:** Project managers in the field of civil engineering are aware that things might not always go as planned and that their project schedules could alter over time. Accepting and embracing risks can help employees work more efficiently because they are aware of the obstacles and unknowns that lie ahead, which facilitates their ability to adjust to changing conditions. Their ability to effectively provide the most practical solutions to issues is another benefit of having risk management skills. Their ability to experiment with various procedures to discover fresh approaches to swiftly completing high-quality projects also contributes to their comfort level.
- **Negotiation:** There is a lot of negotiating involved in a project manager's work. They frequently bargain over contracts with customers and other parties involved, so it's critical to know how to haggle over important issues like time and money. They may also engage in negotiations to come to a consensus amongst different parties, such as suppliers and vendors, on issues like ordering modifications, staff scheduling, and obtaining building supplies. Being a skilled negotiator also means knowing when to make concessions to clients.

- **Flexibility:** Project managers in civil engineering may want to be flexible individuals who recognize that things can change and that they might need to modify their original plans to suit the needs of their clients. These experts frequently exhibit flexibility in accommodating evolving project requirements and making appropriate preparations. In this context, flexibility also refers to adjusting to new technologies or methods of doing tasks in order to maximize productivity.
- **Industry Knowledge:** Civil project managers are well-versed in both their field and the kinds of projects they create. Even though these experts are trained and knowledgeable about current market trends, the construction sector is a large and ever-evolving one where new methods and procedures are created on a daily basis. Effective construction managers know how to apply creative approaches to optimize project efficiency in addition to having a solid understanding of project management fundamentals. These professionals can work more effectively if they have relevant industry knowledge in addition to project management skills.
- **Financial Management:** Project managers need a well-organized and detailed budget plan in order to carry out a construction project effectively. These professionals may be able to forecast economic conditions and adjust their activities and budgets accordingly if they possess strong financial management skills. Civil engineers who oversee projects may have an awareness that even minor modifications can have an impact on how well the project is carried out and drive costs over budget. Effective financial literacy and management abilities can aid project managers in averting undesirable circumstances by monitoring project expenditures and investigating alternative funding options.

10.2. CONSTRUCTION PROJECT MANAGEMENT

From project conception to project completion, construction project management entails planning and coordinating every stage of the project life cycle. The aim of this all-encompassing approach is to complete projects on schedule and within budget. Cost control, scheduling, procurement, risk assessment, and other critical issues are just a few of the many complex issues that must be addressed in the complicated field of construction project management. Project managers communicate with architects, owners, contractors, and other team members when working on a construction project.

Avoiding costly errors takes precedence over all other considerations when millions are at stake in a construction project. You have limited resources and time, so you have to plan ahead and consider every possibility. Whether you're building power plants, modern infrastructure, or oil and gas refineries, construction project management enables you to accomplish all of these goals. Construction project management assists you in keeping the entire construction process effective and on schedule, from planning and design to resource management, budget allocation, and more.

Managing a construction project entail supervising its planning, design, and construction from start to finish. It's a complicated process that needs resource coordination, stakeholder management, regulatory compliance, and risk mitigation in order to deliver projects on schedule, within budget, and to the necessary quality standards.

- **Initiation and Planning:** Determining the goals, parameters, and viability of the project is the first step in the construction project management initiation phase. This entails carrying out preliminary cost estimates, site evaluations, and feasibility studies. To establish the needs and objectives of a project, project managers collaborate closely with clients, architects, engineers, and other stakeholders. To prevent scope creep and guarantee that the project is in line with client expectations, clear communication and agreement on the scope are crucial.

  Detailed project plans are created in the planning stage. This includes defining tasks and dependencies, allocating resources like labor, materials, and equipment, and developing a project schedule with milestones. The critical path—the order of tasks that must be finished on time for the project to be completed on schedule—is identified by using techniques like the Work Breakdown Structure (WBS) and Critical Path Method (CPM) to sequence activities.
- **Execution and Control:** During the execution phase, subcontractors, resources, and construction activities are managed in accordance with the project plan.

Overseeing daily activities on the construction site, project managers make sure that work is completed safely, effectively, and in accordance with legal and regulatory requirements. They keep an eye on the timetable, keep tabs on expenses, and handle any problems or adjustments that come up while the project is being built.

To coordinate activities among various teams and stakeholders during the execution phase, effective communication is essential. Stakeholders are kept informed of project developments and transparency is preserved through regular meetings, progress reports, and site inspections. In order to guarantee that supplies and services are acquired on schedule and within budget, project managers also oversee contracts and procurement.

- **Monitoring and Control:** The phase of monitoring and control entails ongoing performance evaluation of the project. Key performance indicators (KPIs) that project managers monitor include schedule deviation, cost performance, and quality metrics. To keep the project on track, they put corrective measures in place to deal with plan deviations like budget overruns or schedule delays.

  Risk control is essential to a project's entire lifecycle. Early risk identification, impact, and likelihood analysis, and strategy development are the tasks of project managers. This proactive strategy reduces interruptions and strengthens the project's resistance to unanticipated circumstances.

- **Closure and Evaluation:** The construction phase comes to an end during the closure phase, and the client receives the project. Project managers make sure that all deliverables adhere to legal and quality standards. They carry out last-minute checks, take care of any lingering problems, and get the project approved by the client. To determine areas for improvement and to evaluate lessons learned, post-project evaluations are carried out. Project managers collect input from interested parties, assess how well the project is doing in comparison to its original objectives, and record achievements and difficulties. This data is useful for process improvement and better project execution in the future.

  Strong leadership abilities, technical proficiency, and a thorough understanding of construction procedures are necessary for construction project management. Construction projects must be completed successfully, which requires careful planning, diligent work, proactive risk management, and strict monitoring and control. Project managers help ensure that infrastructure that improves communities and satisfies societal needs is completed successfully by allocating resources wisely, keeping lines of communication open, and placing a high priority on safety and quality.

10.2.1. Project Management

The process of overseeing a team's work to accomplish all project objectives within the specified parameters is known as project management. Project documentation, which is produced at the start of the development process, typically contains descriptions of this information. The three main limitations are financial, schedule, and scope. The second challenge is to apply and allocate resources as efficiently as possible in order to achieve predetermined goals (Phillips Joseph, 2004).

Producing a finished project that meets the goals of the client is the aim of project management. Project managers frequently have the dual goal of reshaping or changing the client's brief in order to practically accomplish their goals. Following their establishment, the client's goals ought to inform every choice made by all parties involved in the project, including designers, contractors, project managers, and subcontractors. Making decisions is hampered by project management objectives that are poorly stated or overly restrictive.

A project is a brief, one-of-a-kind endeavor with a defined start and end that is usually time-constrained and frequently confined by funding or staffing. Projects are usually undertaken to meet specific goals and objectives, usually to bring about positive change or added value. Projects' transient nature contrasts with business operations, which are routine, ongoing, or semi-continuous functional activities that result in goods or services. In actuality, managing such diverse production approaches necessitates the development of unique management techniques and technical know-how.

10.2.2. Project Management Types

Any project can benefit from the use of project management techniques. It is frequently customized to a particular project type depending on the size, nature, industry, or sector of the project. For instance, the construction sector, which specializes in constructing buildings, roads, and bridges, has created a unique type of project management known as construction project management, for which project managers can receive certification and training. The information technology sector has also developed its own type of project management, known as IT project management, which focuses on providing technical resources and services needed to go through different stages of the lifecycle, including planning, design, development, testing, and deployment. The complexities of biotechnology research and development are the main focus of biotechnology project management. Although many view this kind of management as a completely distinct discipline, localization project management involves applying many common project management techniques to translation tasks. Project managers, for instance, can significantly improve the translation even if they do not speak the target language because they are well-versed in the study's objectives and can therefore make wise decisions. In a similar vein, project management can also be used in research study management. All government-funded public works are covered by public project management, which can be handled internally by government agencies or externally by contractors. The hard (physical) or soft (non-physical) types of project management are the basis for another classification.

The three main objectives of project management, which are time, quality, and cost, are shared by all of them. Successful projects are finished on time, on budget, and in compliance with predetermined quality standards i.e., satisfying the Iron Triangle, also known as the Triple Constraint, before a project can be deemed successful or unsuccessful. Project managers create and use repeatable templates tailored to the industry they work in for each type of project management. With the express purpose of improving quality, reducing delivery costs, and shortening the time required to produce project results, this enables project plans to become extremely detailed and highly repeatable.

10.2.3. Approaches of Project Management

A 2017 study suggested that the success of any project depends on how well four key aspects are aligned with the contextual dynamics affecting the project, these are referred to as the *four P's*:

- **Plan:** The planning and forecasting activities.
- **Process:** The overall approach to all activities and project governance.
- **People:** Including dynamics of how they collaborate and communicate.
- **Power:** Lines of authority, decision-makers, organograms, policies for implementation, and the like.

There are several ways to plan and finish project tasks, such as incremental, lean, phased, and iterative methods. Project planning has various extensions as well, such as those based on activities (process-based) or outcomes (product-based). Irrespective of the approach taken, the project's overall goals, budget, schedule, and participant and stakeholder roles and responsibilities need to be carefully considered.

10.2.3.1. Benefits Realization Management

By concentrating on the results (benefits) of a project rather than its outputs or products and then monitoring the extent to which that is occurring to keep a project on track, benefits realization management (BRM) improves traditional project management techniques. By meeting predetermined needs (outputs), this can lessen the chance that a project won't succeed when it's finished, i.e., project success but not meeting those requirements' benefits (outcomes) i.e., product achievement. Keep in mind that effective requirements management will guarantee that these advantages are recorded as project requirements and that their fulfillment is tracked throughout the project.

Furthermore, the goal of BRM practices is to guarantee that project outcomes and business strategies are strategically aligned. Recent research showing BRM practices influence project success from a strategic perspective across different countries and industries lends credence to the efficacy of these practices. We refer to these broader impacts as the strategic impact.

Delivering a computer system that will process employee data and manage payroll, holidays, and staff personnel records in less time and with fewer errors is an example of meeting project requirements. When compared without the system, the agreement under BRM may be to achieve a specific reduction in staff hours and errors needed to process and maintain staff data after the system installation.

10.2.3.2. Critical Path Method

An algorithm called the critical path method (CPM) is used to schedule project activities. It is the conventional method for project planning that is predictive in nature. To ascertain the necessary project duration, the CPM method assesses the order of activities, the amount of work required, the interdependencies, and the resulting float time per line sequence. The network diagram's critical path, then, is the set of tasks that have no additional time available, or very little extra time.

10.2.3.3. Critical Chain Project Management

The theory of constraints (TOC) is applied to the planning and management of projects in critical chain project management (CCPM), which is intended to address the uncertainties that arise in project management while accounting for the limited availability of resources (manufacturing and human capital, as well as management and support capacity) required to carry out projects.

Increasing the flow of projects within an organization is the aim (throughput). By using the first three of the TOC's five focusing steps, the resources and system constraints for each project are identified. Tasks on the critical chain are prioritized over all other activities in order to take advantage of the constraint.

10.2.3.4. Earned Value Management

Project management is extended by earned value management (EVM), which offers methods to enhance project monitoring. It shows how the project is progressing in terms of work and value (cost) towards completion. An expansion of EVM theory and application is the Earned Schedule.

10.2.3.5. Iterative and Incremental Project Management

Phased approaches are not recommended for large-scale, multi-company projects with unclear, fast-changing requirements, high levels of risk, dependency, or rapidly evolving technologies, according to critical studies on project management. Some of this can be explained by the uncertainty cone since the project's early planning was fraught with a great deal of uncertainty. This is especially true since the creation of software is frequently the realization of an original or novel product.

10.2.3.6. Lean Project Management

Lean project management uses the principles of lean manufacturing to focus on delivering value with less waste and reduced time.

10.2.3.7. Project Lifecycle

There are five phases to a project lifecycle; known as process groups. Each process group represents a series of interrelated processes to manage the work through a series of distinct steps to be completed. This type of project approach is often referred to as "traditional" or "waterfall." The five process groups are (Figure 10.1):

Typical development phases of an engineering project

1. Initiating
2. Planning
3. Executing
4. Monitoring and Controlling
5. Closing

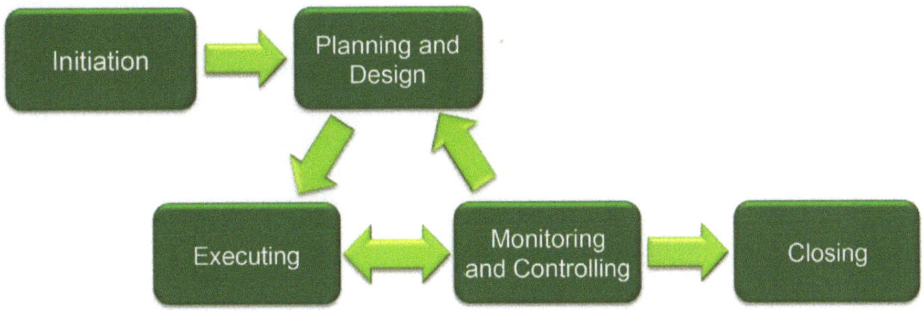

Figure 10.1. *Typical development phases.*

Source: https://upload.wikimedia.org/wikipedia/commons/thumb/b/bb/Project_Management_%28phases%29.png/800px-Project_Management_%28phases%29.png.

Certain project stages may be renamed to better fit the needs of the organization and used in different ways by different industries. Projects will normally go through phases like pre-planning, conceptual design, schematic design, design development, construction drawings (or contract documents), and construction administration when working on a brick-and-mortar design and construction, for instance. The phased approach is useful for small, clearly defined projects, but it can be difficult or even disastrous for larger, more complex projects with a greater number of unknowns, problems, and hazards.

- **Process-Based Management:** The incorporation of process-based management has been driven by the use of maturity models such as the OPM3 and the CMMI (capability maturity model integration.
- **Project Production Management:** The use of operations management in the completion of capital projects is known as project production management. The project as a production system view, which sees a project as converting inputs (raw materials, information, labor, plant, and machinery) into outputs (goods and services), is the foundation of the project production management framework.
- **Product-Based Planning:** A structured method of project management known as "product-based planning" is predicated on the identification of all the products, or project deliverables, that are necessary to meet the project's goals. Therefore, rather than being activity- or task-oriented, it defines a successful project as output-oriented. The most popular application of this methodology is PRINCE2.

10.2.4. Who's Responsible for Construction Project Management?

Team members involved in construction project management include financial planners, C-suite executives, contractors, engineers, and crew members on the ground, among others. But none is as important to the procedure as the project manager. A contractor, an owner who manages projects internally, or even a specialized construction manager can take on the role of project manager. Ensuring the smooth and planned progress of the construction process is the project manager's objective.

They are in charge of assigning resources, avoiding scope creep, keeping a strict timeline, staying within budget, and guaranteeing quality. They notify stakeholders and evaluate risks in real-time. Effective processes that would not be possible without the project manager are made possible by centralizing information and simplifying communication.

10.2.5. What Are the Five Stages of Construction Projects?

Understanding the five major stages of construction projects is a must for successfully managing them.

1. **Planning and Development:** The first and most crucial step in the building process is deciding whether to pursue a project. Project termination is expensive, and the likelihood of losses increases as the project moves forward. This stage is typically characterized by feasibility studies, capital budgeting, pro-con lists, and substantial stakeholder input.

   You can use these practices, and others, to answer key questions about the project:

   - Will it yield a positive ROI?
   - Are the associated risks manageable?
   - Does it fit into your company's portfolio?

   Use a combination of insights from thorough data analysis and input from your team's key stakeholders to address these and other pertinent questions. While diverse team input can help identify potential issues that might otherwise go unnoticed, analytics gives decision-makers an objective perspective on the proposed project.

2. **Design:** It's time to unleash your creative side once you've chosen a project. Creating the project's basic concept and comprehensive blueprints that display the final design are all part of the design phase. From rough sketches to final drawings and specifications, your design will develop; however, each iteration must adhere to the project's specifications while staying within budget and on schedule. After the design has been completed and approved, the preconstruction stage can begin.

3. **Preconstruction:** Making a plan that will direct you through the building process is known as preconstruction. It involves creating a project strategy that outlines everyone's responsibilities, including what needs to be done, when, how, and how much. If everyone follows the plan and does their part flawlessly, they will complete the project on schedule, to the required standard,

and within the allocated budget.

Preconstruction involves a wide variety of critical tasks. Here are a few of the most important:

- Define and allocate resources.
- Set up mini-budgets.
- Create timelines and deadlines.
- Distribute tasks.
- Map out work and operations through work breakdown structures (WBS), organization breakdown structures (OBS), and other tools.

Preconstruction also includes risk assessment and emergency planning in large amounts. Project managers and stakeholders need to be ready for things to go wrong because, during a construction project, things rarely go exactly as planned, frequently because of circumstances beyond their control. The less time, money, and resources you'll waste attempting to get back on track in the event of setbacks, the more proactive you are.

4. **Procurement:** Purchasing, arranging for delivery, and locating the goods and services required to finish a project are all included in procurement. In order to minimize unforeseen cost overruns during this phase, procurement and supply chain managers ought to contribute during the planning stages. However, given that prices are susceptible to changes in the market, some volatility is unavoidable. As much as possible, you should plan ahead for this risk by doing thorough preconstruction planning.

   The procurement of materials from local, regional, or international markets has advantages and disadvantages. While less expensive materials transported over long distances may be more prone to delays and supply chain disruptions, local procurement may be quicker but more expensive. Do a lot of research so that you can make the best decisions and adhere to deadlines while staying within your budget.

   Making a significant decision about when to conduct procurement is another. You can acquire the resources you require as the project moves forward to satisfy changing requirements, as opposed to finishing procurement before the construction project starts. This tactic lowers holding costs, increases flexibility, and maintains liquidity, but it also runs the risk of delayed shipments or shortages, which could cause the project to stall altogether. You run the risk of future price increases as well. Regardless of the strategy you decide on, make an effort to match purchase orders with your construction schedule and make sure you have backup plans in place in case things go off track.

5. **Construction:** You have the tools you need to get going, you've made your plan, and everyone is aware of their responsibilities. Construction can now start. This is the phase where all of your planning and preparation pay off, ensuring a smooth and successful completion of the construction process. Naturally, not even the most elaborate plans can account for every bump in the road, so it's critical to regularly monitor and

assess progress during this phase to ensure that things are moving forward as intended.

Change management is necessary when you need to adapt to new conditions or objectives. Project managers are required to make necessary adjustments while adhering to the project's plan. Seek for a change management tool that can assist in assessing the effects of changes and reducing the project's negative effects.

10.2.5.1. Others Major Stages of a Construction Project

After learning about the main phases of a construction project, let's examine the role that construction management plays. The success of a project depends on these construction management procedures being followed from beginning to end.

1. **Ideation and Research:** Front-end loading (FEL), the planning and design phase of the project life cycle, includes ideation. It occurs when stakeholders carefully examine each project's detail to determine whether it fits into your company's present and future project portfolio.

   The proposed project needs to make sense in terms of risk, return on investment, integration with current projects, and applicable regulations, among other things. Ideas are developed into project proposals during this process by thorough investigation and data-driven analysis. Among the procedures used at this stage are financial breakdowns, brainstorming sessions, feasibility studies, and capital budgeting.

   To make sure the project fully fits within your company's portfolio, actively seek feedback from as many team members as you can while evaluating it. Maintaining a steady flow of high-value projects while controlling risk is one of ideation's main objectives. A portfolio of sound, safe investments may not include a project with substantial risk attached but with the potential for a large return. If it does, more care and attention will be needed for the duration of the project.

2. **Define and Plan the Project:** After deciding on a project, it's important to work out the specifics to avoid scope creep and maintain team cohesion when work gets underway. At this point, it's crucial to discuss the project's scope, a workable schedule, the resources needed, a fair and accurate budget, and key performance indicators (KPIs). During this phase, gathering input from as many pertinent team members as possible and figuring out how many employees the project will need should once again be top priorities.

3. **Determine Roles:** Any construction project involves numerous parties, so it's important to clearly define each one's role. This eliminates uncertainty and delays, makes it easier to hold team members accountable, and makes everyone's roles clear. Redundancies vanish and tasks don't get overlooked when every team member is aware of their responsibilities within the project and how to complete their assigned tasks.

4. **Finalize and Execute Construction Plans:** Project managers should meet with the relevant stakeholders prior to the start of construction in order to go over the plan and make

sure everyone is in agreement. While it's likely that you will encounter unforeseen problems and modifications as the project moves forward, it's nearly a given that you will run into more obstacles if you don't get clear buy-in before work starts. Project managers need to exert every effort to maintain a well-organized and productive process as soon as construction commences. This entails closely monitoring and measuring advancement.

An invaluable insight into the performance and status of the project is provided by comprehensive and relevant KPIs that are updated in real time, something that approximately 56% of project managers lack. Choosing the appropriate KPIs is essential because they offer a quick and easy method to assess if the project is straying from its intended course. You can get the insights you need to keep costs under control and meet deadlines by using data on the present state of your budget, procurement, and other important project aspects.

10.2.6. Handling Construction Project Management Challenges

Construction project management involves its share of challenges. Here are some of the most common and some tips on how to overcome them.

10.2.6.1. Communication and Document Management

It's difficult to keep the lines of communication open between all parties involved in a construction project. It can also be very challenging to maintain an accurate and easily accessible repository for all project documents. However, the dangers of inadequate document management and communication are too great to be disregarded. A misunderstanding that causes even a small delay in a crucial process, such as change orders or ongoing procurement, can derail the project entirely, result in overspending, or fuel tensions between stakeholders.

Use software that centralizes information and communication on a single platform to avoid the difficulties of trying to manage communication through chat, email, the phone, and other channels. Document management is a duty of project managers in addition to encouraging regular communication. Owners, contractors, insurance providers, and other stakeholders require a single, trustworthy source of information.

Examining your company's current document management system, centralizing documents already in existence, standardizing procedures, and utilizing a document management solution with the features you require are all good places to start when developing a successful document management process.

10.2.6.2. Estimating

It is impossible to overstate the difficulty and significance of estimating construction-related costs, resources needed, and schedule. Given the narrow profit margins in the construction sector, even a small estimation error can result in a project loss. When estimating the cost of a large-scale project, such as building a power plant, a large margin of error can even result in the project being abandoned in the middle, incurring enormous sunk costs. There are forecasting tools available to help you take into account all these variables and make your estimates as accurate as possible, even though price fluctuations, currency drops, material shortages, and other factors can make estimation an imprecise science at best.

10.2.6.3. Siloed Data

When information is dispersed and difficult to access, misunderstandings are common and project outcomes are rarely as expected. Data silos may have an impact on risk assessment, project schedule, cost control, and other areas. Your contractor will either have to wait on information that is already available or decide to move forward without it, so you'll have to cross your fingers and hope for the best. Use a data management system that facilitates the easy centralization and integration of pertinent data to eliminate data silos. The best tools enable rapid access to data, wherever it may reside, across all of your systems by your team.

10.2.6.4. Lack of Real-Time Data for Risk Assessment and Change Management

It doesn't matter if the data you share with your team is outdated. Since construction projects are by their very nature ongoing projects, you need access to real-time data to keep track of new developments and resolve issues as they come up. Once more, the solution is the appropriate software package. The platform you select should offer the tools you need to visualize workflows and assess the effects of your decisions, making it simple to dive into the data with up-to-date dashboards and reports.

10.3. COST ESTIMATION AND BUDGETING

Budgeting and cost estimation are essential components of civil engineering projects that are critical to their viability and success. These procedures entail projecting the amount of money needed to finish a project and making sure that resources are distributed effectively and efficiently to meet goals within predetermined parameters.

Cost estimating in civil engineering typically starts in the feasibility and planning phases of a project. It entails a methodical evaluation of numerous elements, including labor, supplies, machinery, and overhead expenses. To generate precise cost estimates, estimators sometimes use sophisticated software tools in addition to historical data and market trends. Cost estimates are heavily influenced by variables like project scope, complexity, location, and regulatory requirements.

For example, land acquisition and environmental regulations may drive up the cost of projects in urban areas. To prevent underestimating expenses, which can result in delays and budget overruns, accurate estimation is essential. In order to account for unforeseen events like inflation, supply chain disruptions, or changes in project scope, estimators frequently include contingency funds. Throughout a project's lifecycle, contingency planning helps to maintain financial viability by reducing risks.

The budgeting process starts after the estimation stage is finished. Setting up financial controls, tracking expenses, and distributing projected costs among project tasks or phases are all part of budgeting. To create a comprehensive budget that satisfies stakeholder expectations and project goals, estimators and project managers work closely together. Throughout the project's execution, this budget acts as a financial road map, directing resource allocation and decision-making.

For budgeting to be effective, it must be regularly monitored and adjusted to account for shifts in the project's scope, costs, or schedule. This iterative procedure guarantees that projects stay on course for successful completion and aids in the maintenance of financial discipline. Real-time budget monitoring is made possible by contemporary project management tools and software, allowing for prompt intervention and modification when needed.

The scope and complexity of civil engineering projects range greatly, from modest infrastructure repairs to massive building projects like bridges, highways, or dams. Customized cost estimation and budgeting techniques are needed for every project in order to meet its particular needs and challenges. Although they are necessary for the delivery of sustainable projects, factors like community involvement, regulatory compliance, and environmental impact assessments make cost estimation and budgeting even more difficult.

In civil engineering, cost estimation and budgeting are essential procedures that guarantee projects are completed on schedule, within budget, and to the necessary quality standards.

Civil engineers can successfully manage financial resources, reduce risks, and produce project outcomes that satisfy the needs of stakeholders and communities by combining precise estimation techniques with strong budgeting processes. The discipline of civil engineering is changing in response to sustainable practices and technology breakthroughs, and innovation and high standards in infrastructure development around the world depend on ongoing improvements in cost estimation and budgeting.

Predicting a project's cost and then allocating funds appropriately is the process of cost estimation and budgeting. Project managers can identify possible risks, allocate resources, and make backup plans thanks to this part of the planning process.

10.3.1. Why Does Cost Estimation and Budgeting Matter?

How important is it to make informed cost estimates for projects?

Consider what can happen if you underestimate the cost of a project:

You'll likely end up going over budget, which will put your project at risk of being canceled or delayed.

Your team will be forced to work longer hours (and maybe even weekends) to make up for the underestimated budget, which can lead to burnout.

On the other hand, if you overestimate the cost of a project, you may:

- Waste valuable resources that could be better used elsewhere
- Hurt your company's bottom line
- Damage your reputation as a project manager

Budgeting and cost estimation are crucial because they enable companies to decide whether or not to move forward with a particular project after receiving accurate information. The company may choose to reduce the project's scope or even scrap it completely if the projected cost of the project exceeds the available funding. The company may decide to allocate more funds in order to boost the chance of success if the project's estimated cost is less than the available budget.

In either case, having accurate cost estimates is essential for making sound decisions about projects (Figure 10.2).

Figure 10.2. *Cost estimation and budgeting.*

Source: https://www.teamly.com/blog/wp-content/uploads/2022/10/Key-Elements-of-cost-estimation-in-project-management-768x353.png.

10.3.2. Key Elements of Cost Estimation in Project Management

Several key elements must be considered when estimating the cost of a project. These include:

- Materials
- Labor
- Overhead
- Equipment
- Permits
- Fees
- Taxes
- Contingencies

Accurate cost estimates for each of these elements are essential for developing an accurate overall estimate for the project (Figure 10.3).

Figure 10.3. *Key elements of cost estimation in project management.*

Source: https://www.teamly.com/blog/wp-content/uploads/2022/10/Methods-for-Estimating-Costs-and-Budgets-for-Projects-768x353.png.

10.3.3. Different Methods for Estimating Costs and Budgets for Projects

Businesses can estimate the expenses of their projects using a variety of techniques. The most popular techniques include reserve analysis, expert judgment, cost of quality, three-point, analogous, bottom-up, and top-down estimation. It is imperative to acknowledge that there is no one "correct" method for estimating project costs. The ideal approach to take will change based on the particulars of the project. Having said that, let's examine each of these techniques in more detail so you can understand when and how to apply them.

- **Bottom-Up Estimating:** The process of bottom-up estimating entails estimating the cost of each project component separately and then adding up all of those estimates to determine the project's estimated total cost. This methodology is frequently employed for extensive and intricate projects due to its exceptional precision. It can, however, also be costly and time-consuming.
- **Top-Down Estimating:** Using a technique known as "top-down estimating," project costs are first estimated overall and then logically allocated to individual components. This methodology is frequently employed for small to medium-sized projects due to its lower time and cost compared to bottom-up estimating. It may, however, be less precise.
- **Three-Point Estimating:** The process of estimating a project's most likely cost, best-case cost, and the worst-case cost is known as "three-point estimating." The average of these three estimates is then used to determine the estimated total cost. This method is frequently applied to projects that involve a lot of uncertainty.
- **Analogous Estimating:** Using a technique known as "analogous estimating," one can estimate a new project's cost by contrasting it with similar projects that have already been finished. This method is frequently applied to projects for which the available data is sparse (Figure 10.4).

Figure 10.4. *Analogous estimating.*

Source: https://www.teamly.com/blog/wp-content/uploads/2022/10/Cost-of-Quality-768x353.png.

10.3.4. Cost of Quality

The cost of quality is a technique that involves estimating the cost of ensuring that a project meets all required quality standards. This approach is often used for projects where quality is a critical concern.

- **Expert Judgment:** This approach is exactly what it sounds like – experts are consulted to provide their estimates of the cost of a project. It's often used for projects where there is a limited amount of data available.

- **Reserve Analysis:** Reserve analysis is a technique used to determine how much cash needs to be set aside for unforeseen expenses that might occur during a project. When there is a lot of uncertainty in a large, complex project, this method is frequently employed. Which method is best? The most appropriate method for estimating a project's cost will vary depending on its particulars.

  A few things to think about are the project's scope and complexity, the volume of data at hand, the value of quality, and the degree of uncertainty (Figure 10.5).

Figure 10.5. *Reserve analysis is a technique that involves estimating.*

Source: https://www.teamly.com/blog/wp-content/uploads/2022/10/must-dos-when-planning-your-budget-768x353.png.

- Here Are Some "Must Do's" When Planning Your Budget

No matter which method you use, there are a few best practices that you should follow to ensure accuracy:

- Use historical data whenever possible
- Use multiple methods to cross-check your estimates
- Get input from as many relevant people as possible
- Use software to help with the estimating process
- Continuously monitor and update your estimates

By following these best practices, you can be confident that your estimates are as accurate as possible. This will help you to budget for your projects more effectively and avoid cost overruns.

10.4. RISK MANAGEMENT IN CONSTRUCTION PROJECTS

The process of assessing and putting policies in place to lessen the impact of risks in construction projects is known as construction risk management. To develop a risk management plan that enables project managers to recognize, track, and reduce risks as they materialize, this risk management process requires extensive planning.

Early on in the construction planning process, a construction risk management plan is created. It describes the potential risks to the project and how to address them. This involves assigning a crew member to take responsibility for and resolve the problem.

10.4.1. Construction Risk Management: Identifying and Managing Project Risks

Building is a dangerous industry. Every construction project is different from the next and presents a different set of opportunities and challenges. Risk identification and management in construction projects can be challenging, but it's not impossible with proper planning and implementation. Construction risk management is critical because, should a risk materialize, it has the potential to disrupt and derail a project.

Once hazards are identified, you must be able to appropriately evaluate, manage, and keep an eye on them in order to prevent disaster. Risks don't always have to be bad. Gaining more revenue, building strong client relationships that result in more projects, and growing your company into new markets and industries can all be attained by having effective risk identification and management skills.

10.4.2. Types of Construction Risks

For proper construction risk management, you need to know the types of risks inherent in construction projects. These can be financial, contractual, operational, and environmental and can be caused by both internal and external sources.

Common risks include:

- Safety hazards that lead to worker accidents and injuries
- Managing change orders

- Incomplete drawings and poorly defined scope
- Unknown site conditions
- Poorly written contracts
- Unexpected increases in material costs
- Labor shortages
- Damage or theft to equipment and tools
- Natural disasters
- Issues with subcontractors and suppliers
- Availability of building materials
- Poor project management

When risks materialize, they can have a significant effect on the project's budget, timeline, and performance, which could cause delays and disagreements later on. The good news is that with careful planning and effective project management, the majority of these risks can be controlled and reduced.

10.4.3. Identifying Construction Project Risks

After going over some of the typical risks associated with construction projects, it's time to determine which risks are specific to your project. This needs to be completed as soon as possible, ideally in the project's preconstruction stage. Keep in mind that if you don't recognize and address a project risk, you are essentially accepting the risk if it materializes. To identify risks, have brainstorming sessions with the stakeholders and the project team. You are not trying to solve any issues at this time.

Finding as many scenarios as possible that could harm the project is the aim here. Make sure you rely on your team's knowledge base, experience, and expertise. Examine previous projects you've completed that are comparable in size, scope, and location to gain a better understanding of the construction risks your current project is facing.

Regularly meeting with your project team and stakeholders is a good way to avoid risks from sneaking up on you as the project moves forward. You can use that time to review your current risk management initiatives as well as to identify any other issues that might arise in the near future.

10.4.4. Construction Risk Management Process

After you have determined which risks could affect your project, you must take a seat and evaluate each one according to its likelihood of happening and the effect it would have on the project should it do so. Sort each risk's impact and likelihood into three categories: high, medium, and low. Risks that have a high probability and high impact should be addressed first; low-probability and low-impact risks can be addressed last. Consider how much money, effort, and time each risk will require to be managed successfully. After you've ranked each risk, carefully go over each one to see if there's anything you can do to avoid, reduce, eliminate, transfer, or accept it.

- **Avoid the Construction Project Risk:** This may mean turning down a project or negotiating the contract to remove the project risks. There's no shame in walking away from a project if the risks outweigh the potential rewards.
- **Transfer the Project Risk:** It's possible that your business is not the best choice to handle a specific risk. Determine who on the project team is most qualified to take on

each risk by consulting with the other stakeholders. Talk to the client about the risks they will take on and the ones you will be in charge of handling. Together with other options for safeguarding your business against risks, work with your insurance provider to ascertain which risks are covered under your current policies.

- **Mitigate the Project Risk:** Risks must be eliminated, reduced, and accepted with careful planning. Divide each risk into manageable steps. Avoid devoting too much of your resources to managing several risks. To properly manage all of your risks, you might need to commit more resources, like hiring more staff or renting more machinery.
- **Accept the Project Risk:** It's important to make a thoughtful decision before deciding to accept a risk. Accepting a few low-probability, low-impact risks might be acceptable. Accepting a high-probability, high-impact risk without any kind of mitigation or management could be bad for your business and the project as a whole.

10.4.5. Legal and Regulatory Aspects of Construction Projects

Construction projects necessitate adherence to laws, regulations, and standards in order to protect public safety, environmental sustainability, and contractual obligations. This is where legal and regulatory aspects come into play. These factors cover a wide range of factors that affect every stage of a construction project, from planning and design to construction and upkeep.

1. **Building Codes and Standards**: Building codes are regulations that specify minimum requirements for the design, construction, and occupancy of buildings. They are enforced by local governments to ensure structural integrity, fire safety, accessibility, and energy efficiency. Compliance with building codes is mandatory and typically verified through inspections during construction.
2. **Permitting and Zoning Regulations**: Before commencing construction, developers must obtain permits from local authorities. These permits ensure that proposed buildings or renovations comply with zoning regulations, land use restrictions, environmental protections, and health and safety requirements. Zoning regulations dictate where and how buildings can be constructed based on factors such as density, height restrictions, and land use.
3. **Contractual Obligations**: Construction projects involve numerous contracts between stakeholders such as owners, contractors, subcontractors, and suppliers. Legal aspects of contracts include terms and conditions, payment schedules, scope of work, warranties, and dispute resolution mechanisms. Contracts must be carefully drafted to mitigate risks and ensure all parties understand their rights and obligations.
4. **Environmental Regulations**: Construction activities can impact the environment through land disturbance, emissions, noise, and waste generation. Environmental regulations enforce measures to

minimize these impacts, including permits for handling hazardous materials, erosion control plans, wildlife protection, and adherence to sustainable building practices. Compliance ensures projects mitigate environmental harm and contribute to long-term sustainability.

5. **Labor Laws and Safety Regulations**: Construction sites are inherently risky environments, making worker safety a priority. Occupational health and safety regulations set forth requirements for site conditions, personal protective equipment (PPE), training, and emergency procedures. Compliance with these laws minimizes accidents, injuries, and fatalities, while also protecting workers' rights related to wages, hours, and working conditions.
6. **Government Contracts and Procurement Laws**: Public construction projects often involve government contracts subject to specific procurement laws and regulations. These laws ensure fair competition, transparency, and accountability in awarding contracts. Compliance with procurement processes is essential to avoid legal challenges and ensure projects receive public funding or approvals.
7. **Dispute Resolution and Legal Remedies**: Despite meticulous planning, disputes may arise during construction projects related to design changes, delays, quality issues, or contractual disagreements. Legal remedies include mediation, arbitration, or litigation to resolve disputes and enforce contractual rights. Parties often include dispute resolution clauses in contracts to streamline resolution processes and minimize disruptions.
8. **Insurance and Liability**: Construction projects carry inherent risks of property damage, bodily injury, and financial loss. Insurance coverage such as builder's risk insurance, liability insurance, and performance bonds protect stakeholders against unforeseen losses and liabilities. Understanding insurance requirements and ensuring adequate coverage is crucial for managing project risks effectively.

Construction projects have a complex legal and regulatory environment that must be navigated with thorough planning, proactive compliance efforts, and thorough knowledge. Construction stakeholders can minimize risks, guarantee legal compliance, and accomplish successful project outcomes by following building codes, obtaining required permits, carrying out contracts efficiently, and giving priority to safety and environmental stewardship. Incorporating legal knowledge into the planning and implementation of projects promotes responsible construction methods that preserve the built environment's sustainability, quality, and safety standards.

10.4.6. Legal Aspects of Construction Contracts

The foundation of the construction industry is construction contracts, which govern the interactions between the various parties involved in a project. The rights, obligations, and responsibilities of all parties involved—including project owners, contractors, subcontractors, architects, and engineers—are outlined in these agreements. Given that construction projects frequently entail substantial financial outlays and intricate procedures, the legal aspects of construction contracts are essential to guarantee the successful, timely, and cost-effective completion of projects.

- **Formation of Construction Contracts:** The creation of a contract is the first stage in any construction project. The parties must make an offer, accept it, and give it some thought. The contractor accepts the offer by committing to supply the services under the terms and conditions outlined in the contract. Usually, the offer originates from the owner who needs construction services.
- **Key Parties in Construction Contracts:** Construction contracts involve various parties, each with distinct roles:
    1. **Owner:** The party who owns or has an interest in the construction project.
    2. **Contractor:** The entity responsible for executing the construction work.
    3. **Subcontractors**: Individuals or companies hired by the contractor to perform specific portions of the project.
    4. **Architects and Engineers:** Professionals responsible for designing various portions of the project.
    5. **Suppliers:** Entities providing materials and equipment for the project.
- **Contract Documents:** Construction contracts are typically composed of several documents, which include:
    1. **Contract Agreement:** This primary document outlines the scope of work, the project schedule, payment terms, and other essential details.
    2. **Drawings and Specifications:** These documents detail the technical and design requirements for the project.
    3. **Change Orders:** In case of alterations to the original contract, change orders document these changes and their impact on costs and timelines.

10.4.7. Project Scope and Performance Standards

The project's scope and performance requirements must be precisely specified in the contract. This contains a thorough explanation of the tasks that need to be completed, quality requirements, and anticipated outputs. Failing to do so may result in disagreements about the original terms of the agreement.

- **Payment and Compensation:** Contracts specify the payment schedule, total contract price, allowances, and contingencies, among other things. Construction contracts frequently involve payment disputes, so it's important to specify in the contract exactly how and when payments will be made.
- **Timeframes and Delays:** A project timeline is usually specified in construction contracts, and late fees may be assessed. It's critical to discuss potential delays and how they will be handled, such as inclement weather, problems with permits, or unforeseen site conditions.

- **Risk Allocation and Insurance:** Contracts for construction divide up the risks among participants. In order to safeguard the parties involved in the event of accidents, damage, or disputes, insurance requirements, such as worker's compensation and liability, are frequently included.
- **Dispute Resolution:** The contract should specify the dispute resolution procedure in the event of disagreements or disputes. This can entail court cases, arbitration, mediation, or negotiation. Costly legal disputes can be avoided by having a clear dispute resolution procedure.
- **Regulatory Compliance:** Compliance with local, state, and federal regulations is a crucial legal aspect of construction contracts. The contract should outline who is responsible for obtaining permits and meeting legal requirements.
- **Termination and Suspension:** Construction contracts should detail the circumstances under which the contract can be terminated or suspended, such as breaches of contract, insolvency, or force majeure events.

CASE STUDY

CONSTRUCTION MANAGEMENT

Project Overview

The case study involves the construction of a new residential condominium complex in an urban area. The project aims to create high-end apartments with modern amenities, including parking facilities, recreational areas, and green spaces. The construction site is in a densely populated neighborhood with limited space for staging and logistics.

Challenges Faced

1. **Site Logistics:** Managing logistics in a tight urban space posed challenges for material delivery, equipment storage, and worker accessibility.
2. **Regulatory Compliance:** Adhering to local building codes, environmental regulations, and zoning laws requires meticulous planning and timely approvals.
3. **Schedule Management:** Balancing construction progress with deadlines to avoid delays while ensuring quality standards.
4. **Budget Control:** Managing costs effectively amid fluctuating material prices, labor demands, and unforeseen expenses.

Strategies Implemented

1. **Advanced Planning:** Detailed project planning with a focus on logistics, scheduling, and regulatory compliance from the initial stages.
2. **Stakeholder Collaboration:** Regular communication and collaboration with stakeholders including architects, engineers, contractors, and local authorities.
3. **Technology Integration:** Use of construction management software for scheduling, cost tracking, and communication to enhance efficiency and transparency.
4. **Risk Management:** Continuous risk assessment and mitigation strategies to address potential delays, cost overruns, and regulatory issues.

Outcomes

Despite initial challenges, the project successfully navigated regulatory hurdles, managed logistics effectively, and maintained quality and safety standards. The integration of technology and proactive risk management strategies contributed to minimizing delays and cost overruns. The condominium complex was completed within the scheduled timeline and budget, meeting client expectations and regulatory requirements.

CLASS ACTIVITY

For a class activity in construction project management, students could participate in a simulation where they are assigned roles such as project manager, contractor, and client. Each group would receive a hypothetical construction project with defined objectives, constraints, and a budget. Students would collaborate to develop a project plan, including scheduling, budgeting, risk assessment, and stakeholder management. Through this hands-on activity, they would experience real-world challenges such as scope changes, resource allocation, and decision-making under pressure. Debriefing sessions would allow reflection on lessons learned, emphasizing the importance of effective communication, teamwork, and strategic planning in construction project management.

SUMMARY

- Project management in civil engineering is crucial for ensuring the successful completion of large-scale infrastructure projects, from bridges and highways to dams and skyscrapers.
- Project planning and scheduling are fundamental processes in project management, crucial for achieving project objectives efficiently and effectively.
- Civil engineering project managers are professionals who oversee large construction projects. They often work on multiple projects simultaneously, and their job requires them to be physically fit and active to carry construction materials on site.
- Civil engineer project managers design project plans and help devise strategies to accomplish them. They also review compliance with various state and government regulations and ensure that their project or site location follows all guidelines.
- Construction project management involves directing and organizing each part of the project life cycle, from ideation to completion. It's a holistic practice to deliver projects on time and under budget.
- The initiation phase of construction project management begins with defining project objectives, scope, and feasibility. This involves conducting feasibility studies, site assessments, and initial cost estimates.
- Critical path method (CPM) is an algorithm for determining the schedule for project activities. It is the traditional process used for predictive-based project planning.
- Procurement encompasses sourcing, purchasing, and transporting the materials and services you need to complete a project.

- Cost estimation and budgeting are crucial aspects of civil engineering projects, playing a pivotal role in their success and sustainability.
- Construction risk management is the process of evaluating and implementing procedures to reduce the impact of risks in construction projects.

REVIEW QUESTIONS

1. What are the key elements of a project initiation document (PID) in civil engineering?
2. How do you define and document project objectives and scope in a civil engineering project?
3. Explain the significance of the critical path method (CPM) in scheduling civil engineering projects.
4. Discuss the challenges associated with managing project timelines in civil engineering and strategies to mitigate them.
5. What are the main factors considered when estimating costs for civil engineering projects?
6. How do you develop and manage project budgets to ensure financial control and accountability?
7. Identify and classify the types of risks typically encountered in civil engineering projects.
8. Describe the process of risk assessment and mitigation planning in civil engineering project management.
9. How is quality assurance different from quality control in the context of civil engineering projects?
10. What are the tools and techniques used to ensure compliance with quality standards in civil engineering construction?

REFERENCES

1. Becker, J., Kugeler, M., & Rosemann, M. (2003). *Process Management: A Guide for the Design of Business Processes.* Springer.
2. Conforto, E. C., Salum, F., Amaral, D. C., da Silva, S. L., & Magnanini de Almeida, L. F. (2014). Can agile project management be adopted by industries other than software development? *Project Management Journal, 45*(3), 21–34.
3. Daniel, P. A., & Daniel, C. (2018). Complexity, uncertainty and mental models: From a paradigm of regulation to a paradigm of emergence in project management. *International Journal of Project Management, 36*(1), 184–197.
4. Dinsmore, P. C., & Cooke-Davies, T. J. (2005). *Right Projects Done Right: From Business Strategy to Successful Project Implementation.* Wiley.

5. Kwak, Y. H. (2005). A brief history of project management. In E. G. Carayannis (Ed.), *The Story of Managing Projects.* Bloomsbury Academic.
6. Nokes, S., & Kelly, S. (2007). *The Definitive Guide to Project Management: The Fast Track to Getting the Job Done on Time and on Budget.* Pearson Education. Prentice Hall Financial Times.
7. Phillips, J. (2004). *PMP Project Management Professional Study Guide.* McGraw-Hill/Osborne.
8. Sha, M., & Childs, J. H. (2014). Applying a project management approach to survey research projects that use qualitative methods. *Survey Practice,* 7(4), 1–8.

Index

K

L

M

N

O

P

FSC
www.fsc.org
MIX
Papier | Fördert gute Waldnutzung
FSC® C083411

Kolibri 360 GmbH
Ferdinand-Jühlke-Straße 7
99095 Erfurt, Deutschland
produktsicherheit@kolibri360.de